Sustainable Civil Infrastructures

About this Series

Sustainable Infrastructure impacts our well-being and day-to-day lives. The infrastructures we are building today will shape our lives tomorrow. The complex and diverse nature of the impacts due to weather extremes on transportation and civil infrastructures can be seen in our roadways, bridges, and buildings. Extreme summer temperatures, droughts, flash floods, and rising numbers of freeze-thaw cycles pose challenges for civil infrastructure and can endanger public safety. We constantly hear how civil infrastructures need constant attention, preservation, and upgrading. Such improvements and developments would obviously benefit from our desired book series that provide sustainable engineering materials and designs. The economic impact is huge and much research has been conducted worldwide. The future holds many opportunities, not only for researchers in a given country, but also for the worldwide field engineers who apply and implement these technologies. We believe that no approach can succeed if it does not unite the efforts of various engineering disciplines from all over the world under one umbrella to offer a beacon of modern solutions to the global infrastructure. Experts from the various engineering disciplines around the globe will participate in this series, including: Geotechnical, Geological, Geoscience, Petroleum, Structural, Transportation, Bridge, Infrastructure, Energy, Architectural, Chemical and Materials, and other related Engineering disciplines.

More information about this series at http://www.springer.com/series/15140

Behzad Fatahi · Aron Mwanza
Dave T. T. Chang
Editors

Sustainable Design and Construction for Geomaterials and Geostructures

Proceedings of the 5th GeoChina International Conference 2018 – Civil Infrastructures Confronting Severe Weathers and Climate Changes: From Failure to Sustainability, held on July 23 to 25, 2018 in HangZhou, China

Springer

Editors
Behzad Fatahi
Civil and Environmental Engineering
University of Technology Sydney
Sydney, NSW, Australia

Dave T. T. Chang
Chung Yuan Christian University
Taoyuan City, Taiwan

Aron Mwanza
University of Zambia
Lusaka, Zambia

ISSN 2366-3405 ISSN 2366-3413 (electronic)
Sustainable Civil Infrastructures
ISBN 978-3-319-95752-4 ISBN 978-3-319-95753-1 (eBook)
https://doi.org/10.1007/978-3-319-95753-1

Library of Congress Control Number: 2018948646

This Springer imprint is published by the registered company Springer Nature Switzerland AG
The registered company address is: Gewerbestrasse 11, 6330 Cham, Switzerland

Contents

Introduction

Due to the scarcity of land, structures and infrastructures cannot always be built at locations with favourable geotechnical conditions. Therefore, there has been a high demand for innovative designs and construction techniques to achieve the required level of performance considering both construction and maintenance costs as well as sustainability issues. The research studies presented in this edited volume present recent research findings and critically review the existing literature related to assessment of geotechnical structures under complex and extreme loading conditions such as cyclic, seismic and blast loads. New experimental investigation on behaviour of various complex geomaterials and new testing methods has been presented, while advanced material models and numerical modelling techniques used to simulate behaviour of various geotechnical systems have been presented. Practicing civil, geotechnical, structural, earthquake and construction engineers can use the research methodologies and outcomes presented here in real-life projects. This volume is part of the proceedings of the 5th GeoChina International Conference on Civil Infrastructures Confronting Severe Weathers and Climate Changes: From Failure to Sustainability, HangZhou, China 2018.

Effect of Pre-deformation by Cyclic Loadings on Seepage Flow-Induced Failure of Volcanic Embankments

Shima Kawamura(✉) and Hieu Minh Dao

Graduate School of Engineering, Muroran Institute of Technology, Muroran, Japan
{skawamur,15096011}@mmm.muroran-it.ac.jp

Abstract. The purposes of this study are to investigate piping mechanisms of volcanic coarse grained soils and to clarify seepage performance of embankments constructed by volcanic coarse grained soils. A series of the 1 g model tests was performed on model embankments constructed by a typical volcanic soil in Hokkaido Japan. In the model tests, the effect of pre-deformation (shear strain) generated by cyclic loadings on seepage flow-induced failure was investigated in detail. Finally, stability evaluation of volcanic embankments was discussed based on the results of a series of seepage flow tests.

1 Introduction

Volcanic soil has been utilized as a useful construction material in Japan, especially man-made earth structures (embankments, dykes or cut slopes, etc.). However, earthquake- and rainfall-induced failures of slopes and grounds have been frequently reported in Hokkaido, Japan (e.g., JGS 2004). It has been said that the failures are caused by the specific performance and topography of a volcanic soil defined as a local or an unusual soil in Japan.

In previous studies, rainfall- and wave-induced failures of volcanic embankments subjected to cyclic loadings (Kawamura and Miura 2014; Kawamura et al. 2016; Kawamura and Tsuchida 2017) and piping phenomenon of volcanic coarse grained soils (Dao et al. 2016, 2017) were investigated in detail. In this study, the authors focus on the feature of "seepage performance" of volcanic embankments with pre-deformation generated by cyclic loadings. In particular, the effect of magnitude of pre-deformation (shear strain) on failure due to seepage flow was elucidated using 1 g model test apparatus.

In order to clarify seepage performance of embankments constructed by volcanic coarse grained soils, field observation was firstly performed on full scale embankment, and then a series of piping tests and seepage flow tests was conducted on the same volcanic soil as the soil material of the embankment under the same compaction conditions. In the consideration of the results, stability of evaluation of volcanic embankments was discussed in detail.

B. Fatahi et al. (eds.), *Sustainable Design and Construction for Geomaterials and Geostructures*, Sustainable Civil Infrastructures,
https://doi.org/10.1007/978-3-319-95753-1_1

2 Field Observation of Full Scale Embankment

In previous studies (e.g., Kawamura et al. 2015), full-scale embankment by volcanic coarse grained soil was constructed to clarify failure mechanisms due to rainfalls and seepage flow and to grasp internal erosion in embankment in Sapporo city, Japan and field observation was conducted from November 01, 2012 to November 15, 2013. The embankment was constructed using a road roller (weight of 5.88 kN) by compacting a typical volcanic soil described later so as to be more than the degree of compaction Dc of 85% (the desired dry density is more than $\rho_d = 0.9$ g/cm^3) for each layer of 0.25 m.

For the embankment, both sides were confined to wooden boards, back and bottom are covered by plastic sheets. The number of roller compactions was 3 times for each layer. According to the in situ tests, the degree of compaction and water content was Dc = 95.9% (average) and w = 42.5% (average), respectively. The slope size was 5 m in height, 4 m in length, 2.7 m in width and the angle of 45° (see Fig. 1). The water supply pipes depicted in Fig. 2 have holes at certain locations and intervals. The water

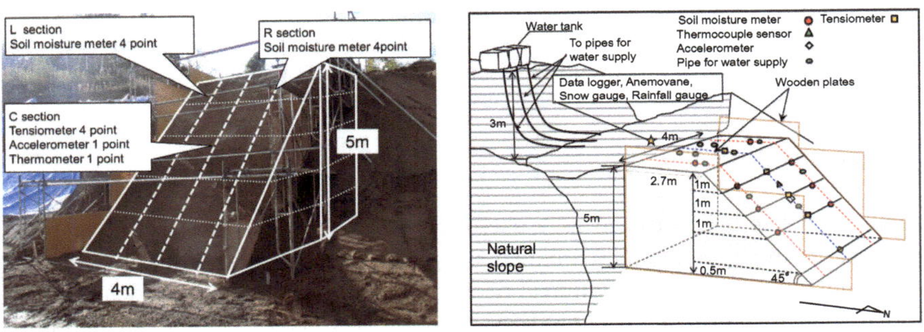

Fig. 1. The whole view of full-scale embankment (Kawamura et al. 2015)

was reserved in three tanks and was provided through each pipe into the embankment. Water supply through the surface of embankment was started from May 7, 2013. The amount of water supply was 1000–3000 l/day. The details of construction procedures and test results were reported by Matsumura (2014), Kawamura et al. (2015) and Dao et al. (2016).

During field monitoring, exudation routes of water in the full-scale embankment were observed. For example, Fig. 3 shows traces of exudation routes of water during field monitoring (in July, 2013). The seepage phenomenon occurred along embankment boundaries. According to the in situ observation, fine particles were flowing out. This means that piping and exudation routes might be easily generated in volcanic man-made soil structures, although the embankment was actually compacted by more than the degree of compaction Dc = 95%. Figure 4 shows the aspects of cross-section after dismantling of the embankment. It can be seen from the photos that piping routes from holes of water supply pipe extend laterally. In the consideration of field

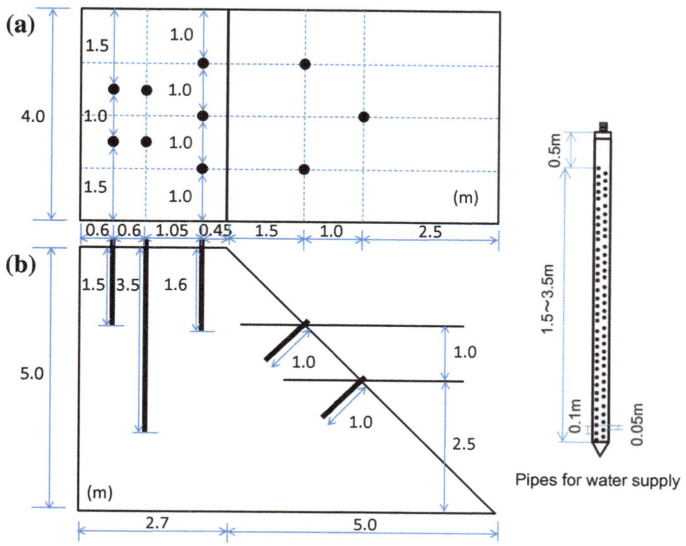

Fig. 2. Locations and specification of pipes for water supply: **a** plane view, **b** side view (Kawamura et al. 2015)

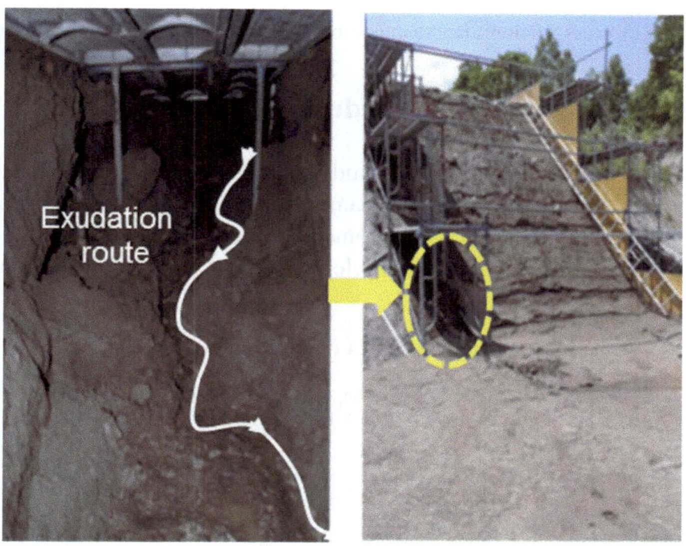

Fig. 3. Trace of exudation routes during field monitoring in July, 2013 (Dao et al. 2016)

observation, it was pointed out that the evaluation of seepage performance is also significant for volcanic man-made structures containing a significant amount of fine particles.

Fig. 4. Aspects of cross-section after dismantling of full scale embankment (Dao et al. 2016)

3 Test Material and Test Procedures

Volcanic coarse-grained soil used in this study was sampled from the ejecta of Shikotsu calderas (*Spfl*) in Hokkaido and was the same materials as the full scale embankment. The sample is hereafter referred to as Komaoka volcanic soil (or K soil). The index properties of the samples are shown in Table 1, and are compared to the characteristics

Table 1. Index properties of Komaoka volcanic soil

Sample name	ρ_s (g/cm^3)	$\rho_{d\ max}$ (g/cm^3)	$\rho_{d\ min}$ (g/cm^3)	w_n (%)	D_{50} (mm)	U_c	F_c (%)
Komaoka	2.50	1.12	0.76	43	0.27	43.0	26–42.6
Toyoura sand	2.68	1.63	1.37	–	0.18	1.5	0

w_n: Natural water content, D_{50}: Mean grain size, U_c: Coefficient of uniformity, F_c: Finer content

of Toyoura sand. As shown in Table 1, the finer contents range from 26.0 to 42.6%. The fines of Komaoka volcanic soil was classified into non plastic material (NP) according to Atterberg limits (liquid limit is 44.5%, plastic limit is NP). In the past, volcanic soils, including Komaoka volcanic soil (*Spfl*), which have been utilized for residential embankments have repeatedly liquefied due to the strong earthquakes; examples include Hokkaido Nansei-oki (1993), Hokkaido Toho-oki (1994) and Tokachi-oki (1968 and 2003). Mechanical behavior of Komaoka volcanic soil has been described by JGS, Hokkaido Branch (2011) and Kawamura and Miura (2013).

The authors have also elucidated hydraulic mechanical behavior of Komaoka volcanic soils subjected to seepage flow by using piping test apparatus and an X-ray CT scanner (Dao et al. 2016, 2017). The test results showed that destabilization of soil structures due to seepage flow is changed depending on an increase of amount of finer soil particles, and that internal erosion is induced by loss of fine particles with changes in void ratio. As a result, a significant variation in hydraulic conductivity was generated. Additionally, internal stability of volcanic soils under several conditions was elucidated by empirical criterion (e.g., Kedzi 1979; Kenny and Lau 1985). It was pointed out that volcanic coarse grained soil including pumice particles with a low specific gravity was internally unstable.

In this study, a series of seepage flow tests was performed on model embankments either subjected to pre-deformation due to cyclic loadings to or not. For the 1 g model testing, it has been assumed that the strength of soil varies proportionally by the normalization of the effective confining pressure in the embankment. Specifically, a conventional approach to element testing is taken in terms of the 1 g model testing (Kawamura and Miura 2013). Based on this technique, the mechanical behavior in the elements of the embankment for the 1 g model corresponds with those for the real field, if seepage speed in the model is consistent with those in the prototype. Owing to this, seepage flow which was thought to be present in the real field was directly applied to the embankment, and volcanic coarse grained soil (NP material) was adopted as test material. As a result, the similarity of time dependency regarding the development of pore water pressure and the consolidation phenomenon may not be required theoretically. Additionally, note that the pore water pressure value obtained from the model test can be similarly evaluated as a true value. Consequently, reproduction by model testing is theoretically enabled, although the assumption that a model and a natural physical phenomenon occur in the same mechanism is required. The correlation between model and prototype in this study is summarized in Table 2.

Table 2. Index properties of Komaoka volcanic soil

	Scale (Prototype/Model)
Length	λ
Stress ratio in element in slope	1
Pore water pressure ratio	1
Stress in element of slope	1
Deformation	λ

Figure 5 depicts a complete view of the apparatus used in the seepage flow tests, which is capable of changes in seepage flow with an increase of groundwater, and the shaking table test. The soil container was 2000 mm long, 700 mm deep, and 600 mm wide, and its front wall was made of reinforced glass to enable observation of

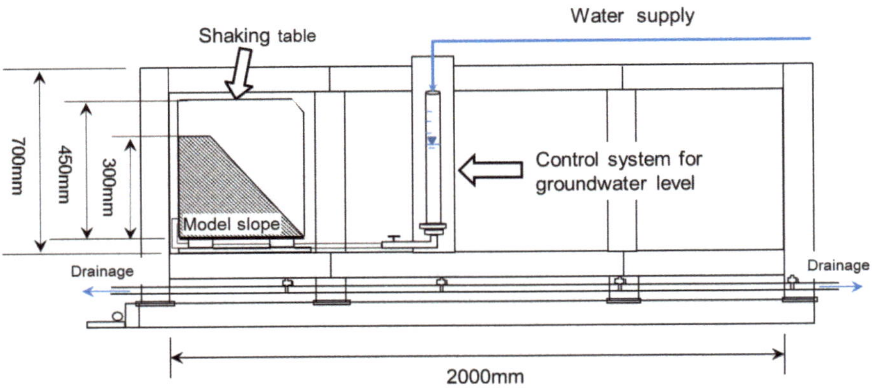

Fig. 5. Whole view of apparatus

deformation when failure occurred. In order to reveal the mechanical behavior of volcanic embankments subjected to pre-deformation due to cyclic loadings, a shaking table was also equipped in the soil container. The shaking table was 400 mm long, 450 mm deep, and 580 mm wide. The details of specifications were reported by Kawamura and Miura (2014).

Model embankments were constructed by compacting to attain the desired value (see Table 3, variation of single model in dry density is within 5%) where constituent particles were not broken by compaction under the initial water contents of w_0 = 37%. Figure 6 depicts compaction curves of the Komaoka volcanic soil which were obtained by the A-c method of JGS (2009) and by a tamping method for model tests. After construction of embankments, the surface was carefully cut to the angle of 45° (relative to horizontal) using a straight edge to eliminate surface disturbance.

Table 3. Test conditions for this study

Slope angle (°)	45
Length of base, B (mm)	400
Initial water content, w_0 (%)	37
Dry density ρ_d (g/cm^3) [degree of compaction D_c (%)]	0.90[85]
Acceleration (gal) [desired shear strain (%)]	200[1.0], 250[2.0], 280[3.0], 550 [4.0–6.0]
Number of cycles	20

For the cases of volcanic model embankments using a shaking table, after the desired shearing deformations (shear strain, γ = 1.0, 2.0, 3.0 and 4.0%) for model embankments of w_0 = 37% generated by cyclic loading of 200, 250, 280 or 550 gal, a series of seepage flow tests was conducted. The number of cycles was 20 times for all tests.

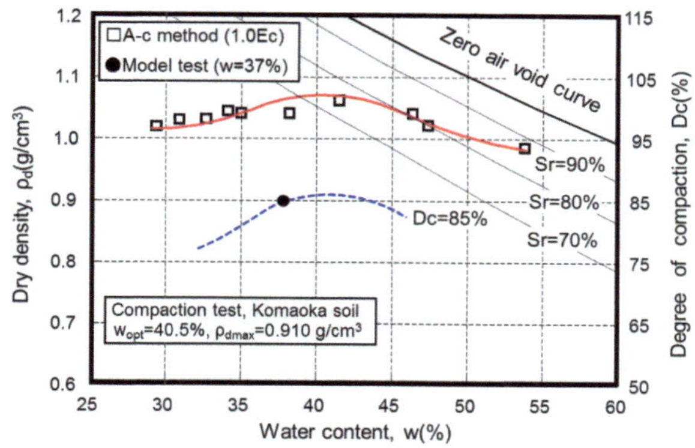

Fig. 6. Compaction curves of the Komaoka volcanic soil by A-c method of JGS (2009) and by a tamping method for model tests

For piping tests, variation in water head within the specimens was measured by three stand pipes at three different depths during upward seepage flow. In seepage flow tests with model embankments, a suitable hydraulic gradient and flow speed to simulate seepage flow was firstly determined based on the piping test results. After that, a series of seepage flow tests was continued until failure. Seepage flow was generated from the back side of model embankment through three pipes under a constant of hydraulic gradient (i = 2.0) and a flow speed (=1 * 10^{-3} cm/s), as described later. During the seepage tests, pore water pressure behavior, deformation behavior and variation in saturation degree were monitored. Figure 7 shows the slope shape (45° slope) and the setting positions of measurement devices.

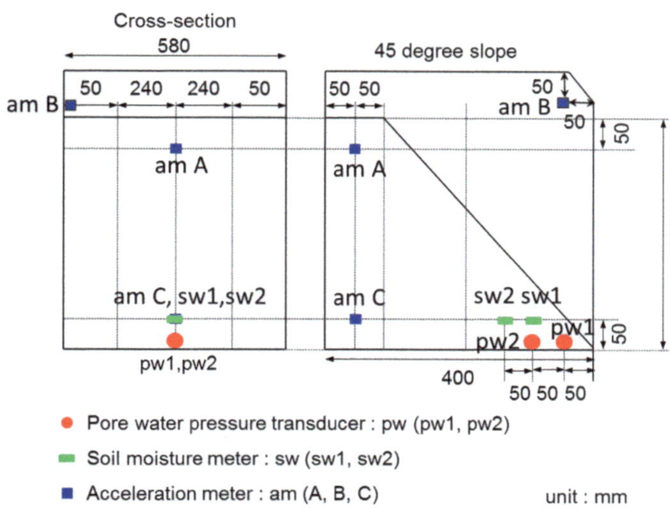

Fig. 7. Model shapes and locations of measurement devices

In this study, mechanical behavior at a shear strain γ' of 6% after seepage flow test, as estimated by the deformations of seven strings of a kite, was defined as failure. In the subsequent examination, pore water pressure normalized by over burden pressure was used as a parameter due to the conventional 1 g model testing approach.

4 Test Results and Discussions

4.1 Piping Phenomenon of Komaoka Volcanic Soil

Before commencing discussions on seepage flow-induced failure of model embankments, mechanisms of piping and boiling of the compacted Komaoka volcanic soils will be examined.

Figure 8 shows the relationship between hydraulic gradient and flow velocity of Komaoka volcanic soil with the different degree of compactions are Dc = 83, 88 and 95%. Piping phenomena in Komaoka volcanic soils occur in early stage (ranging from a hydraulic gradient $i = 0.1$ to $i = 0.2$). Fine particles were washed out by seepage flow. The relationships are also changed depending on the difference of the degree of compaction (see Fig. 8a, b). After reaching the critical hydraulic gradient ($i_c = 0.58$) based on the Terzaghi's equation, boiling phenomenon was observed in the specimen ($i = 0.9 - 1.4$), as shown in Fig. 8c. The difference between values by Terzaghi's equation and by seepage flow tests was caused due to the characteristics of void structures in volcanic coarse grained soils. The details of seepage behavior of volcanic coarse grained soils were presented by Dao et al. (2017).

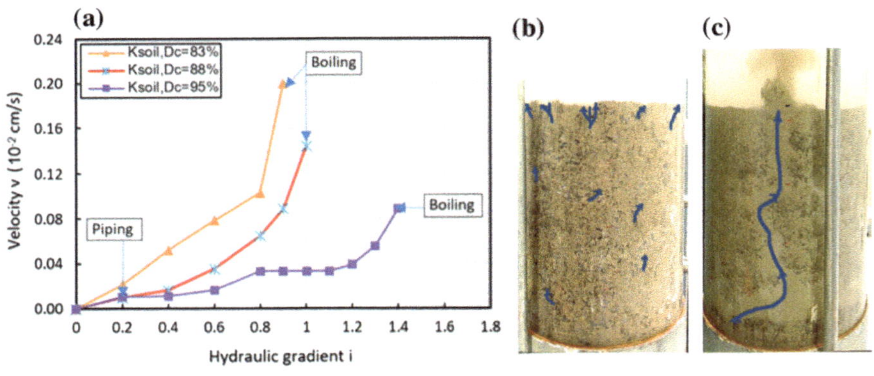

Fig. 8. Piping test results of Komaoka volcanic soils (K soil): **a** relationship between hydraulic gradient and flow velocity, **b** piping, **c** boiling (Dao et al. 2017)

In this study, so the purposes of this study to clarify the seepage flow-induced failure of embankment with pre-deformation, therefore, a hydraulic gradient $i = 2.0$ and a flow speed ($=1 * 10^{-3}$ cm/s) are adopted for the seepage model tests.

4.2 Seepage Flow Tests for Model Embankments with Pre-deformation Generated by Cyclic Loadings

Figure 9 illustrates typical variation of accelerations at basement and crown for initial water content $w_0 = 37\%$ ($\gamma = 1.0\%$ at 200 gal), compared with that of the shaking table. As shown in the figure, the maximum acceleration at the crown of embankment (A point) is larger than those at other points (B and C points). This reason is that the difference in rigidity due to the shape (geometric condition) of embankment affects the mechanical behavior during cyclic loadings. Many cracks on the surface were actually observed. Therefore, the crown of embankment might be more susceptible to damage due to cyclic loadings. A similar tendency was also obtained for other test cases for $w_0 = 37\%$.

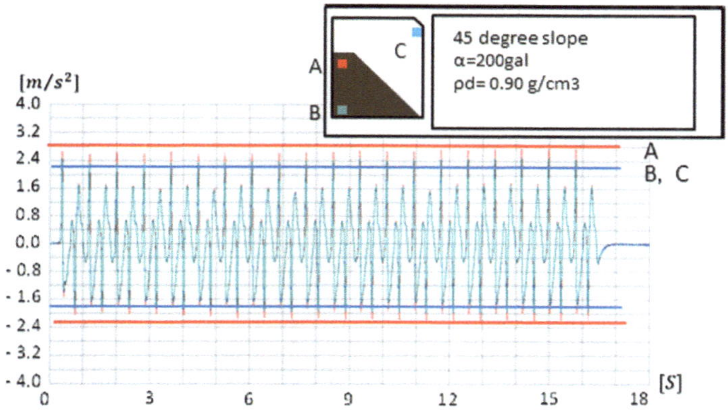

Fig. 9. Variation of accelerations at basement, crown of embankment and shaking table for $w_0 = 37\%$, 200 gal, $\gamma = 1.0\%$

Figure 10 shows the slope shape after the seepage flow tests with cyclic loadings and the behavior of pore water pressure and the changes in degree of saturation for $\gamma = 1.0\%$. From the figures, it is apparent that slope failure is generated at the toe of the embankment (a red line) with the increase of pore water pressure (see Fig. 10a, b). On the other hand, there was little change in degree of saturation (see Fig. 10c). As a result, the embankment failed at 493.5 s. Failure mechanism was the same as other test cases.

Figure 11 shows the comparisons of the developments of shear strain γ' during seepage flow tests around the toe of the embankments for the cases of $\gamma = 0$ and 1.0%. As can be seen in the figures, shear strain γ' during seepage model tests suddenly increases with seepage flow (with elapsed time) for both cases. This is due to the decrease of effective stress with the increase of pore water pressure, as aforementioned above. On the other hand, the elapsed time until failure for the case of $\gamma = 0\%$ (see Fig. 11a) was faster than that for the case of $\gamma = 1.0\%$ (see Fig. 11b). It is interesting that the time required to failure is changed depending on the magnitude of shear strain γ generated by cyclic loadings in advance. Similar tendencies were confirmed for

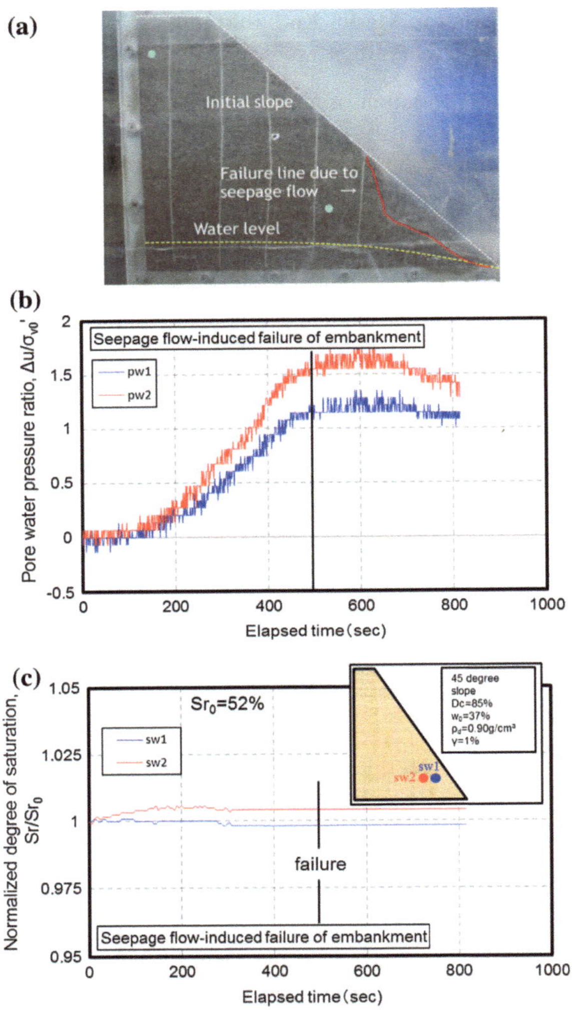

Fig. 10. Results of seepage flow tests with cyclic loadings for w_0 = 37%, γ = 1.0 % : **a** slope shape after seepage flow test, **b** behavior of pore water pressure, **c** changes in degree of saturation

model embankments subjected to seepage flow induced by wave erosion (Kawamura and Tsuchida 2017) or by rainfall (Kawamura et al. 2016). In addition, Kawamura et al. (2016) indicated that the degree of compaction D_c was increased by cyclic loadings for smaller range of shear strain (less than around γ = 2% for w_0 = 43%) in rainfall model tests. Therefore, it can be said that densification in embankment and rearrangement of soil particles due to cyclic loadings have an influence on the stability of embankments. As a result, it might be pointed out that compaction is useful as an improvement technique to reduce piping phenomenon of soil materials.

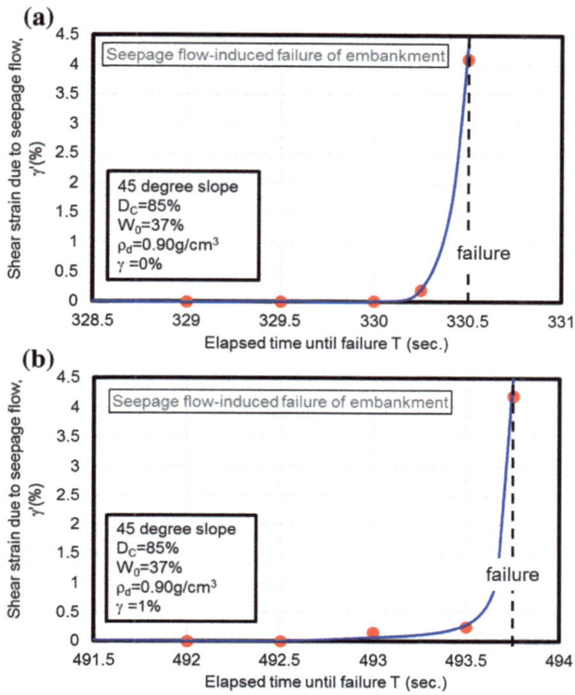

Fig. 11. Comparisons of developments of shear strain at neighborhood of toe of embankments: **a** $\gamma = 0\%$, **b** $\gamma = 1.0\%$

4.3 Stability Evaluation of Embankments Subjected to Various Seepage Flow

In previous studies, the stability of embankments has been discussed on other patterns of seepage flow. For example, Kawamura and Tsuchida (2017) has clarified wave-induced failure of volcanic embankments subjected to cyclic loadings. In particular, the effects of compaction conditions and wave conditions on failure of embankments were investigated. Similarly, Kawamura et al. (2016) elucidated the effect of cyclic loading history on rainfall-induced failure of embankments. In this study, the effect of difference in seepage patterns (due to wave erosion and rainfall and ground water) on slope failure was investigated under the same conditions of this model embankments.

Figure 12 summarized the relationship between shear strain γ due to cyclic loadings and elapsed time until failure normalized by that without cyclic loadings for each test. As shown in this figure, the data for wave-induced failure and rainfall-induced failure is decreasing with the increase of shear strain. In the case where seepage force acted directly on the surface having damage, the time decreases with the magnitude of the shear strain. On the other hand, the data in this study (for the infiltration of ground water from back side of embankment) is increasing until $\gamma = 1.0\%$ and then is decreasing. In particular, cyclic loading history up to $\gamma = 1.6\%$ indicates a good performance against slope failure. This is due to that the influence of the load history varies depending on the place in the embankment, as described in Fig. 9. Similar relationship was obtained in those at wetter side of compaction curve for wave or rainfall-induced failures, for example $w_0 = 43\%$ in this study (Kawamura et al. 2016; Kawamura and Tsuchida 2017).

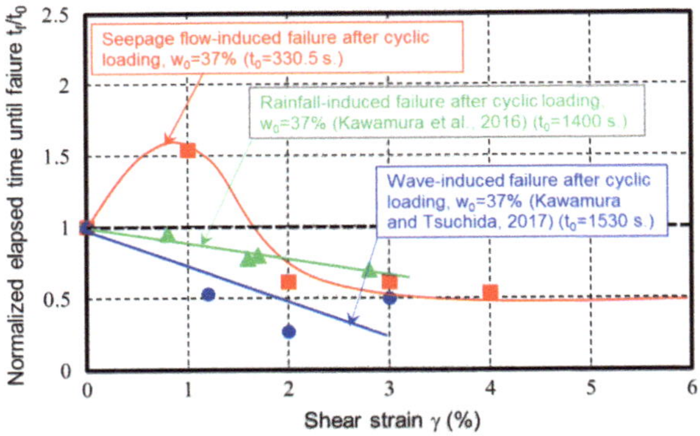

Fig. 12. Relationship between elapsed time until failure and shear strain γ due to cyclic loadings

Therefore, it will be important to evaluate the effect of loading history on failure mechanisms of embankment subjected to seepage flow. However, the above results may change with variations in water content, soil material and inherent error such as scale effect. In any case, further discussions in this direction will be required.

5 Conclusions

The conclusions obtained in this study were as follows;

(1) Piping phenomena in Komaoka volcanic soils occur in early stage (ranging from a hydraulic gradient $i = 0.1$ to $i = 0.2$). After the onset of internal erosion (piping phenomenon), movements of fine particles in pore network of particles are observed for Komaoka volcanic soils.
(2) The crown of embankment might be more susceptible to damage due to cyclic loadings. Many cracks on the surface were actually observed.
(3) Slope failure due to seepage flow from the back side of embankments was induced with the increase of pore water pressure at the toe of the embankment.
(4) Mechanical behavior of model embankments is changed depending on magnitudes of shear strain due to cyclic loadings. Therefore, the effect of pre-deformation due to cyclic loadings on seepage flow-induced failure is significant to discuss the stability evaluation of volcanic embankments.

Acknowledgements. The authors wish to express their sincere gratitude to Messrs. S. Horiike and S. Tanno (Muroran Institute of Technology, Japan) who conducted a major part of experiments. This study was undertaken with the financial supports of KAKENHI (Grant-in-Aid for Science Research (C) No. 15K06205), Japan Society for the Promotion Science.

References

Dao, M.H., Kawamura, S., Matsumura, S., Lan, Y.: Piping phenomenon of embankments constructed by volcanic soils and its evaluation. Geotechnical Special Publication, No.280, ASCE, pp. 688–698 (2016)

Dao, M.H., Kawamura, S., Matsumura, S.: Internal erosion of volcanic coarse grained soils and its evaluation. Int. J. Geomate **13**(38), 165–173 (2017)

Japanese Geotechnical Society, Hokkaido Branch.: Volcanic soils for engineers—characteristics, design, construction and disasters. The Research Committee in Hokkaido Branch, JGS, 1–130. (in Japanese) (2011)

Japanese Geotechnical Society.: Test method for soil compaction using a rammer (JGS 0711-2009). Standards of Japanese Geotechnical Society, JGS (2009). (in Japanese)

Japanese Geotechnical Society.: Reconnaissance report of the 2003 Tokachi-oki Earthquake. The Reconnaissance Committee of 2003 Tokachi-oki Earthquake, JGS (2004). (in Japanese)

Kawamura, S., Miura, S.: Rainfall-induced failures of volcanic slopes subjected to freezing and thawing. Soils Found. **53**(3), 443–461 (2013)

Kawamura, S., Miura, S.: Stability of volcanic slopes in cold regions. J. Geogr. Geol. **6**(3), 34–54 (2014)

Kawamura, S., Tsuchida, Y.: Wave-induced failure of volcanic embankments subjected to cyclic loadings such as seismic loadings and its evaluation. J. Coast. Eng. JSCE **73**(2/B2) (2017) (in Japanese) (in Press)

Kawamura, S., Miura, S., Matsumura, S.: Stability evaluation of full-scale embankment constructed by volcanic soil in cold regions. Japanese Geotechnical Society Special Publication, Vol. 2, No. 26, pp. 971–976 (2015)

Kawamura, S., Miura, S., Dao. M.H, Yamada, R.: Rainfall-induced failure of volcanic embankments subjected to cyclic loadings in cold regions. Geotechnical Special Publication, No. 257, ASCE, pp. 116–123 (2016)

Kenney, T.C., Lau, D.: Internal stability of Granular Filters. Can. Geotech. J. **22**(2), 215–225 (1985)

Kezdi, A.: Soil Physics-Selected Topics, p. 160p. Elsevier Scientific Publishing Co., Amsterdam (1979)

Matsumura, S.: Laboratory and in-situ studies on mechanical properties of volcanic soil embankment in cold region. Doctoral Dissertation, Graduate school of Engineering, Hokkaido University, Japan (2014). http://eprints.lib.hokudai.ac.jp/dspace/handle/2115/55883

In Situ Determination of Void Ratio and Compactness in Saturated Soils Using a Partially Automated Measuring System Based on Microwaves

Maren Katterbach[1(✉)] and Samuel Poretti[2]

[1] Lombardi Engineering Ltd, Minusio, Switzerland
maren.katterbach@lombardi.group
[2] SUPSI University, Manno, Switzerland
samuel.poretti@supsi.ch

Abstract. Void ratio and compactness are two of the principal parameters used for the geotechnical characterization of soils. Amongst others, they determine the deformation, settlement or liquefaction potential, which in turn might significantly impair the quality as well as the performance of any soil foundation. Direct measurement of the natural void ratio, i.e. the parameter typically used as indication for the soil compactness, is particularly difficult since it requires undisturbed samples. Because of the difficulty to obtain such samples in their undisturbed state, indirect methods such as: SPT, NSPT or CPT, to mention only the most commonly used ones, have been developed and are widely used. However, such indirect methods have their limitations and often fail, especially with heterogeneous soils also containing coarser materials. Additionally, results interpretation is done by indirect, empirical correlations, and is often affected by uncertainty. This situation gave rise to develop a more reliable and flexible approach, allowing a direct quantitative measure of the in situ void ratio of saturated soils. In this article a newly developed, partially automated, measuring system based on microwaves, is presented. The measurement method and an application example will be illustrated.

1 Introduction

One of the most important geotechnical properties of granular soils is their degree of compaction. This property exerts considerable influence on the bearing capacity, deformability characteristics, shear strength, and last but not least, permeability. It is expressed in terms of the void ratio, which is defined as the volumetric ratio of void spaces in a soil. As such, it has two limiting values, e_{min}, e_{max}, representing the densest and loosest state of the soil, respectively.

Given the void index of a defined soil, the ratio of the in situ void ratio to the abovementioned limit values, which are determined in the laboratory, is known as relative density, D_R. The relative density is the parameter that indicates the soil compaction level. Amongst others, this is a determining parameter assessing the risk of

B. Fatahi et al. (eds.), *Sustainable Design and Construction for Geomaterials and Geostructures*, Sustainable Civil Infrastructures,
https://doi.org/10.1007/978-3-319-95753-1_2

liquefaction of loose materials subjected to dynamic loads. The relative density is expressed by the relation:

$$D_R = \frac{e_{\max} - e}{e_{\max} - e_{\min}} \tag{1}$$

where,

e_{max} void ratio of the soil in its loosest state
e_{min} void ratio of the soil in its densest state
e in situ void ratio.

The relative density, D_r, varies between 0% for a very loose soil and 100% for a very dense soil. A corresponding classification of granular soils based on values of D_r as suggested by Lamb and Whitman (1969) is indicated in Table 1. Note that the relative density is a characteristic property of coarse grained soils having no fine particles, or just of insufficient quantity to impart any measureable cohesion to the soil.

Table 1. Classification of the relative density

Description	Relative density, D_r [%]
Very loose	0–15
Loose	15–35
Medium	35–65
Dense	65–85
Very dense	85–100

Accordingly, in order to solve Eq. 1 for determining the relative density of a soil, the three previously mentioned void ratios, i.e. e_{min}, e_{max} and e, should be known. While the first two parameters, i.e. the densest and loosest possible state are obtained by means of rather simple laboratory tests on representative samples of the corresponding soil, the determination of the void ratio of the soil in its natural state on the other hand is very challenging. This is especially true for the recovery of undisturbed soil samples from deep-lying strata. Even though appropriate equipment might permit obtaining undisturbed samples, such an expense would nevertheless not be justifiable with respect to the overall project schedule and costs. If a sufficient budget is available, the in situ void ratio can also be determined from laboratory analysis of frozen samples, or from neutron or gamma-ray density logging. There are also tables available that provide corresponding soil property correlations. However, these values should be referred to only in exceptional cases, and should be used only with a certain degree of caution.

This is the reason why the natural void ratio is typically estimated indirectly through empirical correlations with standard-penetration-test (SPT) and conventional cone-penetration-test (CPT) results.

The SPT test requires a sample tube with a hammer and provides information on the sleeve friction resistance. It is quite simple and inexpensive to perform. It is also the best solution for testing soil in areas that are difficult to access with vehicles. The SPT is therefore usually recommended as a quick, easy and low-cost soil testing solution. Its main drawback is that it is inaccurate, especially when sampling coarse sands, or clays. Furthermore, when gravels are encountered, the SPT results get unreliable and mostly unusable.

The CPT, which provides a result in terms of the tip resistance, is considered to be more reliable than the SPT, since it is less subject to errors. CPT is becoming increasingly more popular for site investigations and geotechnical designs. For many construction projects, it is common to use SPT for the preliminary soil investigation, whereas CPT is used for detailed soil investigations and construction quality controls. However, this method requires estimating the content of fines, which presents one of the weaknesses of the method, should no soil samples be available.

In summary, penetration resistance testing is generally subject to both operator and mechanical errors, which impairs results finding and as a consequence, limits the applicability of these methods to certain soil conditions. As such, they finally also affect the reliability of the overall geotechnical analysis based on these tests. In this context it is emphasized that, especially for the determination of the liquefaction risk for instance, proper knowledge of the relative density is fundamental.

2 Development of a Reliable Soil Compaction Test

The limitations of the currently available testing methods and the necessity for more reliable measurements arising from actual project experience, have called for the development of a new device. The main requirements are to have a user-friendly apparatus, which provides a quantitative measure of the natural, thus in situ void ratio. Such apparatus shall be applicable in a wider range of soil conditions than current instruments, and shall be easily integrated in the general project workflow, thereby not compromising it.

2.1 Description and Physical Background

The basic idea of the elaborated soil compaction testing is to determine the water content in saturated granular soil, and thereby to obtain a direct measure for the void ratio. As such, it represents a procedure already known to other industries (i.e. ceramics and cement), where it is used to control the moisture content during manufacturing processes.

The fundamental concept is based on electromagnetic waves of radio frequency, with which the average dielectric constant of the soil can be derived. With this dielectric constant being directly related to the soil water content the void ratio in a saturated soil is obtained immediately.

To verify the effective applicability of this concept for different soil conditions, as a first step, it was tested in how much the dielectric constant depends on the soil granulometry and humidity. Some of the test results for different soil granulometries in not saturated conditions are shown in Fig. 1. It appears that, despite of the different composition, particle size and grading, the initial dielectric constant is with a value of

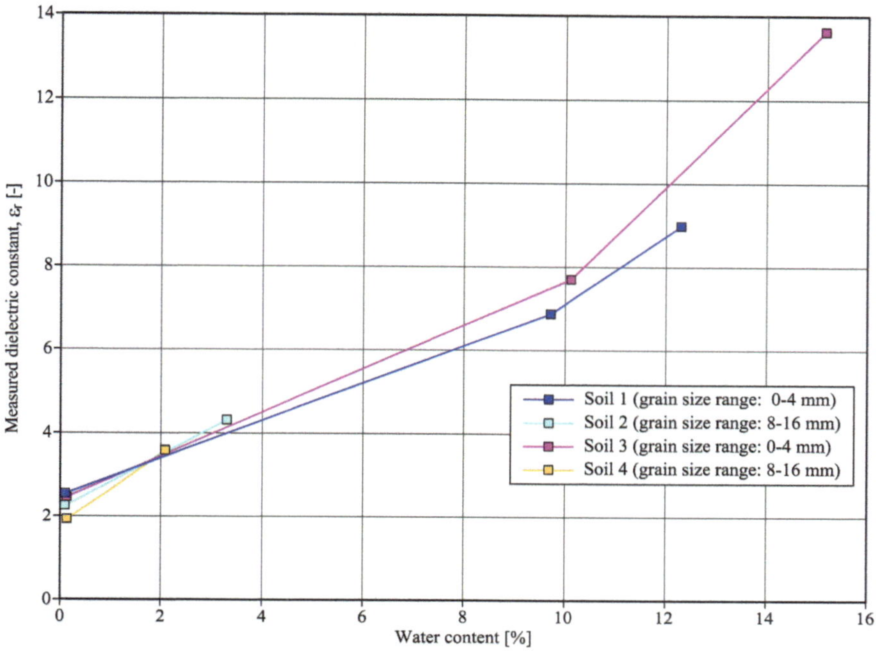

Fig. 1. Results of first laboratory tests. Percentage of water in function of measured dielectric constant for different soil granulometries in not saturated conditions

around 3 very similar for all samples. The same applies for the increase of the dielectric constant in function of the water content, which turns out to be similar for all samples.

Consequently, the average dielectric constant depends mainly on the volumetric ratio between water and soil particles. With the dielectric constant of water being about 30 times greater than that of the soil particles, the soil water content can thus be determined in a good approximation by evaluating the propagation speed, which is related to the dielectric constant by

$$v = \frac{c}{\sqrt{\varepsilon_r}} \tag{2}$$

where

c is the speed of light

ε_r is the relative dielectric constant.

The actual wave propagation speed is additionally affected by various delays caused by cables, antennas, and air that have to be taken into account. Knowing the distance between the sender and receiver, and calibrating the measurements with respect to the mentioned delays, the propagation velocity of the electromagnetic waves can be determined. Provided that the test is performed in saturated soil conditions, the electromagnetic signal propagates through two material types, i.e. water and soil.

Thus, the dielectric constant measured by the SCT device is a combination of these two values and can therefore directly be correlated with the soil-water content. Hereafter, the corresponding analytical conversion is briefly presented.

By knowing the values for the dielectric constant of dry soil, $\varepsilon_{\text{soil}}$, that of water, $\varepsilon_{\text{water}}$ (i.e. ~ 80) as well as the borehole distance, it is possible to determine the soil-water percentage. The analytical solution for the determination of the water content is:

$$\varepsilon_{measured} = \frac{(x_{soil} \cdot \varepsilon_{soil} + x_{water} \cdot \varepsilon_{water})}{d} \tag{3}$$

with d being the borehole distance, defined by

$$d = x_{soil} + x_{water} \tag{4}$$

and with x being the corresponding content of the soil and water, respectively, defined by

$$x_{soil} = \frac{d \cdot (\varepsilon_{measured} - \varepsilon_{water})}{\varepsilon_{soil} - \varepsilon_{water}} \tag{5}$$

$$x_{water} = d - x_{soil} \tag{6}$$

expressed in percentage:

$$\%_{water} = \frac{x_{water} \cdot 100}{d} \tag{7}$$

$$\%_{soil} = 100 - \%_{water} \tag{8}$$

Consequently, as given by Eq. (3) a simple linear relation between the mean dielectric constant, $\varepsilon_{\text{measured}}$, and the water content can be considered. This last-mentioned equation presents the analytical description of the measured dielectric constant, according to which the dielectric constant of dry soil can be considered the same for different soil compositions.

2.2 Prototype and Pilot Tests

With the objective of determining the still unknown dielectric constant of dry soil, $\varepsilon_{\text{soil}}$, required for Eq. (3), as well as to test the functionality of the new measurement system, a series of pilot tests were performed. To do so, tests using a box filled with different soil types were prepared. As shown in Fig. 2, the emitter and receiver were simultaneously lowered within PVC tubes at a distance of 0.76 m. The actual water content was determined by oven-drying test samples taken from the box. These values were then compared with the water content measured by the SCT instrument.

These previous considerations and tests showed that the developed instrument is generally applicable and can be used for the following two measurements:

Fig. 2. Laboratory tests. Left: set-up with holes at 67 cm distance in sand. Right: same box filled with gravel and coarse material

(1) Dielectric constant and water content for non-saturated soils, respectively,
(2) Percentage of water content in saturated soils in order to determine the soil compactness.

With the pilot tests being performed in controlled conditions regarding soil composition and set up, it was also possible to quantify the instrumentation accuracy. Accordingly, the water content can be determined with a precision of about 5%.

After proving the functionality of the measuring system, a first prototype version for in situ tests was created. A schematic illustration of this device is shown in Fig. 3. It

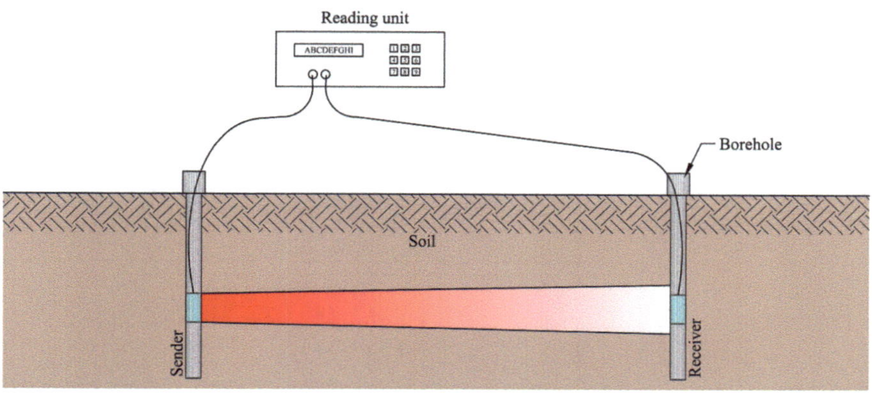

Fig. 3. Schematic illustration of measuring device and test configuration

consists of a portable main unit and two sensors, i.e. a sender and a receiver, respectively. The equipment can be placed in boreholes down to 75 mm (3″) in diameter. For a reasonable and economically feasible device capacity, which in turn determines the signal strength, a testing range distance between the boreholes of up to 20 m can be reached. The tests are carried out in common PVC tubes or similar, as steel casings would impede signal propagation, rendering tests impossible. As mentioned above, the medium to be tested should be a saturated granular soil—only in this way can the assumption of a two phase medium rightly be applied. This applies for example for most dam foundations, with a major part being below the ground water table.

Before starting with the main measurement, an air-calibration is required. To do so, the sensors are located on the surface, right above the two boreholes to be tested, and a single measurement is done. In this way the signal passes exclusively through the air medium—an important constellation which is required for the automatic post-processing. Subsequently, the location of the ground water level is determined. This is necessary to ensure that the readings start in the saturated conditions. The sensors are then stepwise lowered in the two boreholes, while for each depth interval a corresponding measurement is performed. The increment size is selected based on the necessary detail and accuracy required for the project. In most of the cases a range of 0.5 and 1 m appears reasonable. The readings are then performed along the entire depth of the borehole. For a consistency check, it is recommended to measure every borehole couple in both directions, i.e. by alternatively switching the sender and receiver in the holes. Since the signal attenuation and delay is independent of the travel direction, the inverted readings should correspond.

After a general control of the measurements, the data is processed and archived. A specially developed PC-tool allows the user to read off the required information directly. An intuitive user-interface provides the measurement results, in the form of a dielectric constant profile over the measurement depth and an estimation of the water percentage (bar diagram) based on the previously explained analytical method. A screen shot of the user interface with an example measurement is shown in Fig. 4.

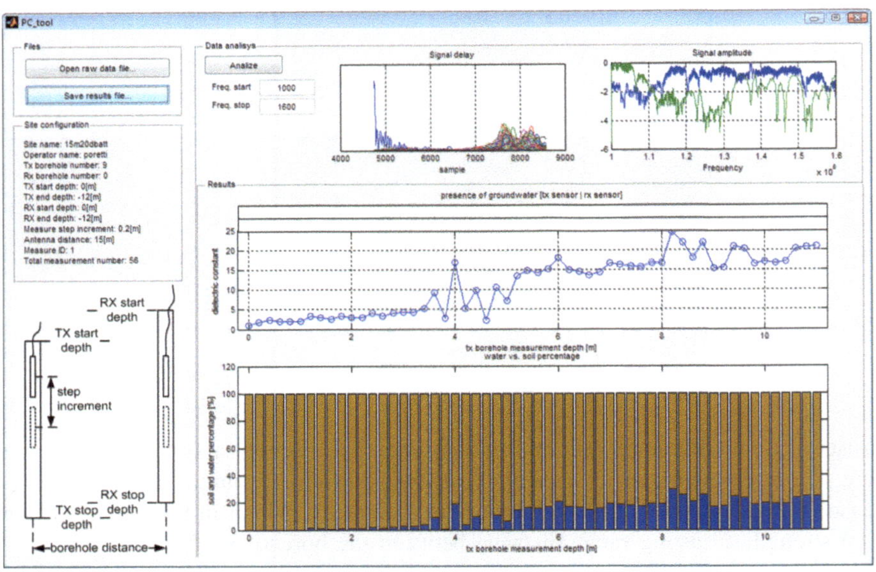

Fig. 4. Screen capture with example of measurement in a 12 m deep borehole

Further parameters, such as the signal delay are also illustrated. They give the user the possibility to perform verifications and to confirm the correctness, consistency and repeatability of the measurements.

3 Application of New Soil Compaction Test

After various laboratory tests as well as on specially prepared field sites, the general functionality of the SCT device was verified on real sites. The main objectives of these tests were to (1) verify again the general functionality of the device (2) have a comparison with conventional measuring devices and (3) to optimize the general system operation, accounting for realistic site conditions.

3.1 Site Description, Need and Execution of Tests

To do so the Nenskra HPP construction site in northwestern Georgia was chosen. In order to ensure the safety of the 130 m high rockfill dam, a careful evaluation of the foundation characteristics was of great importance already at the early design stage. The valley of glacial origin is filled with glacial and alluvial soils, reaching down to around 170 m at the dam site. These soils comprise a wide range of grain sizes, including silts, sands, as well as gravels up to several decimeters in diameter, all the way up to big boulders. In Fig. 5, an example of a typical borehole log indicating lithology, permeabilities as well as photos of the core boxes obtained from the dam foundation are presented. With these soil conditions, special attention was given to the assessment of expected settlements, as well as the degree of compactness in general.

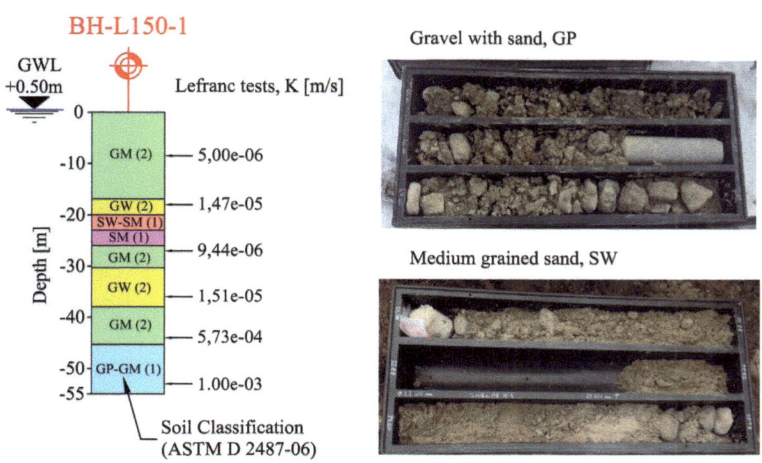

Fig. 5. Left: typical borehole log indicating lithology and permeabilities. Right: photos of the core boxes obtained from the same part of the dam foundation

It turns out that due to the presence of coarser material and boulders, the initially performed penetration resistance tests (SPT) proved to be unusable for any further elaboration. The SPT revealed mostly blow counts indicating a withdrawal already at a shallow depth once coarser material was encountered. Given this situation, measurements by means of the new SCT device were performed in order to provide a better and complete data basis as required to evaluate the in situ soil compactness. Considering the importance of accurate information especially in the central foundation part of the future dam, the SCT campaign focused on the available, accessible boreholes in this part. The exemplary boreholes presented here BH-L-150-1, BH-L-150-1a and BH-L-150b have a distance of between 8 and 15 m. The foundation in this part is characterized by artesian ground water conditions, so that generally saturated soil conditions could be assumed already at shallow depth. With these conditions, it was therefore possible to perform the measurements basically from the borehole mouth down to a depth of 35–40 m below the surface. In view of the required accuracy and resolution of the void ratio profile with depth, the sensors were lowered in increments of 1 m. For consistency, every borehole couple was measured in both directions, exchanging both antennas in the holes. The general test set up is shown in the photographs (Fig. 6).

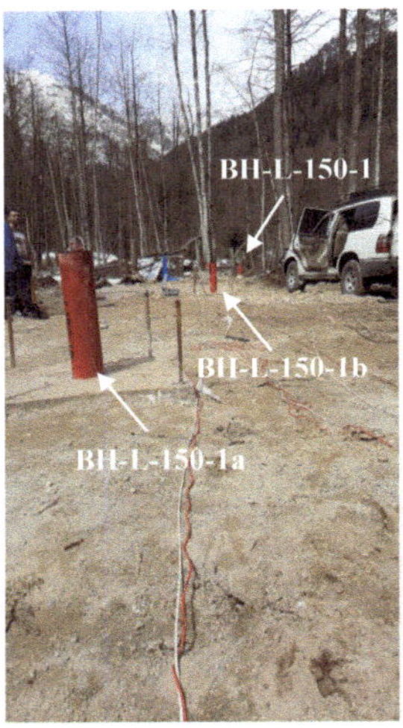

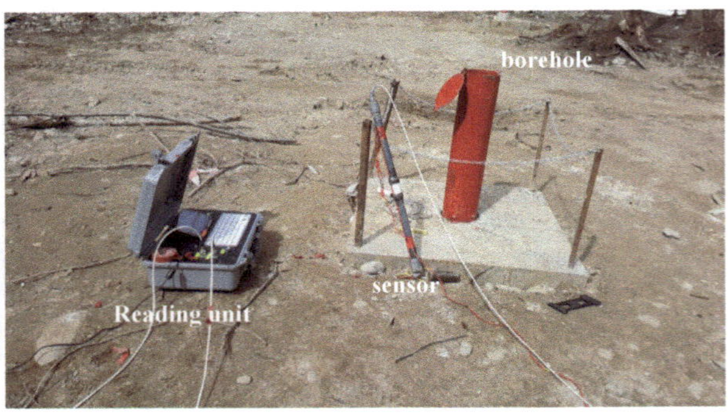

Fig. 6. Photo of SCT campaign in BH-L-150-1, BH-L-150-1a and BH-L-150-1b

3.2 Results of Site Tests

The natural void ratio and water content can be directly read of the SCT PC user-interface. For the present case, special interest was in the relative density values of the foundation soil. This information was essential in order to evaluate as exactly as possible the expected foundation behavior once being exposed to the effective project loads. To create a corresponding relative density profile, the available information from the core logs was used to get the void ratio of the foundation soil in its loosest state,

e_{max} on one hand, and the void ratio of the soil in its densest state, e_{min}, on the other hand. Using Eq. 1, these values were then combined with the in situ void ratio, e, obtained by means of the SCT device to calculate the relative density. The obtained density profile is shown in Fig. 7.

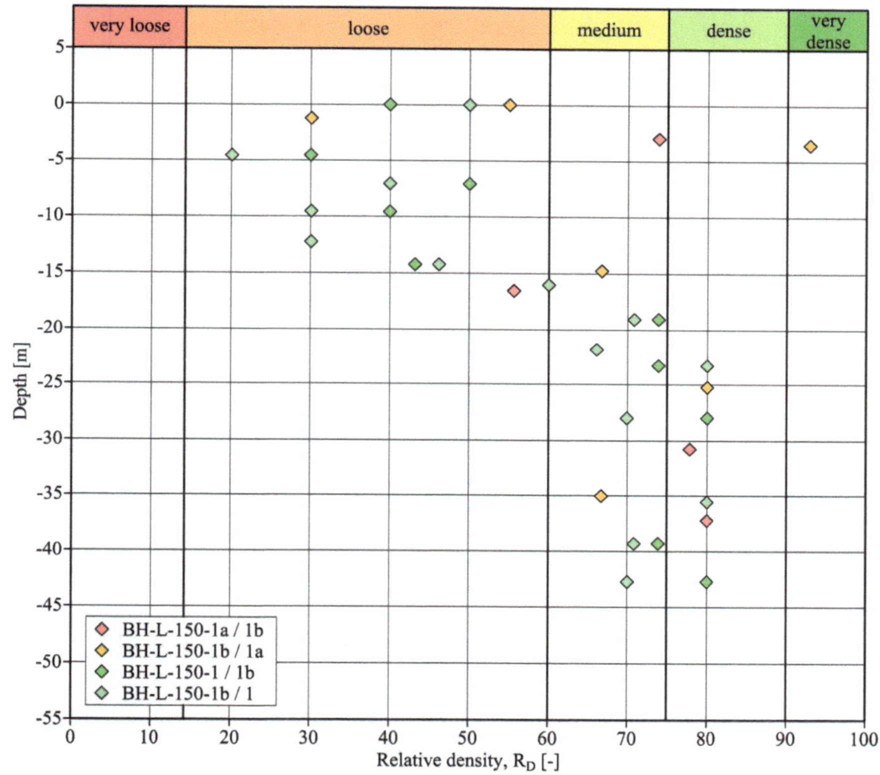

Fig. 7. Relative density in function of depth obtained by means of SCT device (boreholes BH-L-150-1, BH-L-150-1a and BH-L-150-1b)

For comparison in Fig. 8, the corresponding results of the standard penetration tests in the same study area are shown. SPT values which, due to the presence of boulders, were out of range (i.e. blows >50) are not shown. Due to the large scattering of the SPT results it was not possible to determine a reasonable range of the relative density without further ado. It is at this point noted that, especially in this kind of inhomogeneous soil with the presence of larger boulders, this difficulty is often experienced. This is why it is generally indispensable to combine SPT results with other tests (e.g. geophysical tests) for an appropriate geotechnical characterization. Comparing the results of the penetration resistance testing (SPT, Fig. 8) with those obtained by electromagnetic wave technology (SCT, Fig. 7), it can be clearly observed that using the newly developed SCT device the results show significantly less scattering. They

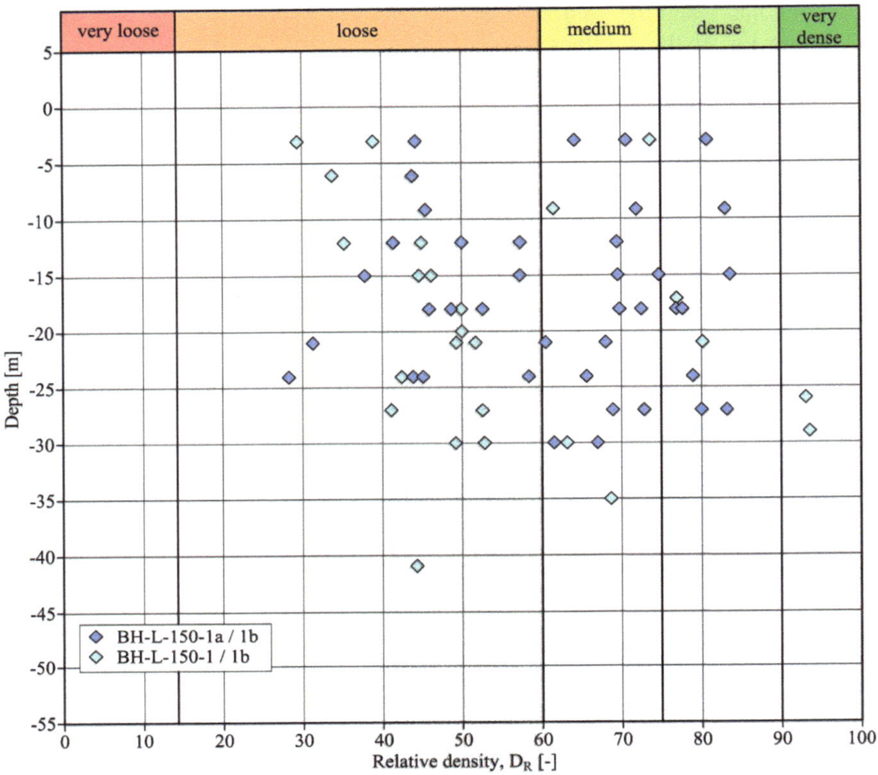

Fig. 8. Relative density in function of depth obtained by means of SPT

thus provide a more accurate insight into the relative density distribution with depth. The upper 10 to 15 m are generally loose to medium dense, while, as expected, the compactness increases with depth. The locally notable decrease is due to the presence of coarser material. The rather high density observed between 3 and 4 m depth can be explained by the presence of very fine grained lacustrine deposits. These findings are in line with the information of the core boxes.

Overall, it can be concluded that the developed SCT device permitted to obtain a complete relative density profile and with it a useful data basis for further engineering evaluations. The measurement campaign using the new device provided continuous and complete information over the entire soil profile under investigation, without any information gap, in contrast to the previously performed SPT, that in most cases failed. It thus provided a key basis for a safe and sound design of the future dam and its support structures.

4 Conclusions

In order to assess the quality and performance of any soil foundation, proper knowledge of the void ratio and compactness is essential. Due to the difficulties associated with obtaining undisturbed soil samples for direct measurements, indirect methods such as SPT, NSPT or CPT are usually used. However, these methods have their limitations and often fail, especially with heterogeneous soils also containing coarser materials. To overcome this limitation, a new device based on microwave technology, referred to as SCT (Soil Compaction Testing) has been developed and tested. By means of the dielectric constant, this method permits to determine both the water content percentage of unsaturated soils (water-soil ratio), and also the void ratio and compactness of saturated soils. As such, the basic concept here is similar to the commonly used cross-holes georadar. The main difference to this widely-known geophysical testing method is the measurement scale, and accuracy of the new SCT device, which have been specifically defined to satisfy typical civil engineering requirements, and to comply with realistic testing conditions. Measurements can be performed between boreholes with a distance of up to 20–25 m. It is possible to determine the water content with an accuracy of about 5%. The measuring device itself is user-friendly, comprising an easy-to-read screen and PC device, with which measurements can be rapidly and conveniently taken, which are also less prone to errors. Moreover, complex pre-/post-processing is not required.

Various tests on actual construction sites performed so far have already proven, not only the general functionality of the SCT device, but also its higher reliability and increased feasibility with respect to common friction based testing methods. This paper presents one simple application example i.e. at the foundation of a 130 m high rock fill dam. By means of the SCT, it was possible to easily and rapidly obtain a sound data basis for further geotechnical assessments such as settlements, without impeding the construction process.

References

Lamb, T.W., Whitman, R.V.: Soil Mechanics. Wiley, New York, Chichester, Brisbane, Toronto, Singapore (1969)

Shroff, A.V., Shah, L.D.: Soil Mechanics and Geotechnical Engineering. A. A. Balkema Publishers, India (2003)

Chen, Q. et al.: The simplified model of soil dielectric constant and soil moisture at the main frequency points of microwave band. In: Geoscience and Remote Sensing Symposium (IGARSS), 2013 IEEE International (2013)

Snowden, D.P., Morris Jr., G.C., van Lint, V.A.J.: Measurement of the dielectric constant of soil. IEEE Trans. Nucl. Sci. **32**(6), 4312–4314 (1985)

University of Applied Sciences and Arts of Southern Switzerland, Telecom-Telemetry-and-High-Frequency Laboratory.: Measurements, Specifications, and device description, (unreleased), Switzerland (2013)

Rapid Compression Test of Cement-treated Sands

Tsutomu Namikawa(✉) and Hayato Usui

Shibaura Institute of Technology, Tokyo, Japan
namikawa@shibaura-it.co.jp

Abstract. During earthquakes, dynamic loads caused by inertia forces are applied to ground improvements by deep mixing method. Therefore, the internal stability of cement-treated soil ground subjected to the seismic loading should be examined in the seismic design of the ground improvement. The speed of the loading caused by the inertia forces during earthquakes is much faster than that in the standard unconfined compression test. It is well known that the loading speed affects the strength of materials. Therefore, the strength under the rapid loading conditions is required to design the ground improvements rationally. The unconfined compression tests of cement-treated soils were conducted under the rapid loading conditions in this study. The rapid load in the shape of a solitary wave is applied to the specimens. The unconfined compression tests with the strain rate of 1%/min were also conducted for comparison. The experimental result indicates that the rapid loading increases the strength by about 20% as compared with the strain controlled loading. The test results were analyzed with the statistical manner to evaluate the variation of the strength of the cement-treated soil specimens prepared in a laboratory. The statistical analysis provides the good understanding of the influence of the data variation on the relationship between the strengths obtained from the rapid loading and strain controlled tests.

1 Introduction

Ground improvements by deep mixing method have been widely used for structure foundations and liquefaction mitigation. During an earthquake, inertia force corresponding to seismic acceleration is induced in ground and stresses due to the inertia force are applied to the ground improvement. Therefore, the internal stability of ground improvement subjected to the seismic loading should be examined in the design of the ground improvements. In current design procedures, unconfined compressive strength of cement-treated soils is adopted for assessing the internal stability of the ground improvements. The strength is normally obtained from unconfined compression test in which the strain rate is controlled. The speed of the loading caused by the inertia forces during earthquakes is much faster than that in the standard unconfined compression test. The loading speed is likely to affect the strength of cement-treated soil. Therefore, the strength under the rapid loading conditions is required to design the ground improvements rationally.

B. Fatahi et al. (eds.), *Sustainable Design and Construction for Geomaterials and Geostructures*, Sustainable Civil Infrastructures,
https://doi.org/10.1007/978-3-319-95753-1_3

It is well known that the loading speed affects the strength of cemented materials. For concrete, several experimental studies have been conducted to investigate the effect of loading rate on the mechanical behavior. The experimental results of past work indicate that the compression strength of concrete increases with strain rate (Bischoff and Perry 1991). Conversely, there are limited studies on the effect of loading rate on the behavior of cement-treated soils. Kongsukprasert and Tatsuoka (2005) conducted a series of drained triaxial compression tests to investigate aging and viscus effects on the strength characteristics of cement-mixed gravelly soils. Noticeable loading rate effects due to viscous property were observed in their test results. Onimaru et al. (2012) conducted a series of undrained compression triaxial tests with different axial strain rates. Their experimental results showed that the strength increases with the strain rate. However, there is little study for investigating the behavior of cement-treated soils under high rate loading induced by earthquakes.

This paper presents the experimental results for examining the strength of cement-treated soils under rapid loading conditions. A series of unconfined compression tests were conducted under the rapid loading conditions. The rapid load in the shape of a solitary wave was applied to cement-treated soil specimens. The loading frequencies are 1 and 2 Hz. The standard unconfined compression tests with the strain rate of 1%/min were also conducted for comparison. The test results are analyzed in the statistical manner to evaluate the variation of the strength of the cement-treated soil specimens prepared in a laboratory. The statistical analysis provides the good understanding of the influence of the data variation on the relationship between the strengths obtained from the rapid loading and strain controlled tests.

2 Experimental Program

Rapid unconfined compression tests were conducted on the cement-treated sand specimens prepared in a laboratory. Standard unconfined compression tests, in which a strain rate is controlled, were also conducted for comparison. The mixing properties of the cement-treated sands used in the tests are described in Table 1. This table shows the weight ratio of the materials composing the specimen.

Table 1. Mixing properties of cement-treated soil

Case	Water	Cement	Sand	Kaolin
1	18.6	5.0	70.9	5.5
2	18.6	6.0	69.9	5.5

Note Weight (%)

Two kinds of mixing property are prepared in this study. All the materials were mixed thoroughly and compacted in a mold by vibration. Thereafter the specimen was cured under wet condition. The curing time is 7 days. A cylindrical specimen of 50 mm in diameter and 100 mm in height is used in the tests. Both the top and bottom ends of the specimen were capped with gypsum. The load in the vertical direction is measured by an inner load cell that is free from the effects of friction at the bearing. The displacement of the loading shaft is measured by a displacement transducer.

The apparatus used for the rapid compression tests is shown in Fig. 1. The rapid axial load is applied by an air-activated Bellofram piston. The schematic picture of the applied stress is shown in Fig. 2. The axial stress of 20 kPa was initially applied to a specimen. Thereafter the rapid load in the shape of a solitary wave was applied to the specimens. Considering the loading during earthquakes, the loading frequencies were selected to be 1 and 2 Hz for Case 1, and 1 Hz for Case 2. In the strain controlled tests, the strain rate is 1%/min, that is, 0.0167%/s. 20 specimens were tested under each condition in the rapid loading and strain controlled tests.

Fig. 1. Rapid loading test apparatus

3 Experimental Results

3.1 Stress–Strain Relationship

Examples of the stress–strain relationship for Case 1 are shown in Fig. 3. This figure shows the comparison between the rapid loading (1 and 2 Hz) and the strain controlled (1%/min) test results. The strain rates calculated from the rapid loading test results shown in Fig. 3 are shown in Fig. 4.

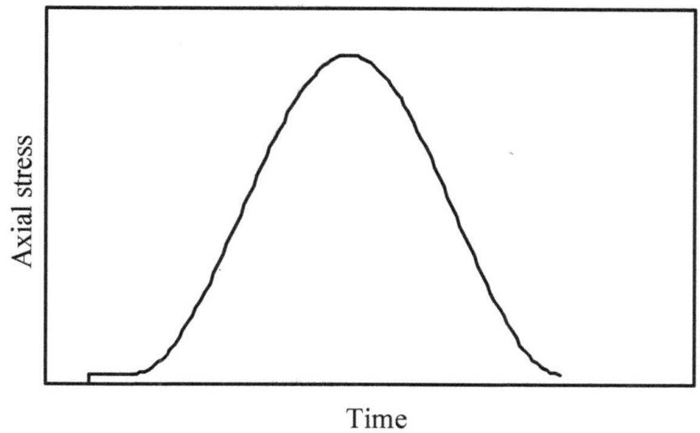

Fig. 2. Schematic picture of rapid load

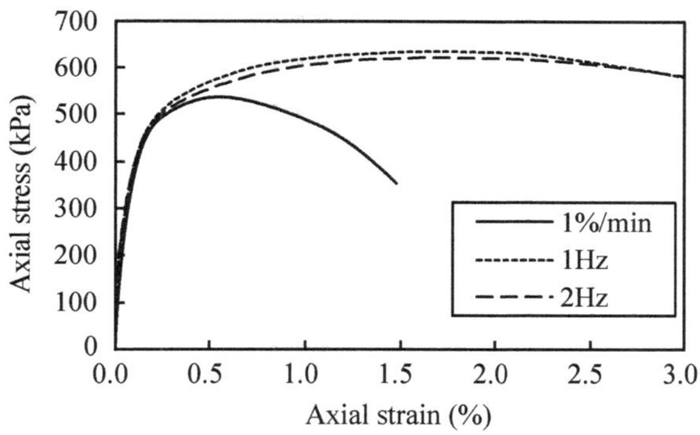

Fig. 3. Typical stress–strain relationship for Case 1

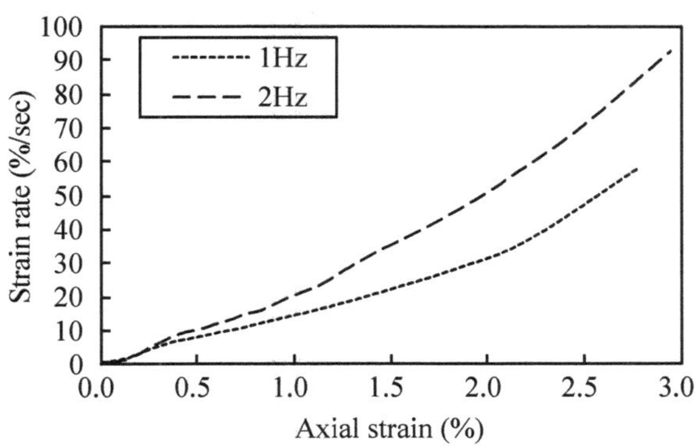

Fig. 4. Axial strain rate in Fig. 3

It can be seen that the stress–strain relationship of the rapid loading tests agree reasonably with that of the strain controlled test until the axial stress reaches about 450 kPa (i.e., about 85% of the peak stress in the strain controlled test). At the axial stress of 450 kPa, the axial strain is about 0.15% and the strain rate is about 2%/s. This strain rate is about 100 times larger than that in the strain controlled test in which the strain rate is 0.0167%/s. The agreement between the stress–strain relationships implies that the loading rate does not considerably affect the stress–strain relationship during the initial loading process. The difference between the rapid loading and the strain controlled test results becomes significant around the peak stress states. The maximum stress in the rapid loading tests is larger than that in the strain controlled test. Moreover, the strain at the peak stress state in the rapid loading tests is larger than that in the strain controlled test. It seems that there is not a significant difference between the stress–strain relationships at 1 and 2 Hz, indicating that the two times difference in frequency does not affect the unconfined compression behavior of the cement-treated soil.

Examples of the stress–strain relationship for Case 2 are shown in Fig. 5. As the results for Case 1, the stress–strain relationship of the rapid loading test agrees reasonably with that of the strain controlled test during the initial loading process. The difference between both the results becomes significant around the peak stress state. The strain at the peak stress state in the rapid loading test is larger than that in the strain controlled test.

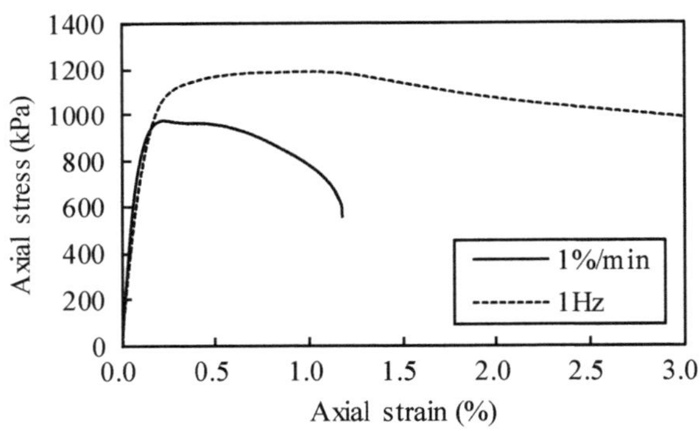

Fig. 5. Typical stress–strain relationship for Case 2

3.2 Strength

The strength obtained from the rapid loading and strain controlled tests is shown in Fig. 6. In both the cases, the strength in the rapid loading is larger by about 20% than that in the strain controlled tests. There is little difference between the strengths at 1 and 2 Hz. In all the cases, the strength varies to some extent.

The influence of the strain rate on the strength is shown in Fig. 7 for Case 1. The strain rate is calculated at the peak stress in the rapid loading tests. Figure 7 shows that the strain rate at the peak stress does not affect the strength in the rapid loading tests.

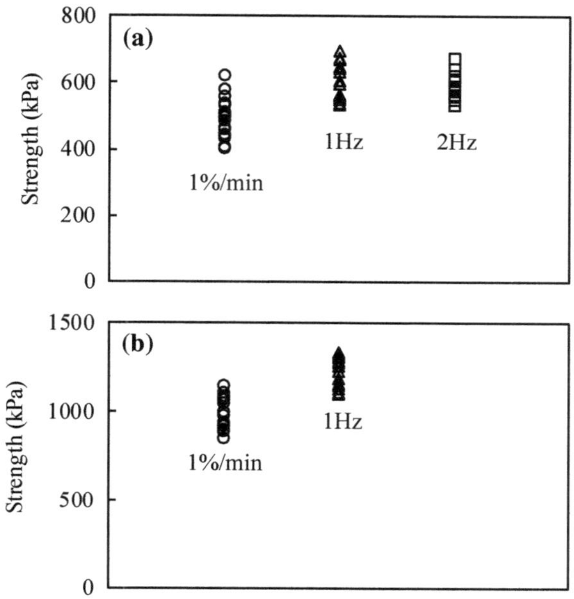

Fig. 6. Summary of strength: **a** Case 1 and **b** Case 2

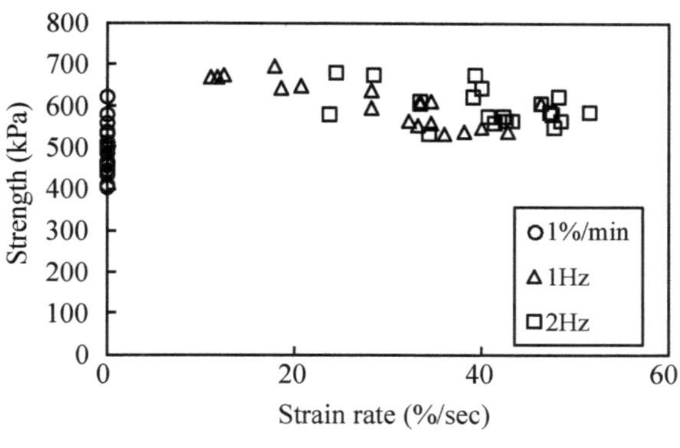

Fig. 7. Strength versus strain rate for Case 1

4 Statistical Analysis of Strength

The experimental results show that there is some variance of the strength of the specimens prepared in the laboratory. The statistical analysis is required for evaluating properly the relationship between the strengths obtained under the rapid and the strain controlled loading conditions.

Summary of the statistical parameters of the strength is shown in Table 2. The mean (m_{qu}) and the standard deviation (s_{qu}) of the strength are calculated from 20 test results for each case. The coefficient of variation V_{qu} is also calculated by dividing s_{qu} by m_{qu}. The values of V_{qu} in the strain loading test are 0.115 and 0.082 for Case 1 and Case 2, respectively. These values of V_{qu} indicate that the specimen strength varies to some extent, even though the specimen is prepared in the same manner in the laboratory.

Table 2. Summary of strength

Case	Loading rate or frequency	n	m_{qu} (kPa)	s_{qu} (kPa)	V_{qu}	$S_{qu}/\sqrt{n}$ (kPa)
1	1%/min	20	494	56.6	0.115	12.7
	1 Hz	20	603	51.9	0.086	11.6
	2 Hz	20	598	42.8	0.072	9.57
2	1%/min	20	986	80.5	0.082	18.0
	1 Hz	20	1223	88.0	0.072	19.7

Note n, number of specimens; m_{qu}, mean of strength; s_{qu}, standard deviation of strength, V_{qu}, coefficient of variation in strength, $S_{qu}/\sqrt{n}$, standard deviation of mean of strength

These values are point estimates for the mean and standard deviation values; no indication is provided as to how closely the sample values estimate the parameters. A more meaningful procedure for estimating the parameters of random variables involves the estimation of the population mean μ_{qu} and standard deviation σ_{qu} with a known degree of uncertainty. The uncertainty of the parameters will be estimated from the sample mean and standard deviation involving errors.

4.1 On Assumption of Normal Distribution

In order to evaluate the degree of uncertainty, the strength is assumed to follow the normal distribution. Then the relationship between m_{qu} and μ_{qu} is given by

$$t = \frac{m_{qu} - \mu_{qu}}{s_{qu}/\sqrt{n}} \tag{1}$$

where n is the number of specimens and t has a Student t distribution with $(n-1)$ degree of freedom. In Eq. (1), $s_{qu}/\sqrt{n}$ denotes the standard deviation of m_{qu}. The values of $s_{qu}/\sqrt{n}$ are shown in Table 2.

The confidence interval for μ_{qu} is calculated from this equation. The 90% confidence intervals of μ_{qu} are plotted in Fig. 8 together with the m_{qu} values. These intervals show a degree of uncertainty of the sample mean value of the strengths of 20 specimens. Figure 8 shows that although some degree of uncertainty is yielded in the estimation of the mean strength, the mean strength obtained from the rapid loading tests is clearly larger than that from the strain controlled tests.

Assuming the strength follows the normal distribution, the sampling distribution of the sample variance s_{qu}^2 is given by

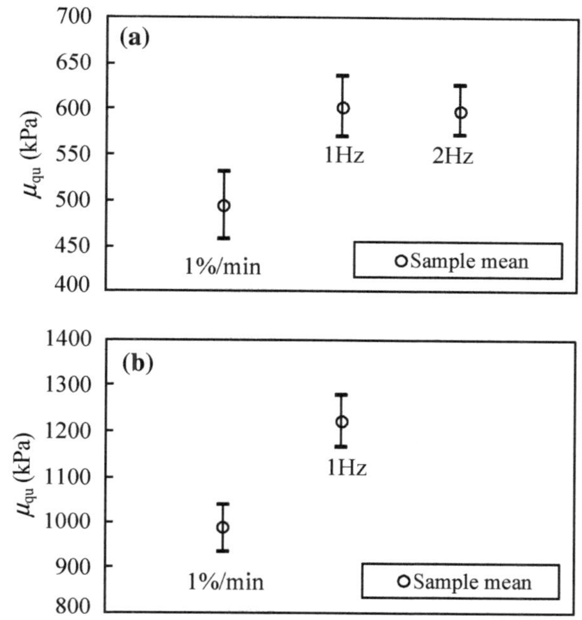

Fig. 8. 90% confidence interval of mean of strength: **a** Case 1 and **b** Case 2

$$\chi^2 = (n-1)s_{qu}^2 \Big/ \sigma_{qu}^2 \quad (2)$$

where χ^2 has a chi-square distribution with $(n - 1)$ degree of freedom. The confidence interval for the variance is calculated from this equation. The 90% confidence intervals of σ_{qu} are plotted in Fig. 9 together with the s_{qu} values. The confidence intervals of σ_{qu} for the rapid loading test agree reasonably with that for the strain controlled test, implying that the variability of the strength in the rapid loading test results is mainly attributed to the variability of the specimen strength. However, the covariance of these strengths cannot be examined based on the experimental results in this study. Further study is required to investigate the covariance between the static and dynamic strengths of the cement-treated soil specimen prepared in the same manner.

4.2 Bootstrap Method

While the normal distribution was assumed for the underlying distribution of the strength in the previous section, the distribution of the parameters is estimated without the assumption of the underlying distribution in this section. Using the bootstrap method (Efron 1979), the distribution of the mean of the strength is evaluated without having to a distributional form for the strength. In the bootstrap method, the data is resampled with the manner of Monte Carlo simulation. The bootstrap sample is obtained by sampling with replacement from the empirical distribution function $\hat{F}(x)$ defined as

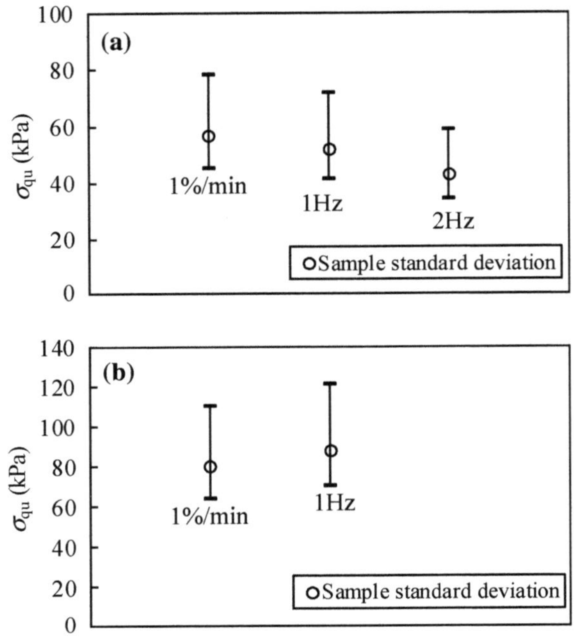

Fig. 9. 90% confidence interval of standard deviation of strength: **a** Case 1 and **b** Case 2

$$\hat{F}(x) = \frac{1}{n}\sum_{i=1}^{n} I(x_i \le x), \quad I(x_i \le x) = \begin{cases} 1 & x_i \le x \\ 0 & x_i > x \end{cases} \tag{3}$$

x_i is the strength of the 20 specimens in each case. A random sample of size n is drawn from $\hat{F}(x)$. The estimate of the mean $\hat{\mu}^*_{qu}$ is calculated from the sampled data. This procedure was repeated 100 times in this study. The variance of $\hat{\mu}^*_{qu}$ is given by

$$\sigma^2_{\mu} = E\left[\left\{\hat{\mu}^*_{\mathrm{qu}} - E\left[\hat{\mu}^*_{\mathrm{qu}}\right]\right\}\right]^2 \tag{4}$$

where σ^2_{μ} is the variance of $\hat{\mu}^*_{qu}$ and E[ ] denotes the mean of a variable. The standard deviation calculated from the variance of $\hat{\mu}^*_{qu}$ is shown in Table 3. σ_{μ} corresponds roughly to $s_{qu}/\sqrt{n}$ shown in Table 2. This correspondence indicates that the normal distribution is allowable as the underlying distribution of the strength.

The degree of uncertainty in the parameters estimated from the data has been discussed. Such estimated uncertainty is useful when transforming from the static strength to the dynamic strength.

Table 3. Summary of Bootstrap sample

Case	Loading rate or frequency	$E\left[\hat{\mu}^*_{qu}\right]$ (kPa)	σ_μ (kPa)
1	1%/min	494	11.0
	1 Hz	604	10.3
	2 Hz	599	8.38
2	1%/min	987	16.5
	1 Hz	1226	16.3

Note $\hat{\mu}^*_{qu}$, mean of strength of bootstrap samples;
σ_μ, standard deviation of $\hat{\mu}^*_{qu}$

5 Conclusions

Rapid unconfined compression tests were conducted on the cement-treated sand specimens prepared in a laboratory to investigate the strength under the dynamic loading applied during an earthquake. The experimental results showed that the strength under the rapid loading is larger by about 20% than the strength at the strain rate of 1%/min. The strengths varied to some extent even though the specimens were prepared in the laboratory. The statistical analysis of the experimental data was conducted to examine the degree of uncertainty in the mean and standard deviation of the strength. The statistical analysis results provide the useful information on the uncertainty that should be considered when transforming the material parameters.

References

Bischoff, P.H., Perry, S.H.: Compressive behaviour of concrete at high strain rate. Mater. Struct. **24**, 425–450 (1991)

Efron, B.: Bootstrap methods: another look at the Jackknife. Ann. Stat. **7**(1), 1–26 (1979)

Kongsukprasert, L., Tatsuoka, F.: Ageing and viscous effects on the deformation and strength characteristics of cement-mixed gravelly soil in triaxial compression. Soils Found. **45**, 55–74 (2005)

Onimaru, S., et al.: Dynamic shear strength of improved soil with B-type blast-furnace cement. J. Soc. Mater. Sci. Jpn **61**(1), 64–67 (2012). (in Japanese)

DEM Study of the Three Dimensional Effect of Soil Arching in Piled-Embankments

Yiyue Ma[1], Xilin Lü[2(✉)], and Maosong Huang[2]

[1] Department of Geotechnical Engineering, Tongji University, Shanghai 200092, China

[2] Key Laboratory of Geotechnical and Underground Engineering of Ministry of Education, Tongji University, Shanghai 200092, China

xilinlu@tongji.edu.cn

Abstract. Soil arching plays an important role in the design of piled-embankments of transportation facilities. This paper aims to study the three dimensional effect of soil arching in piled-embankments by discrete element simulation. The results show the formation of soil arching and the stress transfer mechanism. The evolution of the stress ratio of pile cap and subsoil was obtained, and the overestimation of the arching effect in plane strain condition was revealed. According to the movements of soil particles, the 3D uneven settlement pattern characteristic was plotted and clearly shows the formation of soil arching. The direction of major principal stress in three typical sections clearly shows the 3D effect of stress transfer mechanism. The obtained results are useful for the deformation calculation of high-speed railway piled embankment.

1 Introduction

Efficient deformation control of subgrade in soft ground is essential for the construction and operation of transportation facility. To satisfy the strict requirement of modern transportation facility, i.e. high-speed railway train, piled-embankments are widely used. The installation of rigid piles changes the bearing behaviour of soft subgrade and effectively reduces the settlement. The understanding of soil arching, which is induced by the relative displacement between pile and subsoil, plays a key role in the design of the piled-embankment. Originated from the idea of trapdoor tests by Terzaghi (1943), a variety of physical model tests (Ladanyi and Hoyaux 1969; Chen et al. 2008; Chevalier and Otani 2010; Moradi et al. 2015; Sadrekarimi and Abbasnejad 2010; Vardoulakis et al. 2010; Rui et al. 2016a, b) and theoretical analysis (Hewlett and Randolph 1988; Russell and Pierpoint 1997; Michalowski 2004; Paik and Salgado 2003; Son 2017) were adopted to analyse soil arching. Due to the complex of the soil arching in piled-embankment, previous studies were mostly set as a plane strain problem. Assumption often needs to be introduced into theoretical study. DEM was also adopted to study the soil arching effect; however it was mostly based on 2D case (Nadukuru and Michalowski 2012; Zhang and Jiang 2015; Lai et al. 2016; Rui et al. 2016a, b). The soil arching in piled-embankment is obviously a 3D problem and needs to be studied in 3D case.

B. Fatahi et al. (eds.), *Sustainable Design and Construction for Geomaterials and Geostructures*, Sustainable Civil Infrastructures,
https://doi.org/10.1007/978-3-319-95753-1_4

Although 3D finite element analysis was conducted (Chevalier and Otani 2010; Hello and Villard 2009) on this problem, but the finite element method cannot analyze the mechanical properties between soil particles, such as the contact force chain between particles and cannot be demonstrated. So the 3D deformation pattern and load transfer mechanism during soil arching still need further investigation. The study of the formation and transfer mechanism of soil arching is useful for the deformation calculation and engineering design of piled embankment.

In this 3D discrete element simulation, the formation of soil arching in piled embankment was studied. The evolutions of the earth pressures on the pile top and the surrounding sub-soil were obtained, and the steady value of the efficiency which is defined by the pressure ratio of pile top and soil was obtained. The uneven settlement was calculated by the vertical movements of soil particles, and the evolution of deformation pattern showed the mobilization of soil arching. The force chain at the full mobilized state of soil arching was obtained; the direction of major principal stress in typical sections demonstrates the 3D effect of stress transfer mechanism.

2 Dem Simulation

According to the configuration of piled-embankment (Hello and Villard 2009), a 2×2 pile group, was selected to study the soil arching in a piled-embankment. To model the differential displacement between the pile cap and surrounding sub-soil, the subsoil, which is shown in the shadow area in Fig. 1a, was controlled to moves downward. As shown in Fig. 1a, the spacing between the pile cap is s = 1.2 m, the width of pile cap is a = 0.7 m, the height of soil h is 1 m, and $h/(s - a)$ is 2.0. The sketch map of the soil arching is shown in Fig. 1b. Considering the symmetry, quarter piles were adopted in the model to save computation time. This simplification does not alter the 3D hemisphere representation of the soil arching, and it is able to show the stress transfer mechanism. The embankment was simulated by the multi-layer compaction method (Jiang et al. 2003; Lai et al. 2014); gravity deposition was adopted to ensure homogeneity of the specimen. The generated discrete element model in the simulation is shown in Fig. 2; it was composed of 77,552 soil particles. At the bottom of the model, 4 immovable walls were set up to replace the pile top, and the settlement wall was set in the middle to simulate the subsidence of soil between piles. The sidewall was set under the model to limit the moving area of the soil between piles (Fig. 2). After the initial model generated, the settlement wall decreased and the rest of the wall was kept still.

The grading curve of the soil is shown in Fig. 3; the particle sizes were amplified by 70 times of typical Toyoura sand. The micromechanical properties of the particles are listed in Table 1 (Zhou et al. 2016). The displacement controlled loading was applied to the bottom plane of the subsoil. The loading was divided into 1,000,000 steps, and the velocity was gradually increased to 0.003 m/s at the first 100,000 time steps.

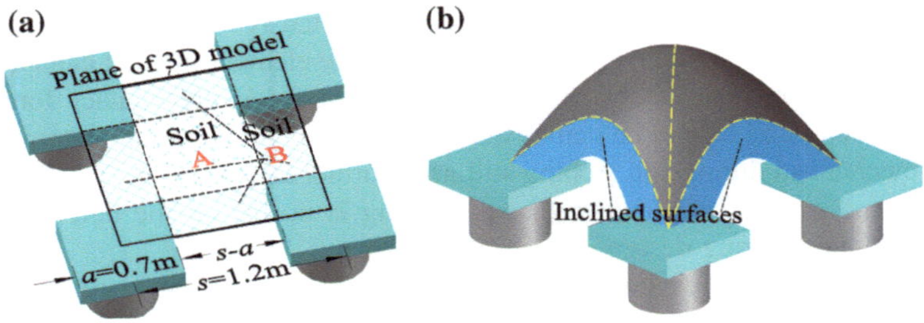

Fig. 1. Soil arching in soft embankment **a** the bottom of the 2 × 2 pile group and **b** sketch-map of soil arching

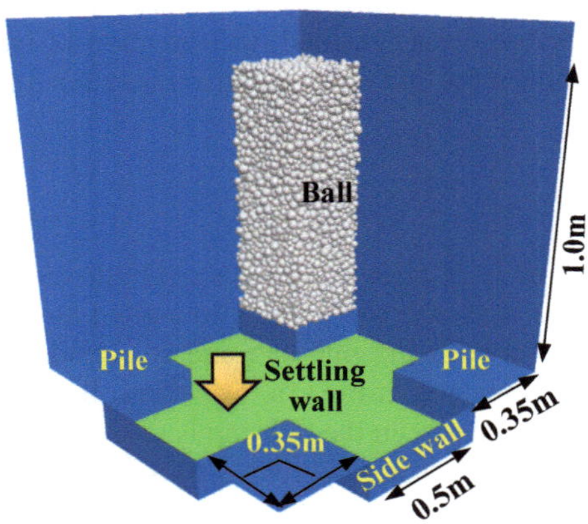

Fig. 2. DEM model in the analysis

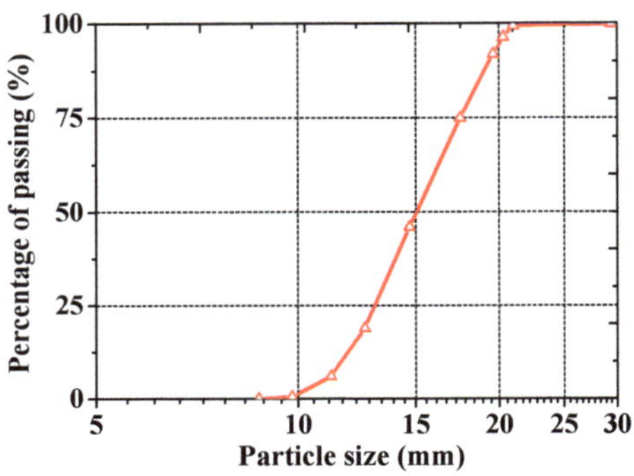

Fig. 3. Particle size distribution

Table 1. Micromechanical properties

Parameters	Value	Parameters	Value
Particle density (kg/m^3)	2650	Normal-to-shear stiffness ratio	1
Young's modulus (Pa)	1.0E + 08	Normal critical damping ratio	0.2
Particle-particle friction	0.05	Shear critical damping ratio	0.2
Wall-particle friction	0	Rolling resistance coefficient	0.1

3 Results and Analysis

3.1 Variation of the Pressures on Pile-Cap and Soil

With the movement of the subsoil, the earth pressure on the top of subsoil gradually decreases and the pressure on the pile cap increases. The obtained earth pressures on pile and subsoil under different embankment height conditions are shown in Fig. 4a. The pressure on the pile cap significantly increases to a stable value. When the embankment is larger than 1 m, the earth pressure on the pile obviously becomes larger due to soil arching effect. Correspondingly, the earth pressure on the subsoil decreases significantly at the initial loading stage, and then decreases slowly. The pressure on zone A + B is greater than that on zone B. The varying trend of the pressures is consistent with the results obtained from plane strain trapdoor tests (Chevalier et al. 2012; Chevalier and Otani 2010; Moradi et al. 2015), but the varying rate is slower. To evaluate the degree of arching effect, the load transfer efficacy E (Hewlett and Randolph 1988) is adopted.

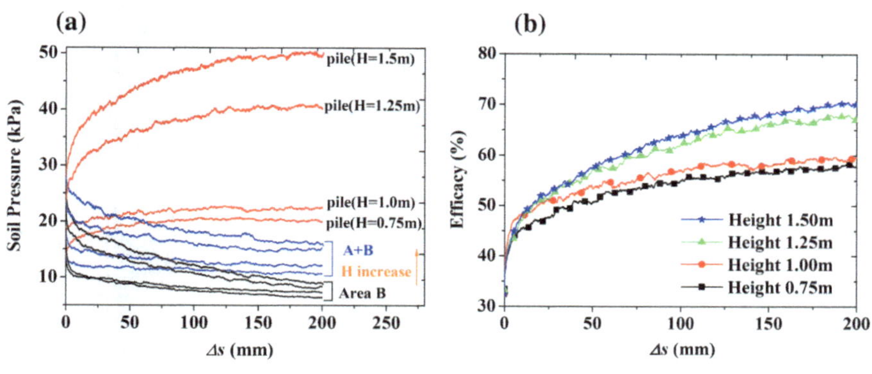

Fig. 4. The obtained results from DEM simulation **a** the soil pressures and **b** the efficacy in the different height of filling material

$$E = F_p / F_p + F_s \quad (1)$$

where F_p and F_s are the average loads applied on the pile and the surrounding subsoil.

The load transfer efficacy E obtained from the discrete element simulation is shown in Fig. 4b. As expected, a higher embankment results in a greater efficacy E and larger displacement required to reach stability. The changing rate of the earth pressure with the differential displacement of subsoil is slower than that of the plane strain trapdoor test (Lai et al. 2016). Due to the ignorance of earth pressure on zone B, the load transfer efficacy will be overestimated in a plane strain trap door test.

3.2 Patter of Settlement

To identify the formation of soil arching, the movement of the particles at different base yielding stage are shown in Fig. 5. With the increase of the relative displacement between pile and subsoil Δs, the settlement of the embankment starts to form. When Δs reach 25 mm (Fig. 5a), an isosceles triangle forms, which shape is similar to the results from the trapdoor model test (Rui et al. 2016a, b). When Δs = 45 mm (Fig. 5b), the triangle develops into an arch. When Δs = 75 mm (Fig. 5c), the height of the soil arch attains 215 mm in height and keep unchanged with the settlement of the bottom soils. When Δs = 135 mm (Fig. 5d), two soil arches form, and the top of the embankment shows uneven settlement. The obtained result is similar to that of the vertical settlement surface in the 2D trapdoor test (Chevalier and Otani 2010; Lai et al. 2016).

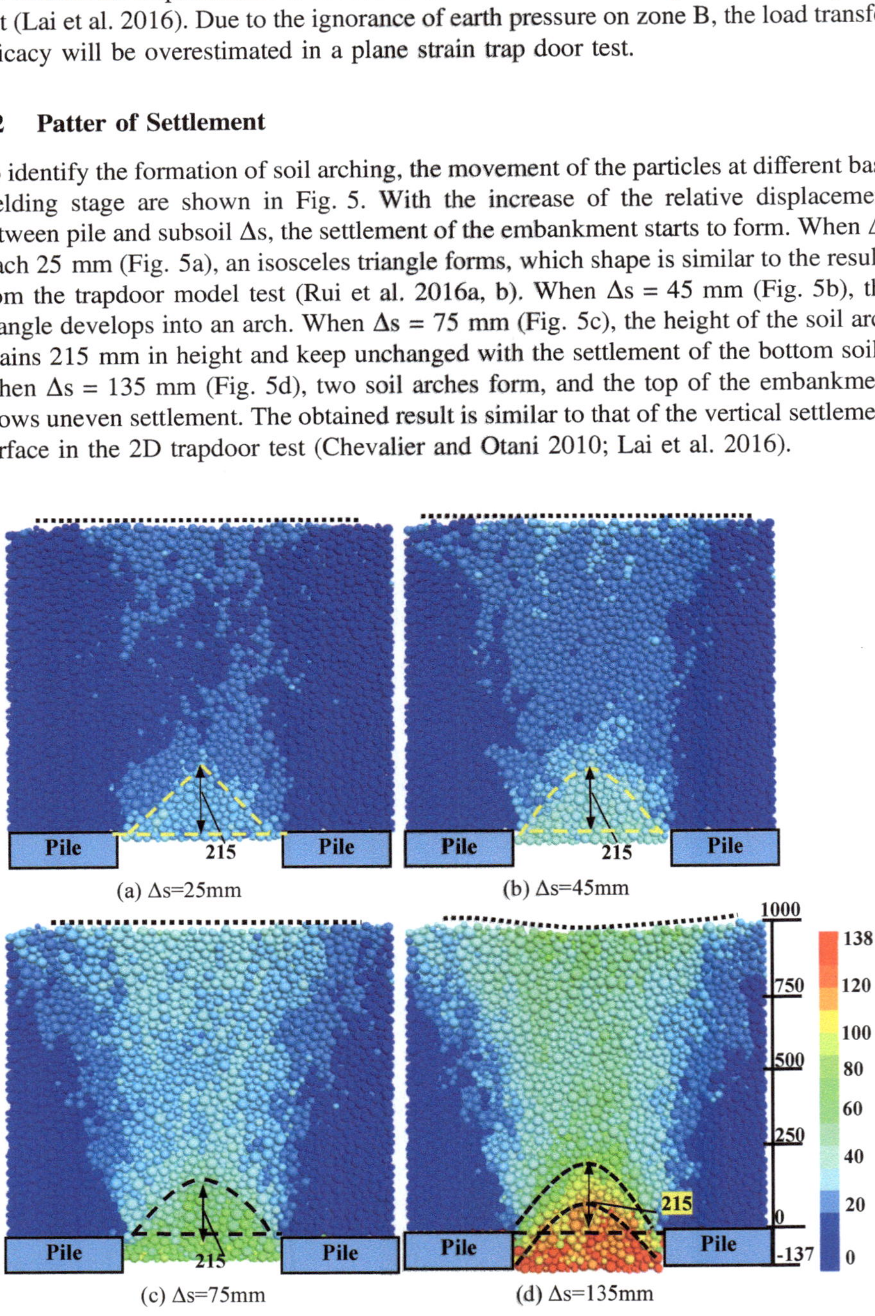

Fig. 5. The evolution of particle displacement under different base yielding stage (r = 2.0)

To clearly show the profile of the settlement, the embankment was divided into 5 layer from bottom to top (i.e. 0, 250, 500, 750, and 1000 mm). As shown in Fig. 6a, puddle-shaped deformation pattern was obtained in each layer. At the height of 250 mm, the boundary of the settlement area demonstrates a cross groove shape, indicating that the uneven settlement at the middle is larger than the other position. At the height of 500 mm, the cross groove shape disappears, and the deformation becomes symmetrical along the vertical centre line. With the increase of the level, the puddle becomes shallower; indicating the influence of the differential displacement in the bottom of the embankment becomes weak. The average uneven settlement of each layer is shown in Fig. 6b; it nonlinearly decreases with the model height and attains a stable value when the model height is large enough. This obtained deformation trend is consistent with the previous analysis.

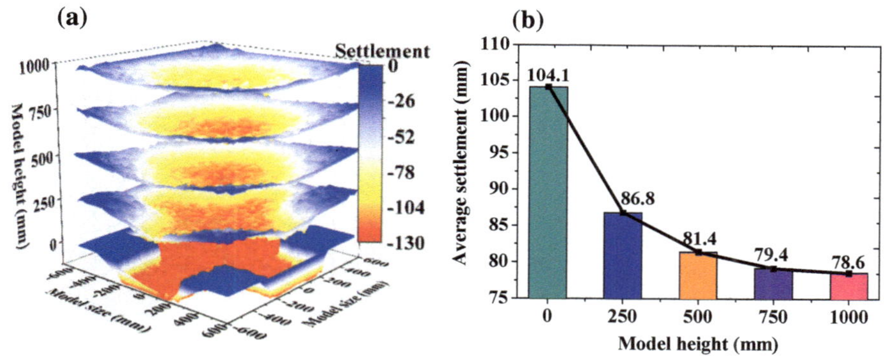

Fig. 6. Settlement of embankment filling at different height **a** pattern of the settlement and **b** average uneven settlement

3.3 Distribution of the Major Principal Stress

In discrete element simulation, the force between particles is transmitted through contact; the features of stress redistribution during soil arching can be visually represented by force chains (Han et al. 2011). During the relative movement of the embankment filling, the magnitudes and direction changes of contacts and forces induced by the particle re-arrangement has been investigated by 2D DEM analysis (Lai et al. 2014, 2016). In this paper, the spatial distribution of stress was analysed to show the internal structure of soil arching in three dimensions. The rolling resistance linear model (Jiang et al. 2015; Zhou et al. 2016) was adopted to calculate the contact force between particles.

As shown in Fig. 7a, three typical cross-sections were selected to analyze the distributions of major principal stresses. Section I is the edge of the soil arching model and the same with the plane strain test if 3D effect is neglected; section II is the middle plane of the model; section III is the diagonal plane of the model. In each section, enough measurement spheres were included to calculate the direction of major principal stress.

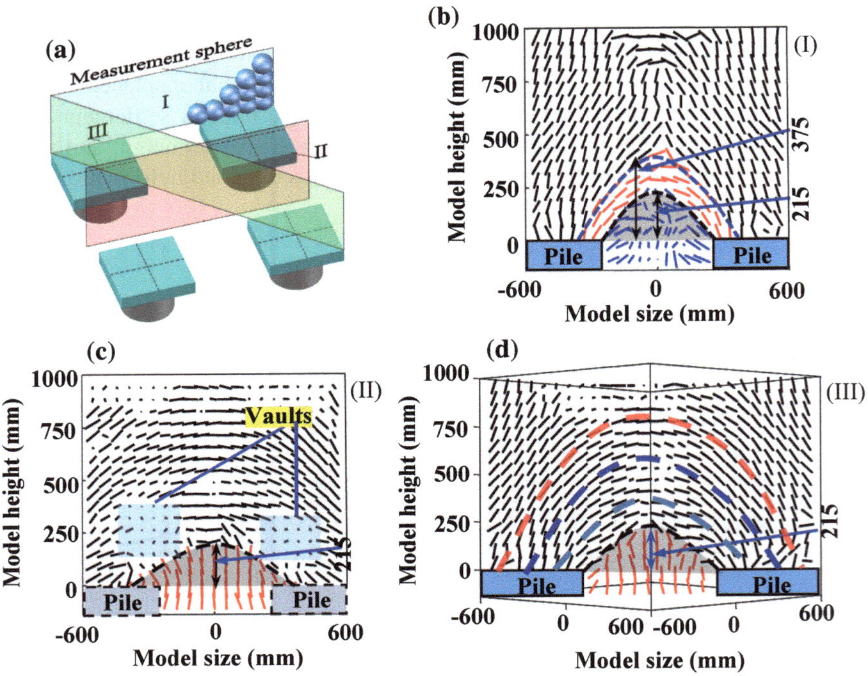

Fig. 7. Distributions of major principal stress direction (section I: the edge of the soil arching model; section II: middle plane of the model; section III: diagonal plane of the model)

The distribution of the major principal stress in section I is shown in Fig. 7a. It is shown that the major principal stress horizontally distributes in two regions, indicating the formation of two arches. The height of the inner arch is about 215 mm from the top of soil arch to the pile cap. The height of the external arch is about 375 mm; it is slightly less than results of 0.8 * (s − a) obtained from plane strain tests (Lai et al. 2016). The distribution of the major principal stress of section II is shown in Fig. 7b. It is vertically distributed in the collapse region, and it turns to be horizontally distributed in the region above the collapse region. The points in the shaded regions indicate that the major principal stress direction is orthogonal to the plane II. The direction of the major principal stress in the section III is shown in Fig. 7d. The difference between the distribution of section III and section I manifest the 3D effect of the soil arching in piled-embankment.

4 Conclusions

Based on 3D discrete element simulation, the 3D effect soil arching in piled-embankment subjected to a differential displacement between pile and surrounding soils was obtained. The deformation pattern and load transfer mechanism were investigated, the main conclusions are as follows.

The ratio of earth pressure on the pile cap and surrounding sub-soil increases with the differential displacement, and the increasing rate is slower than that under plane strain condition. Ignoring the 3D stress state, the steady value of efficiency will be overestimated by plane strain analysis.

The distribution of the movement of embankment filling shows a puddle shape in 3D space. The obtained uneven settlements in different altitude from the bottom to the top of the embankment identify the features of soil arching. The boundary of the settlements of embankment becomes blurry with the soil layer becomes far away from the bottom soil.

The evolution of the force chain clearly shows the 3D force transmission during soil arching, the vertical distribution of the major principal stress direction in the moving area transfers to the horizontal distribution. The 3D effect of soil arching was manifested by the spatial distribution analysis of the principal stress direction in typical planes.

Acknowledgements. The financial supports by National Basic Research Program of China (through grant No. 2014CB049100) and National Science Foundation of China (NSFC through grant No. U1433123) are gratefully acknowledged.

References

Chen, Y.M., et al.: An experimental investigation of soil arching within basal reinforced and unreinforced piled embankments. Geotext. Geomembr. **26**(2), 164–174 (2008)

Chevalier, B., Otani, J.: 3-D arching effect in the trap-door problem: a comparison between X-ray CT scanning and DEM analysis. In: GeoFlorida 2010: Advances in Analysis, Modeling & Design, pp. 570–579 (2010). https://doi.org/10.1061/41095(365)54

Chevalier, B., et al.: Experimental and discrete element modeling studies of the trapdoor problem: influence of the macro-mechanical frictional parameters. Acta Geotech. **7**(1), 15–39 (2012). https://doi.org/10.1007/s11440-011-0152-5

Han, J., et al.: DEM analysis of stresses and deformations of geogrid-reinforced embankments over piles. Int. J. Geomech. **12**(4), 340–350 (2011). https://doi.org/10.1061/(asce)gm.1943-5622.0000050

Hewlett, W.J., Randolph, M.F.: Analysis of piled embankments. Int. J. Rock Mech. Min. Sci. Geomech. Abstr. **25**(6), 297–298 (1988)

Jiang, M.J., et al.: An efficient technique for generating homogeneous specimens for DEM studies. Comput. Geotech. **30**(7), 579–597 (2003). https://doi.org/10.1016/s0266-352x(03)00064-8

Jiang, M., et al.: A novel three-dimensional contact model for granulates incorporating rolling and twisting resistances. Comput. Geotech. **65**, 147–163 (2015). https://doi.org/10.1016/j.compgeo.2014.12.011

Ladanyi, B., Hoyaux, B.: A study of the trap-door problem in a granular mass: discussion. Can. Geotech. J. **6**(4), 1–14 (1969). https://doi.org/10.1139/t69-002

Lai, H.J., et al.: DEM analysis of "soil"-arching within geogrid-reinforced and unreinforced pile-supported embankments. Comput. Geotech. **61**, 13–23 (2014). https://doi.org/10.1016/j.compgeo.2014.04.007

Lai, H.J., et al.: Visualization of the formation and features of soil arching within a piled embankment by discrete element method simulation. J. Zhejiang Univ. Sci. A **17**(10), 803–817 (2016). https://doi.org/10.1631/jzus.a1500302

Le Hello, B., Villard, P.: Embankments reinforced by piles and geosynthetics-numerical and experimental studies dealing with the transfer of load on the soil embankment. Eng. Geol. **106** (1), 78–91 (2009). https://doi.org/10.1016/j.enggeo.2009.03.001

Michalowski, R.L.: Admissible stress fields and arching in piles of sand. Géotechnique **54**(8), 529–538 (2004). https://doi.org/10.1680/geot.54.8.529.52008

Moradi, G., et al.: Experimental modeling and measuring stresses and strains during arching phenomenon. Geosciences **5**(2), 53–61 (2015)

Nadukuru, S.S., Michalowski, R.L.: Arching in distribution of active load on retaining walls. J. Geotech. Geoenviron. Eng. **138**(5), 575–584 (2012). https://doi.org/10.1061/(asce)gt.1943-5606.0000617

Paik, K.H., Salgado, R.: Estimation of active earth pressure against rigid retaining walls considering arching effects. Géotechnique **53**(7), 643–653 (2003). https://doi.org/10.1680/geot.2003.53.7.643

Rui, R., et al.: Evolution of soil arching; 2D DEM simulations. Comput. Geotech. **73**, 199–209 (2016a). https://doi.org/10.1016/j.compgeo.2015.12.006

Rui, R., et al.: Investigation of soil-arching development in dense sand by 2D model tests. Geotech. Test. J. **39**(3), 415–430 (2016b). https://doi.org/10.1520/gtj20150130

Russell, D., Pierpoint, N.: An assessment of design methods for piled embankments. Ground Eng. **30**(10), 39–44 (1997)

Sadrekarimi, J., Abbasnejad, A.: Arching effect in fine sand due to base yielding. Can. Geotech. J. **47**(3), 366–374 (2010). https://doi.org/10.1139/t09-107

Son, M.: Three-dimensional expansion of the terzaghi arching formula considering inclined sliding surfaces and examination of its effects. Int. J. Geomech. **17**(7), 6016043 (2017). https://doi.org/10.1061/(asce)gm.1943-5622.0000844

Terzaghi, K.: Theoretical Soil Mechanics. Wiley (1943)

Vardoulakis, I., et al.: Trap-door problem with dry sand: a statical approach based upon model test kinematics. Int. J. Numer. Anal. Meth. Geomech. **5**(1), 57–78 (2010). https://doi.org/10.1016/0148-9062(81)90991-8

Zhang, R., Jiang, G.: Numerical analyses of soil arching mechanism for passive laterally loaded piles at the mesoscopic scale. Electron. J. Geotech. Eng. **20**(5), 1667–1680 (2015)

Zhou, W., et.al.: Macroscopic and microscopic behaviors of granular materials under proportional strain path: a DEM study. Int. J. Numer. Anal. Methods Geomech. **40**(18), 2450–2467 (2016). https://doi.org/10.1002/nag.2537

Comparative Study Between Mathematical Models for Prediction of Strength Properties of Rocks in United Arab Emirates

Maher Omar(✉)

Department of Civil and Environmental Engineering & Research Institute of Sciences and Engineering, University of Sharjah, P.O. Box 27272, Sharjah, United Arab Emirates
momar@sharjah.ac.ae

Abstract. Identification of rock properties is greatly needed for effective and efficient construction projects, especially in United Arab Emirates and throughout the Arabian Peninsula, where the nature of rocks is generally sedimentary and can be classified on the basis of the most common components: namely sandstone, mudstone, and crystalline gypsum. Unconfined compression strength is a method used frequently, usually by means of laboratory tests to determine it. This paper reports on a comparative study undertaken between a stepwise regression model and an advanced random decision forest model in order to find out which model best predicts the relationship between unconfined compression strength and other properties of rocks. The study was conducted with 210 soil samples collected in an independent study, and these samples were examined for various observations including the Point Load Test and Ultra Sonic Velocity through pulse velocity. Results showed that the regression model, with approximately 95% explanation of variability and 6.67% accuracy of prediction, outperformed the random decision forest model for both Point Load and Ultra Sonic Velocity.

Keywords: Unconfined compressive strength · United Arab Emirates
Stepwise regression · Random decision forest

1 Introduction

1.1 Background

In-depth analysis from a geotechnical perspective is required in a majority of construction projects in United Arab Emirates (UAE), especially in construction projects undertaken during the last decade. Geotechnical investigations vary mainly in terms of completion time and complexities of the investigation and, while many of these procedures are complex, some simpler methods are rather time-consuming as well. Another hurdle is the heavy funding needed for some complex projects. Discovery of an easier, cheaper, and less-sophisticated way of investigation is one of the most significant aspects of studies reported in geotechnical literature. Additionally, engineers and geologists need a simple method that lends itself to straightforward and accurate

B. Fatahi et al. (eds.), *Sustainable Design and Construction for Geomaterials and Geostructures*, Sustainable Civil Infrastructures,
https://doi.org/10.1007/978-3-319-95753-1_5

interpretation. This research aims to identify an analytical solution and related comparative literature that encompasses the strength properties of rocks, considered a vital factor in the majority of geotechnical projects.

The International Society for Rock Mechanics and American Society for Testing and Materials (ASTM 4.08) defined standard procedures for strength measurement of rocks. Since rock samples are usually limited due to the intrinsic nature of sampling, it has become a disadvantage that unconfined compressive strength tests (UCS) require costly and destructive laboratory tests. Conducting Point Load Index, Schmidt hammer rebound tests, ultrasonic tests, and other analyses is a more economic approach, since the equipment of those types of testing is less expensive, and the tests themselves are less complex and minimal sample preparation is required. Additionally, they can be used effectively in investigation of strength properties of rocks.

This study presents an analytical comparison between a Multiple Regression model and a tree-based Random Decision forest model in order to estimate the relationship between Unconfined Compression Strength and various other properties of rocks, as much as possible and appropriate. The comparison is based on Adjusted R^2 (or Coefficient of Determination) and Standard Error of estimates of these models, and the confidence interval of predictions obtained by using the results from these models. In addition, relative importance of variables as assessed from these models is presented. Any correlation existing in this research represents only approximate values. Hence, it is natural to observe a slight deviation.

1.2 Literature Review

Hoek and Bray (1981), Broch and Franklin (1972) suggested that multiplying the point load index by 24 gives the estimate of UCS of rocks. Omar (2017) presented the following predicting equations through nomographic solutions based on regression, for Point Load ($I_{S(50)}$):

$$Ln(UCS) = 4.212\sqrt{I_{s(50)}} - 1.552I_{s(50)} + 0.1Ln(RT) + 0.002\gamma^2 - 1.509 \tag{1}$$

and for Ultra Sonic Velocity (USV),

$$Ln(UCS) = 0.906Ln(USV) + 0.01RT - 0.027\gamma - 5.652 \tag{2}$$

D'Andrea et al. (1965) performed a study on samples of rocks, presenting the regression model:

$$q_u = 16.3 + 15.3I_{S(50)} \tag{3}$$

where q_u = Uniaxial Compressive Strength of rock and $I_{s(50)}$ = Point Load Index for 50 mm diameter core.

Bieniawski (1973) proposed the following model, where D represents the core diameter:

$$UCS = (14 + 0.175D)I_{S(50)} \tag{4}$$

Selçuk and Kayabali (2015) argue that preparing cylindrical rock samples in severely fragile or weak rock blocks is a very complex task. The key benefits of indirect methods of estimation of compression strength are cost effectiveness, portability, and practicability. Sheshde and Cheshomi (2015) proposed a new way of loading small rock samples and assessing UCS in restricted sampling situations. In that study, 510 rock samples were tested, which were all between the size of 3–5 mm and consisted of three types of rocks, namely crystalline limestone, micritic limestone, and sandstone. Their findings presented 89% accuracy in estimating UCS.

Lai et al. (2014) developed a report that estimates UCS as 20–25 times the Point Load index. However, other reports stated the range between 15 and 50 for anisotropic rocks specifically. The correlation model given by Cargill and Shakoor (1990) is shown in the following equation:

$$q_u = 13 + 23I_{S(50)} \tag{5}$$

Fener et al. (2005) described UCS as a function of Point Load Index (I_s) as given in the following equation:

$$q_u = 9.08I_{S(50)} + 39.32 \tag{6}$$

Voznesenskii et al. (2015) argue that acoustic or mechanical properties of the rocks are focused on in most studies. They asserted the need to address the means of evaluation and interrelation, as they are useful in prediction of UCS using Q-factor.

Bursztyn et al. (2015) indicated the lack of comprehensive data for application to surface processes like compressive strength, fracturing, or a combination of both. They identified that another issue in strength measurement of rocks is bias in sampling.

Chau and Wong (1996) from Hong Kong came up with a new formula for estimating UCS based on modified I_S. The index-to-conversion factor is denoted by k. The fitted value of k = 14.9 was close to the experimental value of k = 12.5.

Rusnak and Mark (2000) gave the following relationship for different rocks. This equation was specific to measurement of coal rocks:

$$q_u = 23.62I_{S(50)} - 2.69 \tag{7}$$

In addition, this equation applied to other types of rocks:

$$q_u = 8.41I_{S(50)} + 9.51 \tag{8}$$

The simple mean method proposed by Hack and Huisman (2002) incorporated the use of methods based on the concept of simple mean. Schmidt Hammer, Equotip, and uncompressed strength tests are few of these methods. The primary assessment is done by the evaluation of time of rebound of the piston that hits the rock. Additionally, Analysis of Variance and descriptive analytics are employed for further finding of results. These methods are useful for finding some conclusions, although they may not always guarantee maximum accuracy.

Hoek et al. (2002) failure method for evaluating UCS dealt with underground excavations. The design required for excavations in underground hard rocks is evaluated through their method. Motamedi et al. (2015) proposed the use of artificial intelligence to address this issue. The main issue of the complex mathematical models used for predicting UCS was the long computation time. In this regard, analytical methods based on artificial neural networks can be used. This high-speed computational model has the benefit of compact solutions. An adaptive neuro-fuzzy technique was employed to solve the problem of uncertainty for predicting UCS of pulverized fuel ash sand cement mixtures.

Yilmaz (2009) proposed an advanced method for indirect determination of UCS. A core strangle test is used in this study. This takes care of the limitations of point load index testing. In this test loading a core is accomplished along a circle. Here, the idea is to rely on strangle loading type.

2 Methodology and Data

2.1 Data

A comprehensive dataset consisting 210 experimental test results from one independent study was obtained from the literature to develop the models. This test, reported by Omar (2017) included data on both Point Load ($I_{S(50)}$) and Ultra Sonic Velocity (USV) through its pulse velocity.

The dataset consisted of two separate spreadsheets for Point Load and Ultra Sonic Velocity. Unconfined Compression Strength measurements of 122 observations were recorded along with 8 other features including Rock Type, Point Load (MPa), Modulus of Elasticity (MPa), Moisture Content (%), Bulk Unit Weight and Dry Unit Weight (KN/m^3), and Depth and Water Depth (m). Similarly, there were 88 observations on the same variables for Ultra Sonic Velocity (m) instead of Point Load. Mudstone, sandstone, and crystalline gypsum were the three varieties of rock samples observed and recorded in the dataset.

The objective of this analytical modeling was to evaluate UCS as a function of these variables, or at least some of them, separately for Point Load and Ultra Sonic Velocity. The study used both continuous and categorical variables. Prior to fitting the models, descriptive analytics were performed to acquire more insight into these variables.

2.2 Descriptive Analytics

Basic descriptive statistics for Point Load and corresponding measurements can be found in Table 1. The same for Ultra Sonic Velocity can be found in Table 2. These descriptions include simple arithmetic mean, standard deviation, median, mean absolute deviation, minimum and maximum value, range, skewness, kurtosis, and standard error. Variables were suitably transformed before exploratory data analysis.

Table 1. Properties of rock samples—point load

	Mean	s.d.	Median	Mad	Min	Max	Range	Skew	Kurtosis	s.e.
Unconfined compressive strength	5.16	4.26	3.81	2.92	0.54	22.44	21.90	1.59	2.58	0.39
Point load ($I_{S(50)}$)	0.50	0.40	0.39	0.40	0.02	1.63	1.61	0.90	0.01	0.04
Modulus of elasticity	1057	1150	413	448	38	4621	4583	1.15	0.36	104.13
Moisture content	8.65	7.66	5.20	5.55	0.59	29.20	28.61	0.93	−0.19	0.69
Bulk unit weight	18.70	2.07	18.64	2.42	14.34	23.53	9.19	0.08	−0.84	0.19
Dry unit weight	17.28	2.15	16.76	2.25	13.76	22.84	9.08	0.50	−0.74	0.19
Depth	16.65	7.40	17.00	8.01	2.00	34.60	32.60	0.15	−0.84	0.67
Water depth	2.80	0.77	3.00	0.74	1.50	4.00	2.50	−0.24	−0.91	0.07

2.3 Engineering of Features

The response variable Unconfined Compression Strength was found to be positively skewed for both Point Load ($I_{S(50)}$) and Ultra Sonic Velocity. Also the major explanatory variables were somewhat skewed. In order to increase the normality of the variable values, suitable transformations were applied to these variables; on some of the other explanatory variables, transformation would increase the meaningfulness of the model. Relationships between the transformed variables could be observed even by the naked eye before analytics were performed, once illustrated. Figure 1 illustrates the scatter plots of the new variables of interest in a histogram (Fig. 2).

The type of rocks (as discussed earlier) was noted in the form of a categorical variable, Rock Type. The three different categories were MUDS which stands for samples consisting of mudstone; SANDS for samples with sandstone; and CRGP for samples with crystalline gypsum.

2.3.1 Dummy Coding

The Principle of Parsimony (Occam's Razor) instructs researchers to prefer a linear model to a non-linear one. Moreover, the involvement of a categorical variable along with continuous ones could be difficult to interpret. This brought in the need for dummy coding. For dummy coding of the data in this study, the categorical variable Rock Type was split and transformed into dummy variables that were the same in number as the unique levels in the earlier original variables, and were at the same time pairwise and mutually independent. Thus, three independent dummies were created. Each of these could have a value of either 0 or 1. For instance, the value of the $dummy_m$ variable was 1 if the sample was from the MUDS category and 0 otherwise. Cases were the same for $dummy_s$ and $dummy_c$ for SANDS and CRGP, respectively.

Table 2. Properties of rock samples—ultra sonic velocity

	Mean	s.d.	Median	Mad	Min	Max	Range	Skew	Kurtosis	s.e.
Unconfined compressive strength	7.13	4.35	6.62	4.51	1.52	22.44	20.92	0.98	0.70	0.46
Ultra-sonic velocity	5632.6	3313	4930	3726.5	1133	14,235	13,102	0.63	−0.52	353.2
Modulus of elasticity	12,245	1029	994	1116	105	4000	3895	0.80	−0.33	109.65
Moisture content	6.44	5.46	4.53	4.02	0.59	20.10	19.51	1.05	−0.05	0.58
Bulk unit weight	18.85	1.73	18.59	1.87	14.75	22.60	7.85	0.12	−0.65	0.18
Dry unit weight	17.76	1.89	17.60	2.26	13.86	22.12	8.26	0.21	−0.85	0.20
Depth	17.93	7.96	17.00	9.64	6.00	36.00	30.00	0.35	−0.93	0.85
Water depth	2.89	0.91	3.00	0.74	1.50	5.00	3.50	0.58	−0.18	0.10

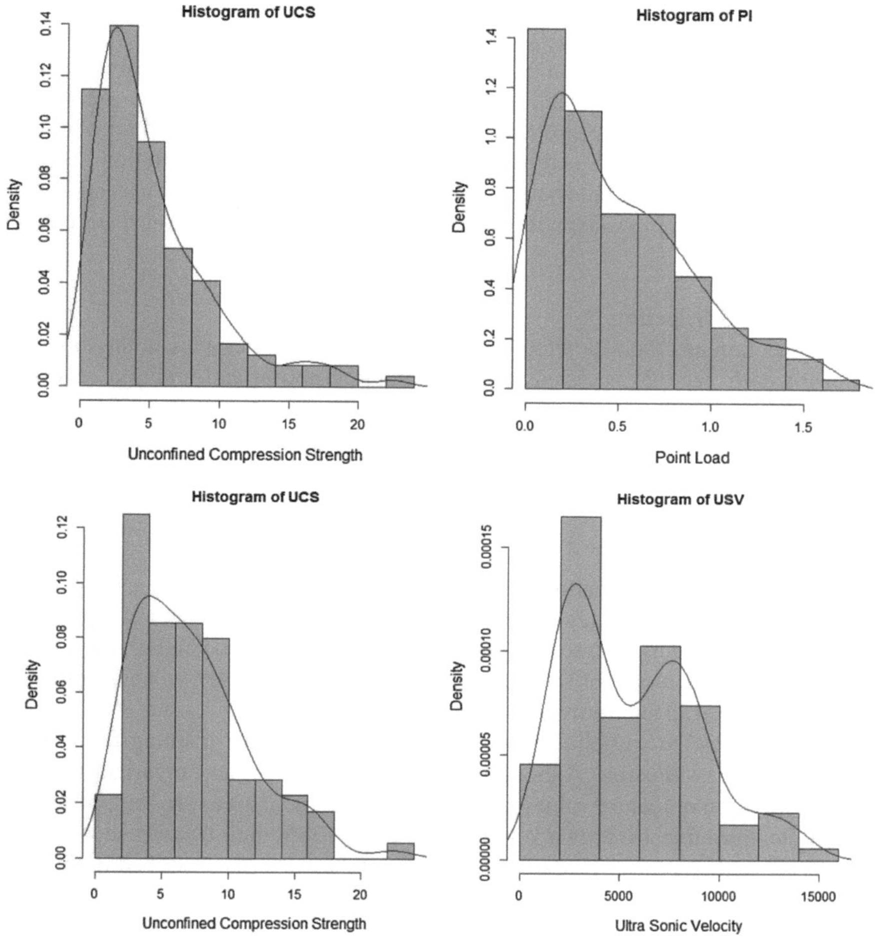

Fig. 1. Histograms of UCS, for both $I_{S(50)}$ and USV

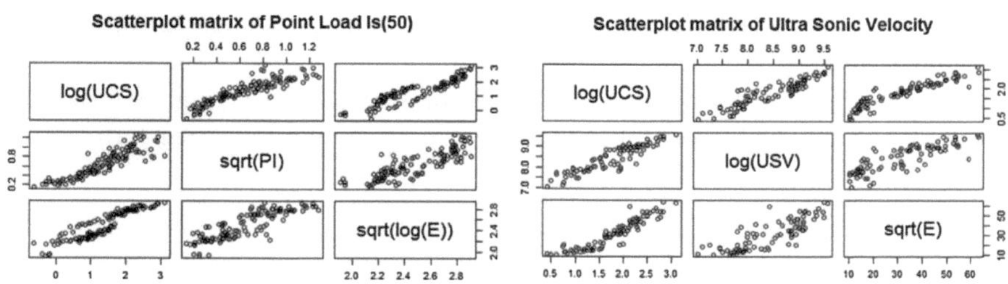

Fig. 2. Scatterplot matrices

2.4 Stepwise Regression

In statistics, stepwise regression is an automatic procedure of choosing predictor variables when fitting regression models. In each step, a criterion is pre-specified according to which one of the explanatory variables is being considered for addition or elimination from the model. Adjusted R^2, Akaike information criterion, Bayesian information criterion, or Mallows's Cp can be used instead of the usual F-tests or t-tests. In order to employ stepwise regression, model uncertainty must be correctly reflected.

2.4.1 Forward Selection

Forward selection has the following steps: Starting with no variables in the model, the addition of each variable is tested using a suitable criterion and addition of variable whose inclusion results in the most statistically significant improvement in fitting. These steps are repeated until no further significant improvement can be seen in the model.

2.5 Random Forests

Using random forests, or random decision forests, is an ensemble learning method popular for tasks such as regression and classification. This method operates by building a forest of decision trees during the model's training time and giving as output the class that is the mode of the classes (in classification case) or mean prediction (in regression case) of the individual trees.

Decision "trees" are a popular method for various machine-learning tasks such as regression and classification. Although this method is robust and invariant to scaling and transformation of feature values, results of some decision trees are seldom accurate. For example, trees that are grown very deep tend to over fit their training sets; with low bias but very high variance. Using random decision "forests" as described in this paper takes care of this issue of overfitting.

2.5.1 Algorithm

Bootstrap aggregating, or bagging, is the basis of the training algorithm used for random decision forests. A random sample with replacement is selected from the training set repeatedly (B times) and trees are fitted to these samples, given a training set X with responses Y.

Predictions for unseen examples (say x') can be evaluated by taking the mean of the predictions of individual trees (for x' in this example) by the following formula:

$$\hat{f} = \frac{1}{B}\sum_{b=1}^{B} f_b(x') \tag{9}$$

or by taking the majority vote in the case of decision trees.

Bootstrapping improves performance of a model since it decreases variance of the model without increasing bias. Since the trees are not correlated, the mean of predictions is not vulnerable to noise. Bootstrap sampling de-correlates the trees by providing them with different training sets. The number of samples/trees, B, is a free parameter. The number of trees depends on the size and nature of the training set and typically, it varies from hundreds to a few thousand.

2.6 Performance Metrics

2.6.1 Coefficient of Determination

Coefficient of determination (R^2) is a measure of the goodness of fit of a regression model. This coefficient lies between 0 and 1. A value of 0 indicates that no relationship exists between the independent variable and the dependent variables; that is, all the regression coefficients β_is are exactly equal to zero. A value of 1 indicates that all the values are predicted exactly through the proposed model. Mathematically, R^2 is given by:

$$R^2 = 1 - \frac{SS_{Res}}{SST} \tag{10}$$

where SS_{Res} is the residual sum of squares and SST is the total sum of squares.

3 Analysis and Results

3.1 Model Development

3.1.1 Stepwise Regression

In order to obtain the governing predictive equations, stepwise multiple regression was employed for both Point Load and Ultra Sonic Velocity. The resulting equation for Point Load is given by:

$$\begin{aligned}\ln(\mathrm{UCS}) = {} & -8.0571 + 0.5355\sqrt{\mathrm{PI}} + 3.1715\sqrt{\ln \mathrm{E}} + 0.0367\mathrm{WD} + 0.0102\,\mathrm{depth} \\ & - 0.0359\,\mathrm{water}_{\mathrm{depth}} + 0.6644\mathrm{dummy}_{\mathrm{M}} + 0.0218\mathrm{dummy}_{\mathrm{S}}\end{aligned} \tag{11}$$

The same for USV is given by:

$$\begin{aligned}\ln(\mathrm{UCS}) = {} & -3.8067 + 0.5752\ln(\mathrm{USV}) + 0.0195\sqrt{\mathrm{E}} \\ & + 0.0041\sqrt{\mathrm{m}} + 0.0573\mathrm{dummy}_{\mathrm{M}} + 0.1976\mathrm{dummy}_{\mathrm{S}}\end{aligned} \tag{12}$$

3.1.2 Random Forest Model

Instead of using coefficients of explanatory variables as in a regression model, random forest models provide an idea about the relative importance of variables that are being used to predict Unconfined Compression Strength. A decrease in the estimate of error is observed when the number of trees in the forest is increased. This also eliminated the issue of variance. For Point Load ($I_{S(50)}$), Fig. 3 illustrates the distribution of error according to the number of trees. The confidence interval of prediction accuracy can be calculated from the residual plot when fitted against values plotted in Fig. 4.

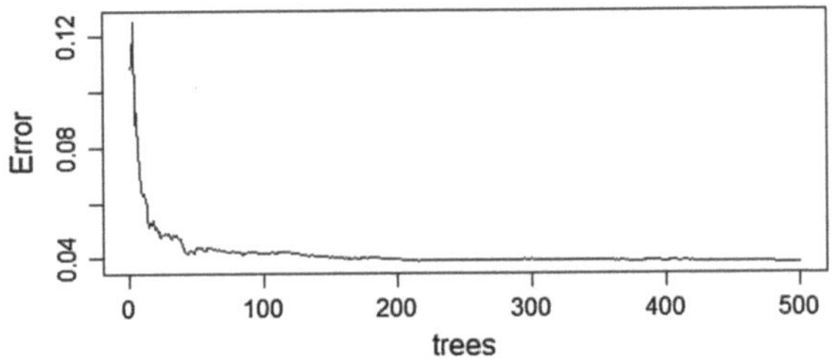

Fig. 3. Error versus no. of trees—point load

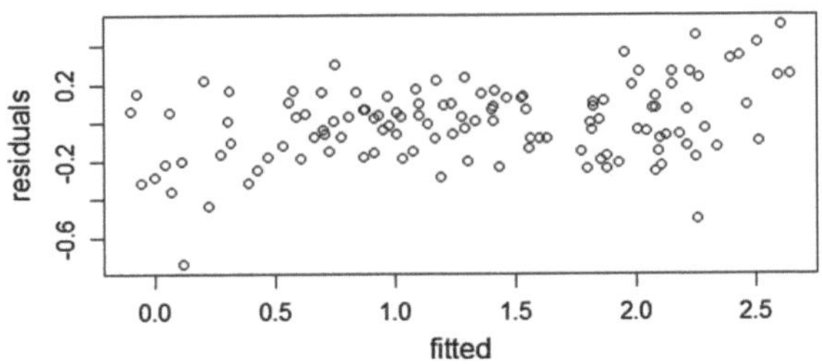

Fig. 4. Residuals versus fitted plot—point load

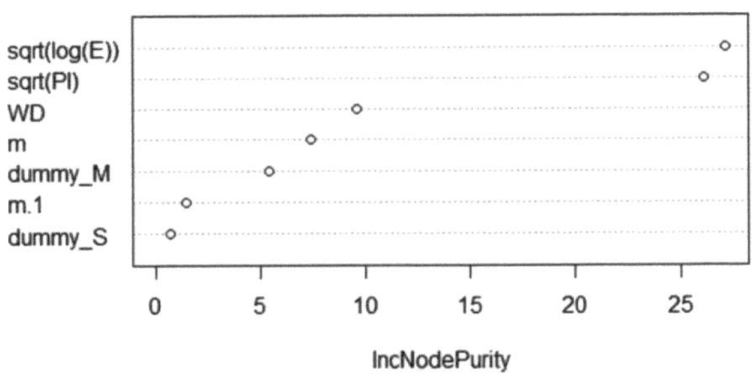

Fig. 5. Variable importance plot for point load

Figure 5 shows the relative importance of explanatory variables in predicting Unconfined Compression Strength for Point Load ($I_{S(50)}$). It indicates that the Modulus of Elasticity (in square-root logarithm scale) and Point Load (in square root scale) are the two most important variables in the prediction. The other variables with important effects in predicting UCS are wet density and depth. The same for Ultra Sonic Velocity can be found in Figs. 6 and 7. Figure 6 illustrates the change in the error according to the increase in number of trees.

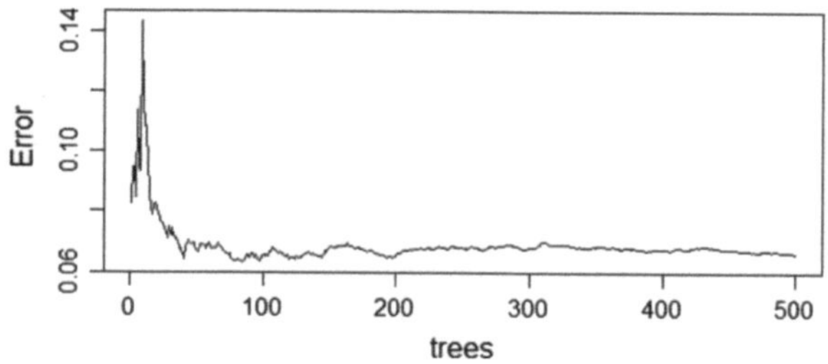

Fig. 6. Error versus no. of trees—ultra sonic velocity

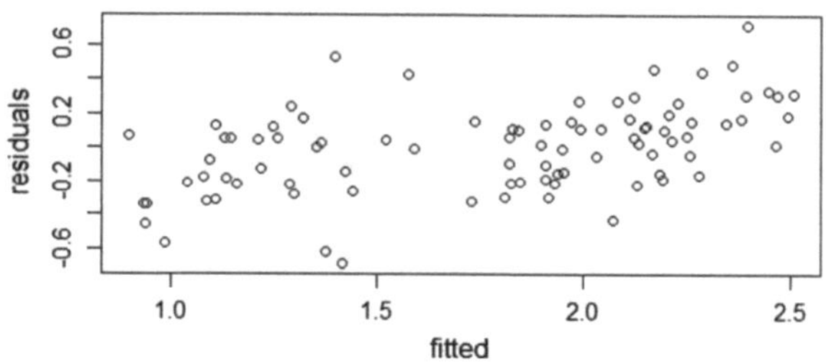

Fig. 7. Residuals versus fitted values—ultra sonic velocity

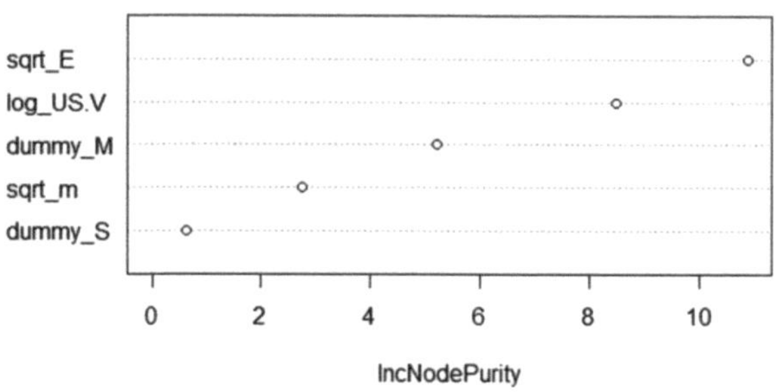

Fig. 8. Variable importance plot for ultra sonic velocity

For Ultra Sonic Velocity, Fig. 8 indicates the Modulus of Elasticity (in square root scale) and Ultra Sonic Velocity (in logarithm scale) are the two most important variables in the purpose of prediction. The other variables that are important in predicting UCS are $dummy_m$ and depth.

3.2 Validation of Model

In order to obtain the best possible results for new rock samples, correctness and applicability of the derived equations was cross-checked using cross-validation data. Keeping in mind the size of each sample, a fourfold cross-validation was performed on each of the derived models. The raw cross-validation estimate of prediction error (denoted by delta) was recorded.

These low values of delta signified that the cross-validation error is low enough to be negligible. Thus, the model under study, although not totally accurate, could be used for prediction purposes with a certain margin of error.

4 Comparison of Models

Keeping in mind the fundamental difference between the proposed models, the notion of comparison is restricted to the applicability, suitability, and accuracy of prediction of strength properties. However, the mechanism behind tree-based Random Forest models does not output coefficients, like regression analysis. Rather, coefficients are utilized for determining the relative importance of variables in prediction. This study aims not to provide a better governing equation, but to establish a model that can better predict with precision.

In this study, the regression model outperformed the random forest model in terms of R^2 and Standard Error of Estimates. Table 3 illustrates the comparison of coefficients of determination for both Point Load ($I_{S(50)}$) and Ultra Sonic Velocity (USV). Table 4 shows the same for Standard Error of Estimates. Comparative study between models led to the finding that stepwise regression could produce an acceptable fit to the data.

Table 3. Comparison of R^2

	R^2—point load ($I_{S(50)}$)	R^2—ultra sonic velocity (USV)
Stepwise regression	0.95	0.93
Random forest	0.94	0.83

Table 4. Comparison of SEE

	SEE—point load ($I_{S(50)}$)	SEE—ultra sonic velocity (USV)
Stepwise regression	0.16	0.15
Random forest	0.20	0.26

In Figs. 9, 10, 4, and 7, accuracy of prediction of the models can be estimated by observing the spread of residuals plotted against the fitted values for each model. Results justify the suitability of stepwise regression as a predictive model for the experimental data under study. Thus Eqs. 11 and 12 derived from the regression model could be used as predicting equations that govern the relationship between Unconfined Compression Strength and various properties of rocks (Table 5).

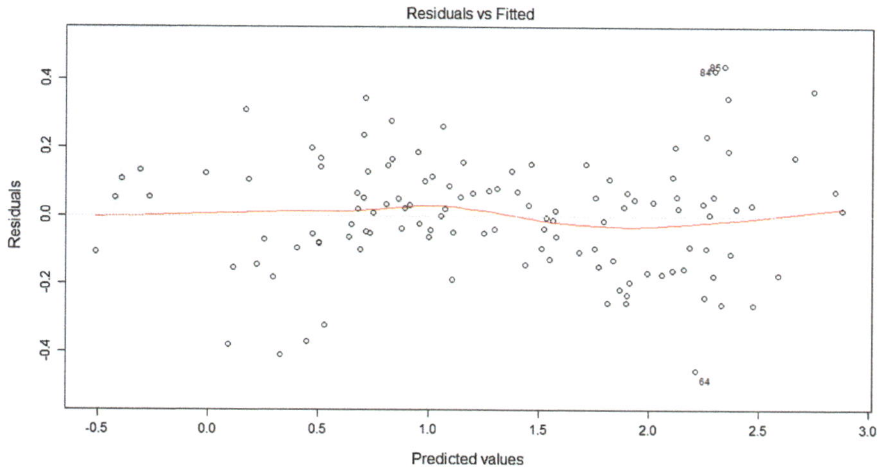

Fig. 9. Residuals versus predicted values plot for $I_{S(50)}$

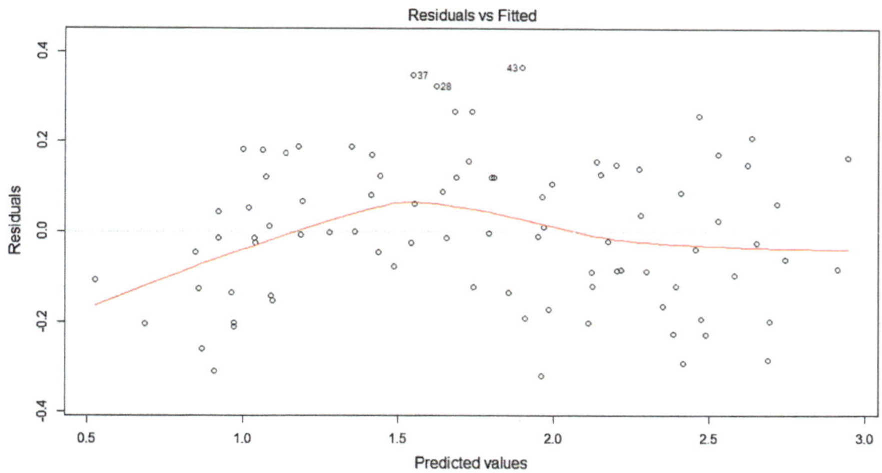

Fig. 10. Residuals versus predicted values plot for USV

Table 5. Comparison of accuracy of prediction

	Conf. interval—point load ($I_{S(50)}$) (%)	Conf. interval—ultra sonic velocity (%)
Stepwise regression	6.67	6.67
Random	8	15

5 Conclusion

The more that engineers and geologists understand mechanical properties of different types of rocks, the better the assessment of rock strength and therefore the better the resulting construction projects. An experimental dataset consisting of 210 rock samples was tested for Point Load ($I_{S(50)}$) or Ultra Sonic Velocity (USV). A statistical model based on stepwise regression and an advanced tree-based Random Decision Forest model were developed. The development of each predictive model was followed by validation of the model using fourfold cross-validation data. A comparative study of the results from these models led to the finding that a stepwise regression model for both Point Load and Ultra Sonic Velocity outperformed the tree-based model (Table 6).

Table 6. Estimate of prediction error—cross validation delta

	Point load ($I_{S(50)}$)	Ultra-sonic velocity
Delta	0.03303	0.02791

It was observed that the regression model could predict Unconfined Compression Strength within ±6.67% confidence interval. For Point Load ($I_{S(50)}$), the adjusted R^2 for the regression model was 0.95 and the Standard Error of Estimates was 0.16. On the other hand, for Ultra Sonic Velocity the measured values were likely to lie within the same confidence interval, with an adjusted R^2 of 0.93 and Standard Error of Estimates of 0.15. The research terminated with the conclusion that the developed models can be used for approximating the Unconfined Compression Strength of rocks in United Arab Emirates, and the models can be seen as reasonably accurate in this regard.

Although the results from the research were satisfactory and could be used for the purpose of benchmarking, there is scope for further initiatives that involve properties of rocks that could not be entertained in this study due to the lack of time and funding.

References

ASTM D2166-06: Standard test method of unconfined compressive strength of cohesive soil. ASTM Int'l

Bieniawski, Z.T.: Engineering classification of jointed rock masses. Trans. S. Afr. Inst. Civ. Eng. **15**(12), 335–344 (1973); and authors reply: 4F, 4T, 39R, Trans. S. Afr. Inst. Civ. Eng. **16**(7), 239–254 (1974)

Broch, E., Franklin, J.A.: The point-load strength test. Int. J. Rock Mech. Min. Sci. Geomech. Abstr. **9**(6), 669–676 (1972). (Pergamon)

Bursztyn, N., Pederson, J.L., Tressler, C., Mackley, R.D., Mitchell, K.J.: Rock strength along a fluvial transect of the Colorado plateau–quantifying a fundamental control on geomorphology. Earth Planet. Sci. Lett. **429**, 90–100 (2015)

Cargill, J.S., Shakoor, A.: Evaluation of empirical methods for measuring the uniaxial compressive strength of rock. Int. J. Rock Mech. Min. Sci. Geomech. Abstr. **27**(6), 495–503 (1990). (Pergamon)
Chau, K.T., Wong, R.H.C.: Uniaxial compressive strength and point load strength of rocks. Int. J. Rock Mech. Min. Sci. Geomech. Abstr. **33**(2), 183–188 (1996). (Pergamon)
D'Andrea, D.V., Fischer, R.L., Fogelson, D.E.: Prediction of Compressive Strength from Other Rock Properties, vol. 6702. U.S. Dept. of the Interior, Bureau of Mines (1965)
Fener, M., Kahraman, S., Bilgil, A., Gunaydin, O.: A comparative evaluation of indirect methods to estimate the compressive strength of rocks. Rock Mech. Rock Eng. **38**(4), 329–343 (2005)
Hack, R., Huisman, M.: Estimating the intact rock strength of a rock mass by simple means. In: Proceedings of the 9th Congress of the International Association for Engineering Geology and the Environment Durban, South Africa, pp. 1971–1977 (2002)
Hoek, E., Bray, J.D.: Rock Slope Engineering. CRC Press (1981)
Hoek, E., Carranza-Torres, C.T, Corkum, B.T.: Hoek-Brown failure criterion. In: Proceeding of the 5th North American Rock Mechanics. Toronto, pp. 267–273. University of Toronto Press (2002)
Lai, G.T., Rafek, A.G., Serasa, A.S., Simon, N., Ern, L.K.: Empirical correlation of uniaxial compressive strength and primary wave velocity of malaysian granites. Electron. J. Geotech. Eng. **19E**, 1063–1072 (2014)
Motamedi, S., Shamshirband, S., Petković, D., Hashim, R.: Application of adaptive neuro-fuzzy technique to predict the unconfined compressive strength of PFA-sand-cement mixture. Powder Technol. **278**, 278–285 (2015)
Omar, M.: Empirical correlations for predicting strength properties of rocks—United Arab Emirates. Int. J. Geotech. Eng. (2017). doi: https://doi.org/10.1080/19386362.2016.1214339
Rusnak, J., Mark, C.: Using the point load test to determine the uniaxial compressive strength of coal measure rock. In: Proceedings of the 19th international conference on ground control in mining, pp. 362–371. West Virginia University, Morgantown, WV, Aug 2000
Selçuk, L., Kayabali, K.: Evaluation of the unconfined compressive strength of rocks using nail guns. Eng. Geol. **195**, 164–171 (2015)
Sheshde, E.A., Cheshomi, A.: New method for estimating unconfined compressive strength (UCS) using small rock samples. J. Petrol. Sci. Eng. **133**, 367–375 (2015)
Voznesenskii, A.S., Kutkin, Y.O., Krasilov, M.N., Komissarov, A.A.: Predicting fatigue strength of rocks by its interrelation with the acoustic quality factor. Int. J. Fatigue **77**, 194–198 (2015)
Yilmaz, I.: A new testing method for indirect determination of the unconfined compressive strength of rocks. Int. J. Rock Mech. Min. Sci. **46**(8), 1349–1357 (2009)

Yield Criterion and Simulation Validation for Different Deformation Modes of Layered Rock Mass Based on Transversely Isotropic Theory

Shuangying Zuo[1(✉)], Xiaoyong Zhao[2], Donglei Zhao[1], and Jing Zhang[1]

[1] College of Resources and Environment Engineering, Guizhou University, Huaxi District, Guiyang 550025, China
shyzuo@163.com, {10760829,1464750356}@qq.com
[2] Guizhou Construction Science Research and Design Institute of CSCEC Co. Ltd., Nanming District, Guiyang 550006, China
33499967@qq.com

Abstract. Due to a distinct layered structure, the layered rock mass has not only manifest anisotropic characteristics in deformation and strength, but also is dissimilar from the homogeneous rock mass in the destruction modes. First, the indoor uniaxial compression tests for 7 groups of rock samples with different dip angles are carried out. The variation of elastic modulus and uniaxial compressive strength of layered rock mass with the dip angle of the layer are analyzed from the experiment results. It is concluded that the layered rock mass has obvious anisotropic characteristics and takes on multiple failure modes: rock block shear, rock block cracking, stratification plane gliding and stratification plane cracking. Second, according to the elastic constitutive relation of the layered rock mass in the local coordinate system, a self-defined model on the transversely isotropic feature could be established by the transformation of the stress coordinate. The yield criteria under various failure modes is built and their corresponding stress recalculation methods are deduced based on the plastic flow rules. Thus, the self-defined constitutive model called as "Layered" is formed. Third, the numerical simulation of uniaxial compression for standard rock samples on Layered model is carried out and compared with the indoor tests. The two results keep a very high consistency. The results show that the self-defined Layered constitutive model is very applicable, correct and superior to the layered rock mass.

1 Introduction

In the long geological historical conditions, the rock mass has formed a variety of primary or secondary structure plane, such as bedding, joint, schistosity, folds, weak surface and so on. These rock with the planes developed along a directional distribution are regarded as a generalized layered rock mass. In 1963, Lekhnitskii started from the generalized Hooke law of continuous media, and derived the general equation of the anisotropic elastic theory, which provides a theoretical basis for the study of anisotropic

B. Fatahi et al. (eds.), *Sustainable Design and Construction for Geomaterials and Geostructures*, Sustainable Civil Infrastructures,
https://doi.org/10.1007/978-3-319-95753-1_6

problems. According to the symmetry surface of the material properties, the anisotropic body can be divided into extreme anisotropic body, monosymmetric anisotropic body, orthotropic body and transversely isotropic body. The layered rock mass studied in this paper belongs to transversely isotropic body, which is formed mainly by sedimentation action, and its physical and mechanical properties in the bedding direction and in the vertical bedding direction are different. Therefore, when we study the mechanical effects caused by force, it is often simplified as transversely isotropic medium.

1.1 Research Status of Constitutive Model of Layered Rock Mass

The constitutive relationship of rock mass is the core content of the theory of rock mechanics. It is the foundation for the establishment of rock mechanical model, numerical simulation and calculation analysis. With the deep study of the mechanical properties of layered rock mass, the following three types of mechanical constitutive models have been proposed by scholars at home and abroad.

1.1.1 Equivalent Isotropic Model

The strength of the stratification plane is incorporated into the strength of the rock block by reduction, without considering the failure characteristics of the structural plane, and the elasto-plastic analysis was done by the isotropic strength criterion. Sitharam et al. (2001) studied the characteristics of equivalent and continuous of jointed rock mass by experiments. Yujun and Yixing (1999) established the equivalent elastic–plastic constitutive model considering the strength anisotropy of layered rock mass. Yin et al. (2017) virtualized the layered composite rock as an equivalent homogeneous rock, and established the failure criterion of lamellar composite rock based on MLC criterion under real triaxial stresses. This method is feasible when the anisotropy of the rock mass is weak, but there are obvious limitations when the stratification plane effect is strong.

1.1.2 Anisotropic Parameter Model

The constitutive model of the layered rock mass is established by introducing various parameters of the anisotropic characteristic in the strength criterion of isotropic model. Yujun and Yiping (2002) used the nonlinear Mohr-Coulomb criterion to study the numerical analysis of elastic–plastic and visco-elasto-plasticity analysis of underground caverns in layered and orthogonal anisotropic rock mass, considering the variation of shear strength parameters with the structural plane, based on the traditional anisotropic constitutive relation. Xu et al. (2013) studied the general method of anisotropic yield criterion based on the isotropic Mohr-Coulomb shear yield criterion, which showed that the yield criterion of orthogonal anisotropy is related to the three principal stress and the yield surface did not always coincide with the maximum shear stress plane. This method takes into account the anisotropy of the strength of the layered rock, but does not involve the failure mode that the structural plane may cause.

1.1.3 Anisotropic Function Model

Based on the test results, the function relation between the strength of the rock mass and the dip angle of stratification plane or the maximum principal stress orientation is derived to establish the yield criterion of layered rock. Zhang and Zhang (2014) put forward an empirical formulae in which shear strength parameters of layered rock mass can change with layer orientation, and try using Mohr-Coulomb yield criteria to search and judge if the stratified rock mass would generate shear failure and the failure position under the three dimensional stresses. Lee (2015), Lee and Pietruszczak (2008) hypothesized that the cracks in isotropic rock mass were controlled by Hoek-Brown's failure criteria, and described the anisotropy of strength by the azimuth dependence of parameters "*m*" and "*s*". Saroglouh (2008) completed the correction of Hoek-Brown's failure criteria by introducing a new parameter "K_b" that reflects the impact of strength anisotropy. Pietruszczak and Pande (2012) applied the second order microstructure tensor to describe the inherent anisotropy of sedimentary rocks, and the distribution function showed exactly the azimuth dependence of strength parameters. This method considers the level effect and the anisotropy of materials, but it has to be studied further that how to determine the parameters in the area where the geological conditions are more complicated or geo-stress change relatively largely.

Due to the complexity of the anisotropy, the research on the anisotropic constitutive relation and yield criterion of rock mass is still placed in theories research stage, and it is rarely applied in practical rock engineering. The main reasons are as follows: (1) The constitutive relation and yield criterion of anisotropic rock mass are very complicated, there are too many parameters and it is very difficult to get them. (2) The anisotropic constitutive relation is difficult to express with "explicit" mathematical formula. The calculation process is too complicated and the calculation is too large.

1.2 Research Ideas

Although the study of mechanical properties and deformations failure mechanism of layered rock mass has made many achievements, the numerical models that can reasonably reflect the anisotropy and integrated failure modes of layered rock are rare at present. The Elastic, Transversely Isotropic Models and the Ubiquitous-Joint Model embedded in FLAC^{3D} can take into account the transversely isotropic and characteristics of structural plane, but they are flawed in application. The former can only be used for simple linear elastic calculation, can not carry out the stress calculation of the plastic phase and the discrimination of plastic zone. In the latter model, the Mohr-Coulomb model can be used for rock blocks and structural planes to establish their own strength yield criteria, but the constitutive model is still isotropic, and does not take into account the different parameters of the rock material in each direction, and can not truly reflect the transversely isotropic deformation characteristics of the layered rock mass (Zuo et al. 2013).

In this paper, the anisotropic characteristics of layered rock mass are analyzed by tests of rock mass mechanical parameters, and the Ubiquitous-Joint Model of FLAC^{3D} will be improved. The failure mode of layered rock mass is divided into 4 kinds: rock block shear failure, rock block cracking, stratification plane gliding and stratification plane cracking. The yield criterions are established and the stress modification method are deduced under various failure modes. The validity, correctness and accuracy of the model are verified by comparing the numerical results of standard rock samples with the indoor test results.

2 Rock Uniaxial Compressive Test

2.1 The Experiment Condition and Process

The samples of layered limestone with thin-medium layer are in the size of 50 mm diameter and 100 mm height. The angle of inclination β is defined as the angle between the bedding plane and the horizontal plane, which is respectively 0°, 15°, 30°, 45°, 60°, 75° and 90°, and the sample number of each group is 6. All 7 groups with different dip angles are shown in Fig. 1.

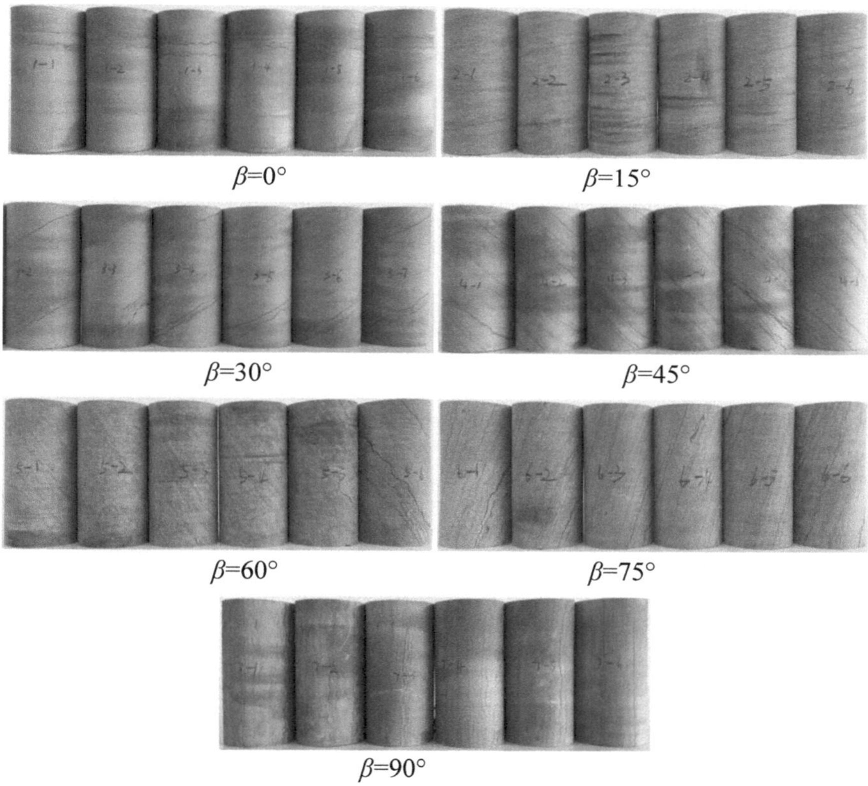

Fig. 1. Photos of rock specimens with different angles β

2.2 The Experiment Results

2.2.1 Axial Stress–Strain Curve

(1) The same rules: Fig. 2 shows that the stress–strain curve is roughly convex shape. The entire compression process can be divided into compaction stage, elastic stage, plastic stage and failure stage.

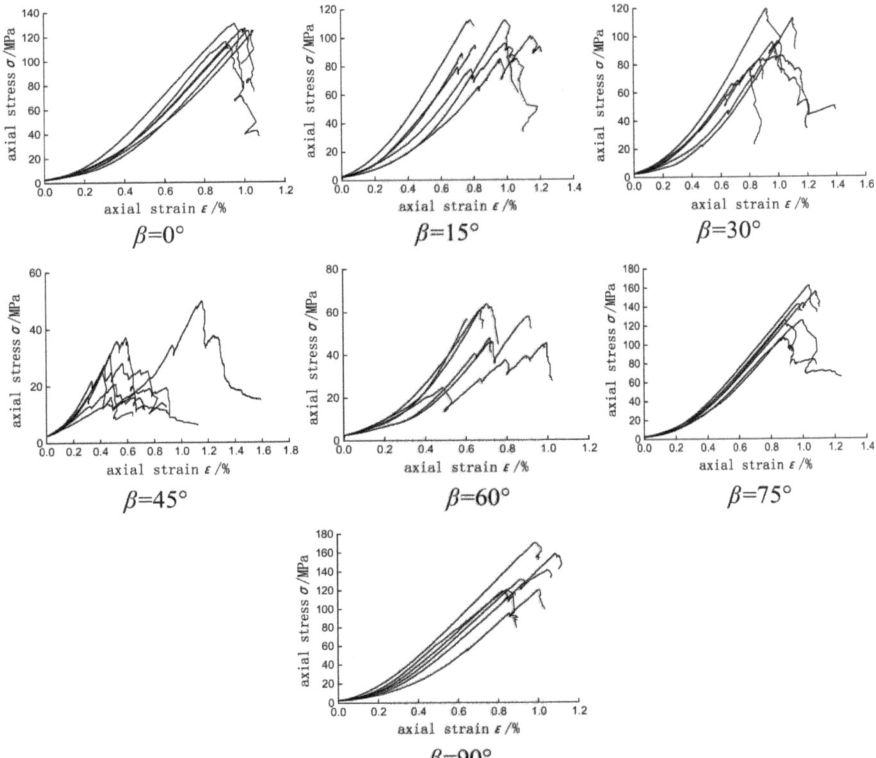

Fig. 2. Axial stress–strain curves for specimens at different layer angles

(i) Compaction stage: At the beginning of the loading, the curve rising is relatively slow, the strain increment decreases with the increasing of stress. The main reason is: when the stress is very small, bedding is compressed or microcracks are closed under the action of stress. Rock sample appears a long obvious compression deformation.
(ii) Elastic stage: With stress increasing, bedding planes and microcracks in rock mass are further compressed and gradually in the closed state. And the curve is steepening, similar to a straight line.
(iii) Plastic stage: After near mid-loading, the phenomenon that the new cracks appear constantly and original cracks continue to expand, makes some irreversible deformation of rocks. The curve fluctuates until it reaches its peak.
(iv) Failure stage: when the stress exceeds its maximum carrying capacity, and the rock sample begins to rupture, and enters the destruction phase. After the peak, the stress decreases rapidly with the increasing of strain, and shows the obvious softening phenomenon.

(2) Different characteristics: When the dip angle of rock samples is different, the peak of the axial strain and the corresponding stress is different. The microcrack, opening width, cementation degree of interface and the failure mode in the rock can also affect the intermediate development of the curve. For example, the rock samples with inclination 45° are repeatedly fluctuating; the 0° and 90° rock samples rise and fall relatively smoothly.

2.2.2 Failure Modes

The deformation and failure of layered rock mass are decided by rock block strength, bedding surface strength and dip angles. According to the experimental results, failure modes can be roughly divided into the following kinds:

(1) Splitting along the axial stress direction (rock block cracking, rock block shear): When β is in 0°–15°, the dip angle is gentle and the load is mainly applied to the bedding structure vertically. As the load increases, the cracks in the rock block are cutting through from both the ends toward the middle of the rock block gradually. The splitting failure is shown in Fig. 3a, b.

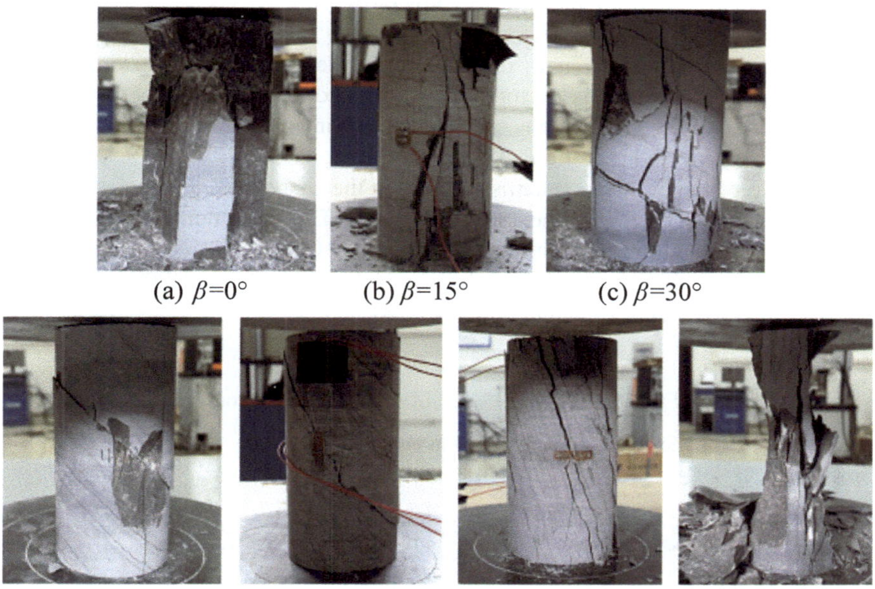

(a) β=0° (b) β=15° (c) β=30°

(d) β=45° (e) β=60° (f) β=75° (g) β=90°

Fig. 3. Fracture morphology of different angled rock samples

(2) The fracturing along the axial stress direction (rock block cracking) and the slip failure of the structural plane along the layer (stratification plane gliding): With the increasing of β, a part of the load acts on the tangential direction of the bedding plane gradually. The deformation of the rock is transformed from the normal to the tangential

of layers and the shear strength of the bedding surface begins to control the deformation and movement of the rock. When the shear stress on the bedding plane exceeds its shear strength, the rock specimens starts to stagger along the bed and come to shear slip failure, accompanied by fracturing along the direction of axial stress, shown in Fig. 3c.

(3) The sliding along the bedding structure (stratification plane sliding): When the β is further increased, most of the load begins to act mainly on the tangential direction of the bedding plane, and the deformation of the rock is completely controlled by the bedding structure, shown in Fig. 3d, e.

(4) The cracking along the bedding structure (stratification plane cracking): when $\beta \geq 75°$, almost all of the load acts on the tangential direction of the bedding structure. The deformation of the rock is mainly controlled by the tension between the layers. When the tensile strength of the bedding is not enough to resist the opening deformation caused by the load, stratification cracking will occur. First, some local microcracks are generated, then connected to form continuous lines as the load increases. The rupture surface is slightly cluttered and poorly regular, shown in Fig. 3f, g.

2.2.3 Anisotropy of Deformation and Strength

(1) Anisotropy of deformation: In order to study the deformation anisotropy of the layered rock mass, the elastic modulus E (defined as the line slope of σ-ε *curve*) obtained from the six specimens of each inclination angle are statistically analyzed, listed in Table 1 and plot the E-β curve as shown in Fig. 4. It can be seen that the E-β curve of the rock specimen is roughly in the shape "U". E decreases first and then increases along with the increasing of β. The maximum of E is 7.3 GPa, appeared at $\beta = 90°$; the minimum of β is 1.9 GPa, appeared at $\beta = 45°$. The corresponding elastic modulus is different when β is 0° and 90°, that is when $\beta = 0°$, $E = 5.9$ GPa, less than when $\beta = 90°$.

Table 1. Elastic modulus E (GPa) of different angled rock specimens

Specimen no.	Inclination angle β						
	0°	15°	30°	45°	60°	75°	90°
1	5.3	4.8	4.5	1.9	2.2	5.7	7.3
2	6.9	4	4.6	2.0	3.1	6	7.6
3	5.4	6.2	3.2	2.4	3.4	5.4	7.4
4	5.4	4	5.4	1.4	3.6	7.3	8.4
5	6.4	5.1	3.8	2	2.5	6.9	5.3
6	5.8	5.8	4.5	0.9	1.9	8.1	7.5
Average	5.9	5.0	4.3	1.9	2.8	6.6	7.3

(2) Anisotropy of strength: The results of the peak strength σ_p (defined as the peak load divided by cross-sectional area) and the residual strength σ_r (defined as the

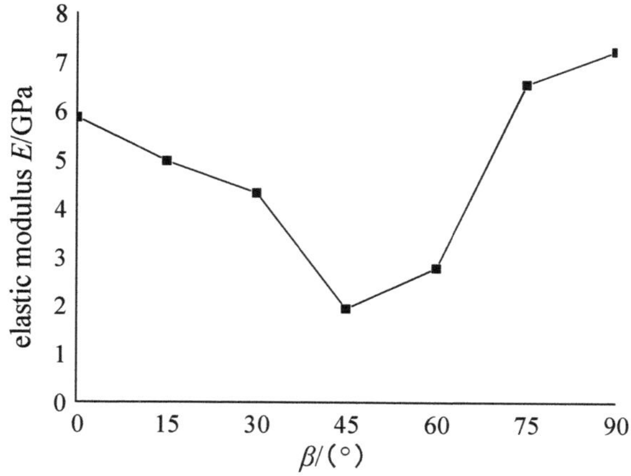

Fig. 4. Relation between elastic modulus and layer angle

residual load divided by cross-sectional area) are listed in Table 2. The relation curves of the peak strength σ_p and the residual strength σ_r of the rock samples are shown in Fig. 5. The curves of σ_p-β, σ_r-β take on approximately "V" style and their laws are the same as that of E-β curve. σ_p and σ_r decrease first and then increase along with the increasing of β. The maximum is 138 and 129 MPa respectively when $\beta = 90°$, the minimum is 27 and 11 MPa respectively when $\beta = 45°$.

Table 2. Peak strength σ_p and residual strength σ_r of different angled rock specimens (Unit: MPa)

Inclination angle β	Peak strength σ_p and residual strength σ_r						
	1	2	3	4	5	6	Average
0°	125	130	125	115	115	125	123
	97	76	76	36	112	117	96
15°	110	110	110	95	90	95	102
	36	87	107	61	87	92	87
30°	75	110	80	120	85	95	94
	25	92	71	71	36	46	63
45°	25	25	35	29	39	50	27
	10	10	10	5	10	15	11
60°	45	60	62	65	55	45	55
	25	51	52	46	56	41	49
75°	125	125	105	155	140	160	135
	66	76	92	138	138	132	115
90°	120	120	140	170	120	160	138
	117	82	132	153	102	143	129

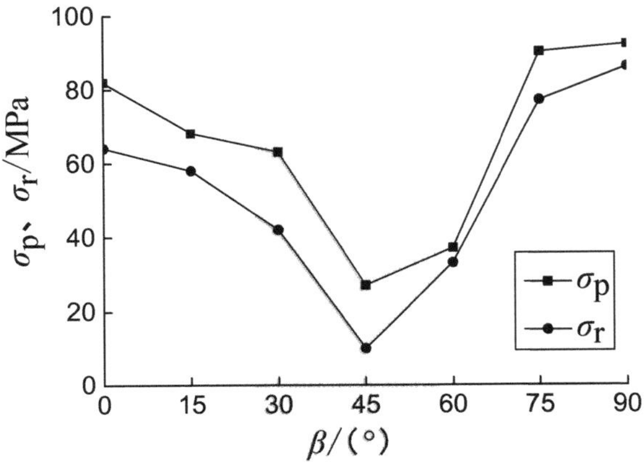

Fig. 5. Relation between peak strength plus residual strength and layer angle

3 Constitutive Model of Layered Rock Mass

3.1 Elastic Constitutive Relation of Layered Rock Mass

3.1.1 Elasticity in Local Coordinate System

The layered anisotropic rock mass is isotropic in the $x'y'$ plane parallel to the layer, and the z' direction perpendicular to the layer is an elastic direction that is different from the layer. According to the elastic theory, the relationship between stress $\{\sigma'\}$ and strain $\{\varepsilon'\}$ in layered normal coordinate system (also called as the local coordinate system) is as follows:

$$\{\sigma'\} = [D'_e]\{\varepsilon'\} \tag{1}$$

where, $[D'_e]$ is the elastic matrix of the layered anisotropic medium in the layered normal coordinate system.

$$[D'_e] = \begin{bmatrix} A & B & C & & & \\ B & A & C & & 0 & \\ C & C & D & & & \\ & & & G_2 & & \\ & 0 & & & G_2 & \\ & & & & & G_1 \end{bmatrix} \tag{2}$$

where, $A = E_1(1 - n\mu_2^2)/m$, $B = E_1(\mu_1 + n\mu_2^2)/m$, $C = E_1\mu_2(1 + \mu_1)/m$, $D = E_2(1 - \mu_1^2)/m$,
$G_1 = \frac{E_1}{2(1+\mu_1)}$, $G_2 = \frac{E_1 E_2}{E_1(1+2\mu_1)+E_2}$, $n = E_1/E_2$, $m = (1 + \mu_1)(1 - \mu_1 - 2n\mu_2^2)$.

$E_1, \mu_1, G_1,\ E_2, \mu_2, G_2$ represent the elastic modulus, Poisson's ratio and shear modulus parallel to the plane and normal to the plane respectively.

3.1.2 Elasticity in Integral Coordinate System

In the practical engineering, the integral coordinate system of calculation is often inconsistent with the normal coordinate system of the stratification plane. So the elastic matrix $[D'_e]$ in the layered normal coordinate system must be transformed into $[D_e]$ in the integral coordinate system. The *x-y* coordinate plane in the integral coordinate system *Oxyz* is used as the reference plane. The angle between the layer and the *x-y* surface is the dip angle. The angle between the projection line of the normal of the layer on the reference plane and the positive direction of the y axis is dip direction (counterclockwise direction is positive). The dip direction and dip angle are α and β respectively shown in Fig. 6.

Hence, the coordinate transformation matrix is obtained

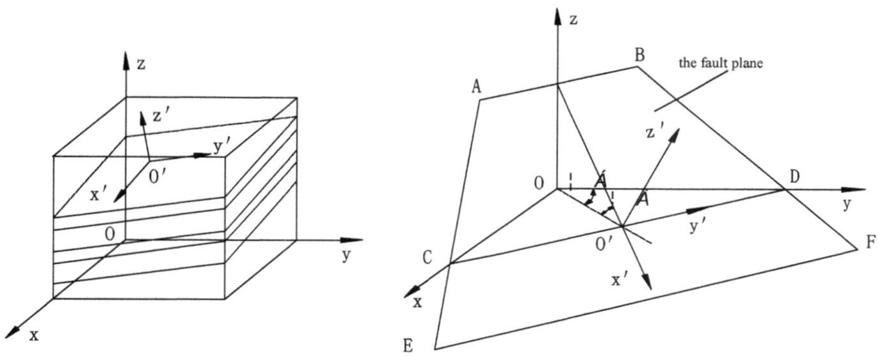

Fig. 6. Spatial relationship between bedded plane and the datum

$$[R] = \begin{bmatrix} \sin\alpha\cos\beta & \cos\alpha\cos\beta & -\sin\beta \\ -\cos\alpha & \sin\alpha & 0 \\ \sin\alpha\sin\beta & \cos\alpha\sin\beta & \cos\beta \end{bmatrix}$$

The stress conversion matrix from the integral coordinate system *Oxyz* to the local coordinate system *Ox'y'z'* can be written as:

$$[S] = \begin{bmatrix} l_1^2 & m_1^2 & n_1^2 & 2m_1n_1 & 2l_1n_1 & 2l_1m_1 \\ l_2^2 & m_2^2 & n_2^2 & 2m_2n_2 & 2l_2n_2 & 2l_2m_2 \\ l_3^2 & m_3^2 & n_3^2 & 2m_3n_3 & 2l_3n_3 & 2l_3m_3 \\ l_2l_3 & m_2m_3 & n_2n_3 & m_2n_3+m_3n_2 & l_2n_3+l_3n_2 & l_2m_3+l_3m_2 \\ l_1l_3 & m_1m_3 & n_1n_3 & m_1n_3+m_3n_1 & l_1n_3+l_3n_1 & l_1m_3+l_3m_1 \\ l_1l_2 & m_1m_2 & n_1n_2 & m_1n_2+m_2n_1 & l_1n_2+l_2n_1 & l_1m_2+l_2m_1 \end{bmatrix}$$

Therefore, the relationship between the stress $\{\sigma\}$, strain $\{\varepsilon\}$ in the integral coordinate and the stress $\{\sigma'\}$ and strain $\{\varepsilon'\}$ in the local coordinate system can be written as:

$$\{\sigma'\} = [S]\{\sigma\}, \{\varepsilon'\} = [S]\{\varepsilon\} \tag{3}$$

We can get the elastic relationship in the integral coordinate system as follows:

$$\{\sigma\} = [S]^{\mathrm{T}}[D'_e][S]\{\varepsilon\} \tag{4}$$

Comparing the stress expression in the integral coordinates with the local coordinates, we can get the elastic matrix under the integral coordinate system as follows:

$$[D_e] = [S]^{\mathrm{T}}[D'_e][S] \tag{5}$$

3.2 The Yield Criterion for Each Failure Modes

3.2.1 Rock Block Damage

According to the results of indoor uniaxial compression test, the yield of the layered rock mass occurs within the rock block when the dip angle is gentle.

(1) Rock block shear failure: The envelope of failure function $f(\sigma_1, \sigma_3) = 0$ follows Mohr-Coulomb criterion, and the yield function is:

$$f^s = \sigma_1 - \sigma_3 N_\phi + 2c\sqrt{N_\phi} \tag{6}$$

where, ϕ is the friction angle, c is the cohesion, $N_\phi = \frac{1+\sin(\phi)}{1-\sin(\phi)}$.

When $f^s > 0$, rock block shear failure occurs, and the shear plastic flow of the rock block is defined by the unassociated flow rule. The plastic potential function g^s is:

$$g^s = \sigma_1 - \sigma_3 N_\psi \tag{7}$$

where, Ψ is the dilatancy angle, $N_\Psi = \frac{1+\sin\Psi}{1-\sin\Psi}$.

For Eq. (7), take the partial derivatives to each stress component:

$$\frac{\partial g^s}{\partial \sigma_x} = 1, \frac{\partial g^s}{\partial \sigma_y} = 0, \frac{\partial g^s}{\partial \sigma_z} = -N_\psi, \frac{\partial g^s}{\partial \tau_{yz}} = 0, \frac{\partial g^s}{\partial \tau_{xz}} = 0, \frac{\partial g^s}{\partial \tau_{xy}} = 0 \tag{8}$$

Put Eq. (8) into the elastic constitutive Eq. (4) of the integral coordinate system, and calculate the imaginary elastic stress increment.

$$S_1\left(\frac{\partial g^s}{\partial \sigma_x},\frac{\partial g^s}{\partial \sigma_y},\frac{\partial g^s}{\partial \sigma_z},\frac{\partial g^s}{\partial \tau_{yz}},\frac{\partial g^s}{\partial \tau_{xz}},\frac{\partial g^s}{\partial \tau_{xy}}\right) = D_{11} - D_{13}N_\psi$$
$$S_2\left(\frac{\partial g^s}{\partial \sigma_x},\frac{\partial g^s}{\partial \sigma_y},\frac{\partial g^s}{\partial \sigma_z},\frac{\partial g^s}{\partial \tau_{yz}},\frac{\partial g^s}{\partial \tau_{xz}},\frac{\partial g^s}{\partial \tau_{xy}}\right) = D_{21} - D_{23}N_\psi$$
$$S_3\left(\frac{\partial g^s}{\partial \sigma_x},\frac{\partial g^s}{\partial \sigma_y},\frac{\partial g^s}{\partial \sigma_z},\frac{\partial g^s}{\partial \tau_{yz}},\frac{\partial g^s}{\partial \tau_{xz}},\frac{\partial g^s}{\partial \tau_{xy}}\right) = D_{31} - D_{33}N_\psi$$
$$S_4\left(\frac{\partial g^s}{\partial \sigma_x},\frac{\partial g^s}{\partial \sigma_y},\frac{\partial g^s}{\partial \sigma_z},\frac{\partial g^s}{\partial \tau_{yz}},\frac{\partial g^s}{\partial \tau_{xz}},\frac{\partial g^s}{\partial \tau_{xy}}\right) = D_{41} - D_{43}N_\psi$$
$$S_5\left(\frac{\partial g^s}{\partial \sigma_x},\frac{\partial g^s}{\partial \sigma_y},\frac{\partial g^s}{\partial \sigma_z},\frac{\partial g^s}{\partial \tau_{yz}},\frac{\partial g^s}{\partial \tau_{xz}},\frac{\partial g^s}{\partial \tau_{xy}}\right) = D_{51} - D_{53}N_\psi$$
$$S_6\left(\frac{\partial g^s}{\partial \sigma_x},\frac{\partial g^s}{\partial \sigma_y},\frac{\partial g^s}{\partial \sigma_z},\frac{\partial g^s}{\partial \tau_{yz}},\frac{\partial g^s}{\partial \tau_{xz}},\frac{\partial g^s}{\partial \tau_{xy}}\right) = D_{61} - D_{63}N_\psi$$

where, $S_i(\bullet)$ is a generalized Hooke function, D_{ij} is the element of the integral elastic matrix $[D_e]$.

The new stress expression is

$$\sigma_i^N = \sigma_i^I - \lambda S_i\left(\frac{\partial g}{\partial \sigma_n}\right) \tag{9}$$

where, σ_i^I is the imaginary elastic stress calculated according to the total strain increment, λ is the stress correction factor, and the expression is:

$$\lambda = \frac{f(\sigma_i^I)}{f[S_n(\frac{\partial g}{\partial \sigma_n})] - f(O_n)} \tag{10}$$

where, $f(O_n)$ denotes a constant term for the yield function.

Substituting S_1, S_2, S_3, S_4, S_5, S_6 into the new stress expression (9), we have:

$$\begin{array}{ll} \sigma_x^N = \sigma_x^I - \lambda^s(D_{11} - D_{13}N_\psi) & \sigma_y^N = \sigma_y^I - \lambda^s(D_{21} - D_{23}N_\psi) \\ \sigma_z^N = \sigma_z^I - \lambda^s(D_{31} - D_{33}N_\psi) & \tau_{yz}^N = \tau_{yz}^I - \lambda^s(D_{41} - D_{43}N_\psi) \\ \tau_{xz}^N = \tau_{xz}^I - \lambda^s(D_{51} - D_{53}N_\psi) & \tau_{xy}^N = \tau_{xy}^I - \lambda^s(D_{61} - D_{63}N_\psi) \end{array}$$

According to Eq. (10), we have:

$$\lambda^s = \frac{f^s(\sigma_1^I, \sigma_3^I)}{(D_{11} - D_{13}N_\psi) - (D_{31} - D_{33}N_\psi)N_\varphi}$$

Therefore, the incremental expression of each stress component is:

$$\Delta\sigma_x = -\lambda^s(D_{11} - D_{13}N_\psi),\ \Delta\sigma_y = -\lambda^s(D_{21} - D_{23}N_\psi),\ \Delta\sigma_z = -\lambda^s(D_{31} - D_{33}N_\psi)$$
$$\Delta\tau_{yz} = -\lambda^s(D_{41} - D_{43}N_\psi),\ \Delta\tau_{xz} = -\lambda^s(D_{51} - D_{53}N_\psi),\ \Delta\tau_{xy} = -\lambda^s(D_{61} - D_{63}N_\psi) \tag{11}$$

(2) Rock block cracking: The failure function of rock block cracking is:

$$f^t = \sigma_3 - \sigma^t \tag{12}$$

where, σ^t is the tensile strength of the rock block.

When $f^t > 0$, it indicates that the tensile stress of the rock block exceeds the tensile strength and the rock block is cracked. The extension of the rock block is defined by the associated flow rule, and its drawability potential function g^s is:

$$g^t = \sigma_3 \tag{13}$$

According to the similar principle of plastic flow law, the incremental expression of each stress component can be obtained as follows:

$$\begin{aligned}\Delta\sigma_x &= -\lambda^t D_{13},\ \Delta\sigma_y = -\lambda^t D_{23},\ \Delta\sigma_z = -\lambda^t D_{33}\\ \Delta\tau_{yz} &= -\lambda^t D_{43},\ \Delta\tau_{xz} = -\lambda^t D_{53},\ \Delta\tau_{xy} = -\lambda^t D_{63}\end{aligned} \tag{14}$$

where, $\lambda^t = \frac{f^t(\sigma_1^I, \sigma_3^I)}{D_{33}}$.

3.2.2 Stratification Plane Damage

(1) Stratification plane cracking: According to the indoor uniaxial compression test of the layered rock block, when the dip angle is more than 75°, the rock tension failure perpendicular to the structural plane occurs. According to the stress component in the integral coordinate system, the stress component in the local coordinate system is calculated by the stress conversion matrix [S]: $\{\sigma'\} = [S]\{\sigma\}$, and then the tensile failure is judged according to the stress component in the local coordinate system. The yield criterion is:

$$f_j^t = \sigma_z' - \sigma_j^t \tag{15}$$

When $f_j^t > 0$, that is, $\sigma_z' > \sigma_j^t$ (σ_j^t is the tensile strength of the layer), indicating that the normal stress of the layer exceeds the tensile strength of the layer, creaking will appear in the perpendicular to the direction of the layer. The stress component is modified according to the flow law associated with tensile stress yield, and the incremental expression of each stress component is obtained as follows:

$$\Delta\sigma'_x = -(\sigma_z^{\prime o} - \sigma_j^t)\frac{C}{D}, \Delta\sigma'_y = -(\sigma_z^{\prime o} - \sigma_j^t)\frac{C}{D}, \Delta\sigma'_z = \sigma_j^t - \sigma_z^{\prime o}, \Delta\tau' = 0 \quad (16)$$

where, C and D are the elements in the local coordinate system elastic matrix $\left[D'_e\right]$, whose expression is shown in Eq. (2).

(2) Stratification plane gliding: According to the uniaxial compression test of stratified rock block, when the dip angle is between 30° and 60°, the rock begins to slip along the bedding faces. The stratification plane gliding is the same as the plastic shear failure, which follows the Mohr-Coulomb criterion:

$$f_j^s = \tau' - \sigma'_z \tan\phi_j - c_j \quad (17)$$

where, c_j, ϕ_j are the cohesion and internal friction angle of the stratification plane respectively, $\tau' = \sqrt{\tau_{yz}^{\prime 2} + \tau_{xz}^{\prime 2}}$.

When $f_j^s > 0$, the stratification plane appears to glide, according to Mohr-Coulomb non-associated plastic flow orthogonal law to correct the stress, the incremental expression of the stress component is:

$$\begin{gathered}\Delta\sigma'_x = -\lambda_j^s C\tan\psi_j, \quad \Delta\sigma'_y = -\lambda_j^s C\tan\psi_j, \quad \Delta\sigma'_z = -\lambda_j^s D\tan\psi_j \\ \Delta\sigma'_{xy} = -\lambda_j^s G_1, \quad \Delta\tau'_{yz} = -\lambda_j^s G_2\frac{\tau_{yz}^{\prime O}}{\tau^O}, \quad \Delta\tau'_{zx} = -\lambda_j^s G_2\frac{\tau_{zx}^{\prime O}}{\tau^O}\end{gathered} \quad (18)$$

where, $\lambda_j^s = \frac{f_j^s(\sigma_z^{\prime O}, \tau^O)}{G_2 + D\tan\psi_j\tan\phi_j}$.

4 Numerical Calculations of the Model

The constitutive model of layered rock mass is developed secondly by VC ++ 6.0 programming language, and the dynamic link library file Layered.dll is generated and embedded in the main program of FLAC^{3D}. The numerical simulation of uniaxial compression test is carried out on the standard rock specimen to verify the rationality and the correctness of this model.

4.1 Computational Model and Physical Mechanics Parameters

The calculation model adopts the bidirectional pressurized cylindrical specimen of the same size as the indoor test, as shown in Fig. 7. The physical and mechanical parameters are shown in Table 3.

4.2 Results and Analysis

4.2.1 The Numerical Stress–Strain Curve

The numerical simulation of the standard rock specimen is carried out by the self-defined constitutive model of layered rock mass. The relationship curve between

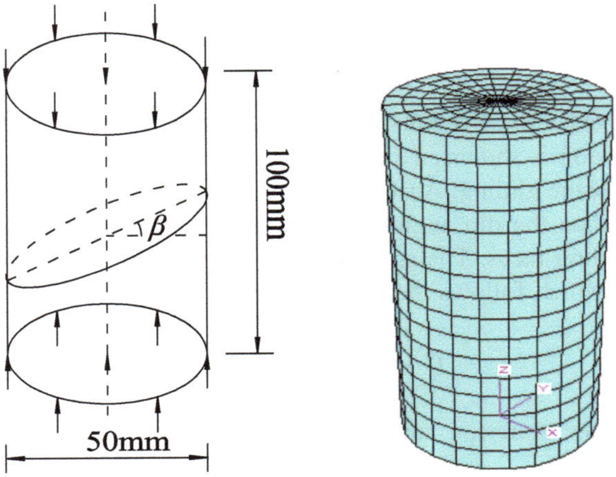

Fig. 7. The diagram of model size and grid

Table 3. The mechanical parameters of numerical experiments

Type	E (GPa)		$\mu_{\parallel}$	$\mu_{\perp}$	c (MPa)	φ (°)	σ_t (MPa)
	0°	90°					
Rock block	5.9	7.3	0.24	0.21	3.50	38	3.30
Stratification plane	–		–	–	0.142	24.8	0.15

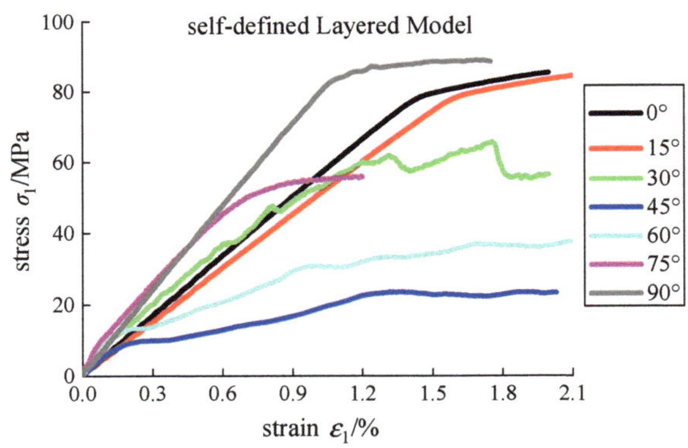

Fig. 8. Stress–strain curve with the change of dip angle

axial stress and strain in each dip angle is obtained, as shown in Fig. 8. The numerical stress–strain curve is roughly folded. The slope of the straight line corresponding to the different inclination is different. When $\beta = 45°$, the slope is the smallest. Elastic ultimate strength and the corresponding axial strain are also different, the peak stress of

45°–60° reaches the minimum, and the corresponding strain is the largest. It is mainly due to the obvious anisotropic characteristics of the deformation and strength.

4.2.2 The Change Rule of Uniaxial Compressive Strength

The relationship curves between the uniaxial compressive strength and the dip angle of rock specimen in each dip angle in two experimental conditions (indoor test, simulation test of self-defined Layered Model) are shown in Fig. 9. The uniaxial compressive strength is defined as the peak stress at the end of elastic stage in simulation curve.

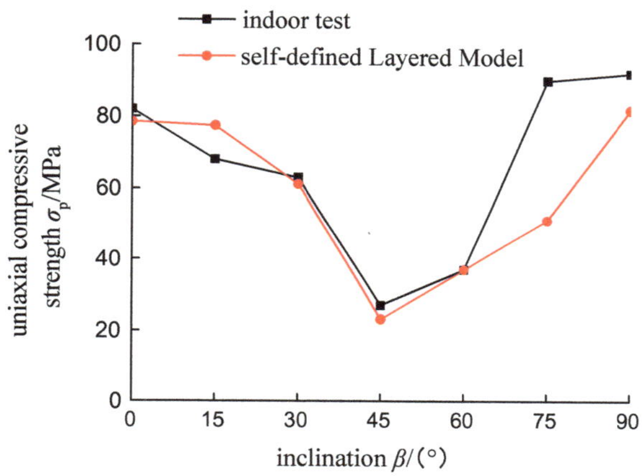

Fig. 9. The relationship between uniaxial compressive strength and dip angle

As can be seen from Fig. 9, the results of the two tests are basically consistent. The uniaxial compressive strength begins to decrease significantly after $\beta > 15°$, and the minimums appeared at $\beta = 45°$, are 23.2 and 27.0 Mpa, respectively. The strength began to increase after $\beta > 45°$, the maximum reached 81.9 and 92 Mpa when $\beta = 90°$. It is proved that the self-defined Layered Model can reflect the strength anisotropy of the layered rock mass well.

4.2.3 The Change Rule of Elastic Modulus

The relationships between the elastic modulus and the dip angle of the rock specimen in each dip angle in the two experimental conditions are shown in Fig. 10. The elastic modulus is the line slope of elastic stage in simulation curve.

Numerical experiments of uniaxial compression of the layered rock mass model show that the elastic modulus of the rock mass decreases first and then increases with the increasing of the dip angle, which is consistent with the indoor test result. It is proved that the self-defined Layered Model can reflect the deformation anisotropy of the layered rock mass well.

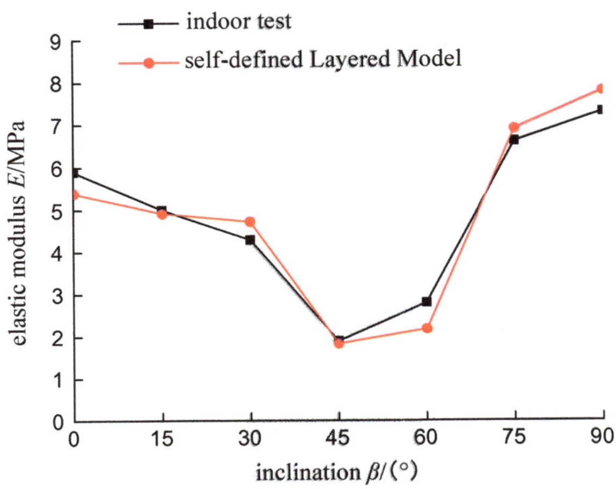

Fig. 10. The relationship between elastic modulus and dip angle

4.3 Advantages of Self-defined Layered Model

(1) The geometric model for layered rock mass is established by the "implicit" way. It does not consider the structure surface when we mesh the model, we just need to input the dip direction α and the dip angle β, the stress on the layer and the integral stress can be automatically converted. It is versatile and time-saving.
(2) The constitutive relation and yield criterion of layered rock mass can hardly be expressed by explicit mathematical formula. The calculation process is complex and the parameters are too many, and it is very difficult to get parameters. The self-defined Layered Model just needs to input the parameters of the horizontal layer and vertical layer in calculation, and it can be easily run in $FLAC^{3D}$.
(3) In the indoor test, the specimen for the horizontal layer and vertical layer is easily made, but the specimen with a certain inclination is very difficult to product, it may cost a lot of time and not easy to shape. Therefore, using this model also greatly reduces the workload of indoor tests.

5 Conclusion

Based on the transversely isotropic characteristics of the layered rock mass, the constitutive model of the layered rock mass is defined according to the elastic–plastic theory. Through the comparison between the numerical simulation of the standard rock specimen and the indoor test,rationality, correctness and precision of the model is verified.

(1) The variation of elastic modulus and uniaxial compressive strength of layered rock mass with the dip angle of the layer is analyzed from the experiment results. The results show that the deformation and strength of the layered rock mass are of

obvious anisotropy, and the modulus and strength of the rock specimen decrease first and then increase with the increasing of the dip angle. When the dip angle is 30°–60°, the uniaxial compression strength and elastic modulus of the layered rock are lower.

(2) According to the elasto-plastic theory, the anisotropy model of layered rock mass which can reflect the transversely isotropic characteristics is established, and the yielding criterion and stress correction method for four different failure modes are proposed. Through the secondary development platform provided by $FLAC^{3D}$, use VC ++ 6.0 programming language, generate Layered.DLL file, and successfully embed it into the $FLAC^{3D}$ main program.

(3) The numerical simulation of uniaxial compression for standard rock samples on self-defined Layered Model is carried out and compared with the indoor tests, and obtained the change rule of mechanical parameters with the dip angle. The results show that the distribution of uniaxial compressive strength and elastic modulus of each layered rock specimen under two different experiments are similar. Thus, it is proved that the self-defined Layered Model is reasonable and feasible.

6 Discussion

(1) The structure surface is implicitly in the solid elements, so the thickness of layered rock is not considered. At the same time, the creep of the geo-material has been ignored in this study.

(2) The numerical simulation curves are not exactly the same as the test curves. The reason that the softening has not appeared after the peak is that the adopted yield criterion is M-C criterion (ideal elastic–plastic body), and the strain softening model is not used in the simulation model.

Acknowledgements. This work is supported by the National Natural Science Foundation of China (51469007), Guizhou Science and Technology Support Project for Social development ([2015]3055), Science and technology research and development project of CSCEC (CSCEC4B-2015-KT-03). They are gratefully acknowledged.

References

Itasca Consulting Group Inc.: Fast Lagrangian analysis of continua in 3 dimension (Version 2.1) user's guide. Itasca Consulting Group Inc., Minneapolis, USA (2003)

Lee, Y.K.: Simulation of polyaxial tests using a failure condition for transversely isotropic rocks. Geosyst. Eng. **18**(1), 29–37 (2015)

Lee, Y.K., Pietruszczak, S.: Application of critical plane approach to the prediction of strength anisotropy in transversely isotropic rock masses. Int. J. Rock Mech. Min. Sci. **45**(3), 513–523 (2008)

Lekhnitskii, S.G.: Theory of Elasticity of an Anisotropic Elastic Body. Holde-Day Inc., San Francisco (1963)

Pietruszczak, S., Pande, G.: Description of hydraulic and strength properties of anisotropic geomaterials. Stud. Geotech. Mech. **34**(3), 3–16 (2012)

Saroglouh, T.G.: A modified Hoek-Brown failure criterion for anisotropic intact rocks. Int. J. Rock Mech. Min. Sci. **45**(3), 223–234 (2008)

Sitharam, T.G., Sridevi, J., Shimizu, N.: Practical equivalent continuum characterization of jointed rock masses. Int. J. Rock Mech. Min. Sci. **38**(3), 437–448 (2001)

Xu, W., Zhang, J., Wang, R.-B., et al.: An elasto-plastic model and its return mapping scheme for anisotropic rocks. In: Constitutive Modeling of Geomaterials, SSGG, pp. 371–380. Springer, Berlin (2013)

Yin, G., Li, X., Lu, J., et al.: A failure criterion for layered composite rock under true triaxial stress conditions. Chin. J. Rock Mech. Eng. **36**(2), 261–269 (2017). https://doi.org/10.13722/j.cnki.jrme.2016.0682. (in Chinese)

Yujun, Z., Yiping, L.: 3D elasto-plastic FEM analysis on layered rock mass. Chin. J. Rock Mech. Eng. **21**(11), 1615–1619 (2002). (in Chinese)

Yujun, Z., Yixing, T.: FEA on underground tunnel with directional strength of laminated rock. Undergr. Space **19**(l), 30–34 (1999). (in Chinese)

Zhang, Y., Zhang, W.: Finite element analyses of computational effects with of different shear strength expressions for layered rock mass. Rock Soil Mech. **35**(Supp. 1), 359–364 (2014). (in Chinese)

Zuo, S., Ye, M.-L., Tang, X.-L., et al.: Numerical model and validation of failure mechanism for layered rock mass of underground caverns. Rock Soil Mech. **34**(Supp. 1), 458–465 (2013). (in Chinese)

Zuo, S., Shi, W.-B., Liang, F., et al (2015). Numerical simulation and engineering application for failure modes and criterion of layered anisotropic rock mass **37**(Supp. 1), 191–196. (in Chinese)

Experimental Study on the Seepage Feature of Natural Loess Joint

Xiaojun Liu(✉), Qing Kong, and Xing Zheng

College of Civil Engineering, Xi'an University of Architecture and Technology, Xi'an, China
liuyu8345@163.com

Abstract. In order to obtain the seepage law of natural loess joint, this paper puts forward a set of methods for getting the natural loess joint and processing standard triaxial specimens to test its permeability coefficient on the basis of a large number of site experiments. By designing contrast experiment, the permeability coefficient of each sample is measured by making use of Global Digital System (GDS) triaxial osmotic instrument. The experiment results are shown as follows: (1) The permeability coefficient of complete sample is about 2.23–3.58 $\times$ 10^{-6} cm/s and it is about 85–360 $\times$ 10^{-6} cm/s when it includes joint, the existence of joint greatly enhances the seepage ability of loess mass. (2) When the loess joint shares same average opening, its permeability coefficient is slightly lower than the artificial joint because the large fluctuation and certain contact area of natural joint. (3) Loess and joints can produce a certain amount of water exchange, which improves the whole penetration ability significantly.

1 Introduction

Loess in China is widely distributed with many formation, covering an area of about 6.4 $\times$ 10^5 km^2, accounting for 6.6% of the total land area. The loess joint is the gap formed by the formation of the loess and through internal and external stress and new structure. It is widely found in the Loess Plateau, and its existence has an important effect on the loess. The joint is the main storage area and flow channel of underground water in the loess area and it's the main factor that affects the destruction of mountain, highway slope and underground cavern in loess area, it is also the basis of soil erosion, landslide and ground fissure. Therefore, it is of great significance to study the stability of loess joints in the evaluation of slope, tunnel and highway engineering of loess area, guide the arid areas to find water sources, control soil erosion, landslides and other disasters, and understand the development laws of loess landforms [1].

Researchers have studied the formation mechanism of loess joints by site investigation, theoretical analysis, simulation experiments and simulation techniques [2–6]. For joint seepage model, the early study simplified the complex natural rock fissures into the gap formed by the parallel plate model and established the classical formula that the flow rate of the joint surface is directly proportional to the third power of the opening, that is, the famous cubic law.

B. Fatahi et al. (eds.), *Sustainable Design and Construction for Geomaterials and Geostructures*, Sustainable Civil Infrastructures,
https://doi.org/10.1007/978-3-319-95753-1_7

Later, some scholars believe that natural joint seepage does not conform to the parallel plate model, but lies in several major channels, and proposed a trench model [7]. But because of the groove distribution of the joint surface under the stress is very complicated, it cannot establish the perfect groove model so that the groove model cannot be widely used. The main direction of the present study is to modify the cubic law obtained by the parallel plate mode. The rough surface of natural joint and the deformation produced by external force has great influence on seepage, the three-dimensional surface morphology of joints could be measured and the software was used to calculate composite morphology parameters [8]. Considering the influence of roughness, an improved formula of fracture seepage is obtained [9]. Considering the coupling of seepage and deformation, the generalized cubic law of seepage is obtained [10]. There are few studies on loess joints at home and abroad, and the main achievements are as follows: (1) The vertical joints and macropores in Loess media play different roles in water infiltration under different water contents [10]. The relationship between internal friction angle of loess joint and moisture content is revealed to be a parabola one [11]. (2) The existence of loess joint will promote the seepage of loess the joint has a great influence on the wetting peak migration of the loess field [12]. The seepage test direction can impact the connection between stress and permeability tests [13]. (3) Rock mass joints can be regarded as one-dimensional seepage in unit width, and two-dimensional seepage occurs in Loess joint. A two-dimensional seepage model of loess joint is established on the premise of reasonable assumption, and deduced the differential equation of fluid element by Euler's variable method, and obtained the seepage differential equation with flow velocity as function [14].

Cubic law is applicable to the model with smooth and contactless parallel plate and crack seepage is considered to be stable laminar flow. But the surface morphology of actual natural joint is very complex, which limits the application of cubic law in natural joints. The measured values of the test are compared with calculated values by cubic law can be found that the gap between the both can be as much as 20 times [15]. The two-dimensional seepage model of loess joint is based on cubic law [14], which is similar to the applicable condition of cubic law and will produce great error when it is directly used for natural joint. There is no calculation method for the seepage of natural loess joints at home and abroad, and few seepage experiments of natural loess joints have been done. In this paper, the natural loess joints are cut and the standard samples are prepared to determine the permeability coefficient of loess joints and study the seepage law of natural loess joint. It is of great significance to establish the calculation method suitable for natural loess joint and evaluate the influence of joint seepage on soil stability.

2 Experiment Scheme

2.1 Sample Preparation

The soil samples were taken from the cliff walls on the two sites near the County Road 309 in Bin County, Shaanxi. The height of the upper soil layer is about 5 m. The selected joints are cut through and emptied for half meter inwards. The width and trend of inner joints are the same as those on the surface. The basic physical properties of soil are listed in Table 1.

Table 1. Basic physical properties of samples

Category	Dry density ρ_d (g/cm^3)	Water content (%)	Porosity ratio e	Liquid limit water content w (%)	Plastic limit water content w_p (%)	Plasticity index I_p
Site I	1.31	2.74	1.09	32.73	17.14	15.59
Site II	1.28	2.89	1.10	33.20	17.92	15.28

In order to reduce the disturbance to soil samples, large machinery is not used during sampling. After trying, we explore a way to cut the natural loess joint. The specific operation is as follows: (1) On the surface containing the joint soil, use a shovel to remove the surface floating soil and flatten the soil surface; (2) Mark the cutting range, and cut it along the cutting range with a saw; (3) Use a shovel, pickaxe and homemade tools to cut soil within a certain distance outside the cutting range; (4) Use cutting tools and other small tools to cut the root of the soil containing joints. During cutting the joint soil should be supported to prevent sudden falling; (5) Pack the soil containing joints with film and wrap it with tapes, put into a sponge bedding box, then seal and load into a car. The sampling photo is shown in Fig. 1.

(a) Cutting the soil (b) Cutting completed

Fig. 1. Sampling on site

The size of the soil with joints is about 50 × 50 × 50 cm^3, 30 cm thick surface soil is discarded, the internal soil is selected for testing, and the internal soil physical index is determined before processing samples.

The steps of processing the natural joint specimen are as follows: (1) Use hand saws, wire saws, cutters and other hand tools to cut soil samples with joints into appropriate size; (2) Use tools including cutting tools and other tools to process the standard sample. In this process, take many reinforcement measures, cut the sample by segments, and finally level it. The processed sample is a cylinder with the diameter of 6.18 cm and the height of 12.5 cm, as shown in Fig. 2.

Fig. 2. The sample of natural loess joint

The steps of manual processing of the sample are as follows: (1) Cut the smooth joints in the complete soil, and then process it as the natural joints; (2) Use the standard filter paper to control the artificial joint opening. The filter paper is slightly shorter than the sample and the width is narrow, which has little influence on seepage, as shown in Fig. 3.

2.2 Experiment Device

The penetration experiment was conducted by using the GDS triaxial penetration test system (as shown in Fig. 4). The system consists of a standard triaxial osmotic pressure chamber, a digital pressure controller, and a software-driven computer. Global Digital System Laboratory (GDSLAB) software containing penetration test module is used to control test and collect data.

2.3 Experiment Principle

The constant head penetration test module of GDS triaxial osmotic instrument can control three pressure controllers. The top of the sample is connected to the upper head pressure controller. The bottom and surrounding of the sample are connected with the

Fig. 3. The sample of artificial loess joint

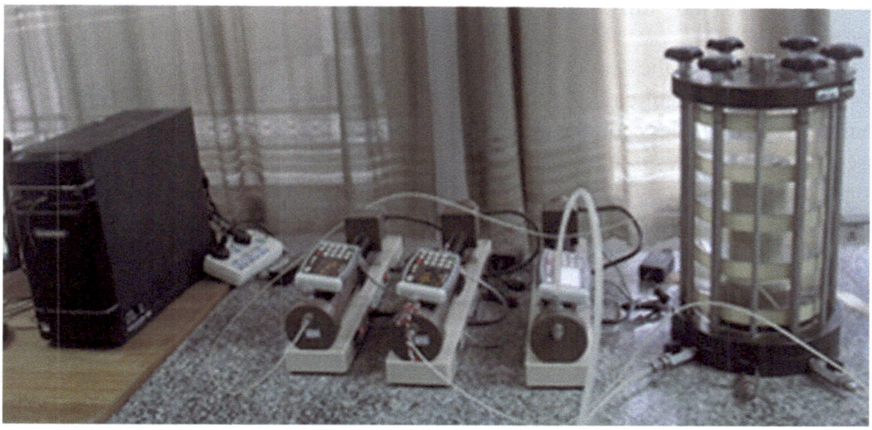

Fig. 4. GDS triaxial osmotic instrument

lower head pressure controller and confining pressure controller. Each pressure controller can precisely control the confining pressure of the specimen, the upper head pressure and the lower head pressure and record the change in the amount of water in the pressure controller. GDSLAB software automatically collects the pressure provided by the three pressure controllers and its inlet and outlet water volume and calculates the permeability coefficient of the sample based on the input sample size and boundary conditions by using Darcy's law and the volume change of the sample.

2.4 Experiment Design

Relevant theory of loess joint seepage is less and no one has done experiment study of natural loess joint seepage. This experiment will complete the effect of joint opening on the permeability of joint soil, the relationship between artificial joint and natural joint seepage and the influence of surrounding soil on joint seepage. The experiment design is as follows: 1. Make some natural joint samples with different openings on the both sites and make some artificial joints with the same opening for comparison, and study effects of joint opening on joint seepage and the relationship between artificial joint and natural joint seepage; 2. Study the influence of the surrounding soil on the seepage by controlling whether the surrounding soil seeps together with the joint.

2.5 Experiment Method and Steps

The experiment method consists of two parts: sample treatment and permeability coefficient determination. In order to eliminate the influence of the initial moisture content on the permeability coefficient, all samples are naturally dried in the room for one month before the determination of the permeability coefficient. After one month, the mass is measured at least once every two days. Mass change less than 0.01 g shows that the sample has fully become dry. Then the basic physical parameters are determined.

The volume of the cavity of the natural loess joint is measured by using self-designed loess joint cavity volume analyzer (precision is 0.1 ml), and the average opening of natural loess joints (precision is 0.01 mm) is obtained, and then the processed chip filters are used to adjust the same average opening of artificial joints as natural joints. Finally, by using Vaseline on the upper and lower surfaces of the jointed sample, the surrounding soil is controlled to be involved in the seepage.

Method and procedure for determination of permeability coefficient: Make vacuum saturation of the sample before test (saturation time is not less than 10 h), The plastic film is covered at the top and bottom of the saturated frame to allow water to saturate through the surrounding soil, so as to prevent the flow of water from impacting the joint surface during seepage. Set up GDS triaxial osmotic instrument, and suck appropriate amount of water into the pressure controller, and set related coefficient of constant head penetration test module in GDSLAB software. And then fully saturate sample and put it into the osmotic pressure chamber, fill the pressure chamber with air-free water (no bubbles occur in the whole process), and start the GDS triaxial osmotic instrument. The software will automatically collect confining pressure, head pressure, permeability coefficient and other parameters of the sample at each time point. Some of the test process is shown in Fig. 5. In permeability coefficient determination test, the too large confining pressure would cause serious deformation, so we chose a smaller value of 3 kPa; The head pressure should be smaller than the confining pressure, and we take it as 2 kPa. The test is completed after the permeability coefficient of the sample becomes stable.

(a) Sample saturated (b) Sample fixed (c) Sample in test

Fig. 5. The process of testing permeability coefficient

3 Experiment Results and Analysis

To facilitate the presentation, the shorthand rules for the following situations are described. (1) Any sample unjointed is called complete sample. (2) Samples with natural/artificial loess joint which surface is treated with vaseline to prevent water from seeping into the soil is called "natural/artificial single-joint sample". (3) Sample with natural/artificial loess joint is not treated with vaseline is called "natural/artificial jointed sample".

3.1 Permeability Coefficient of Complete Sample

The permeability coefficient results of complete samples at two sites are obtained by use of GDS triaxial osmotic instrument, as shown in Tables 2 and 3:

Table 2. The soil permeability coefficient of the first site

SN	Sample	Permeability coefficient (10^{-6} cm/s)	
		Test value	Average
1	Sample I	2.13	2.23
2	Sample II	2.06	
3	Sample III	2.50	

Table 3. The soil permeability coefficient of the second site

SN	Sample	Permeability coefficient (10^{-6} cm/s)	
		Test value	Average
1	Sample I	3.80	3.58
2	Sample II	3.41	
3	Sample III	3.52	

From Tables 2 and 3, it can be seen that the average permeability coefficients of the first and second sites are 2.23×10^{-6} and 3.58×10^{-6} cm/s, and the latter is larger than the former. From Table 1, we can know that the dry density of loess at the second site is less than that at the first site, which means the more the pore ratio is, the higher the permeability capacity of loose loess is.

3.2 Influence of Joint Aperture on Permeability Coefficient

For "natural single-joint sample", the results of the testing for relations between average joint aperture and permeability coefficient are as shown in Fig. 6.

The permeability coefficients of complete samples at the first and second sites are 2.23×10^{-6} and 3.58×10^{-6} cm/s. From Fig. 6, the permeability coefficient of jointed sample is far above that of complete sample, and joints greatly improve the permeability capacity of jointed sample. Besides from the relation curve of average joint aperture and permeability coefficient, it can be known that the permeability coefficient of jointed sample increases nonlinear as increase of joint aperture.

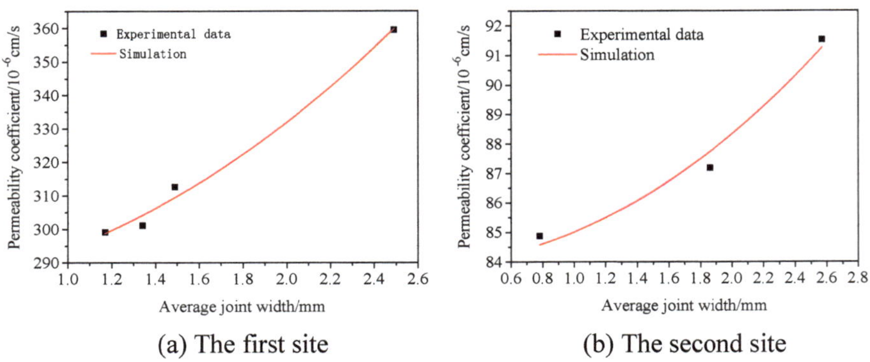

Fig. 6. The relationship of permeability coefficient and width

3.3 Comparison of Permeability of Natural Jointed Sample and Artificial Jointed Sample

The comparison of the permeability coefficients for natural joint and artificial joint is shown in Fig. 7, from which it can be obtained that: (1) For artificial jointed sample and natural jointed sample with the same average aperture at the same site, the permeability coefficients are almost same, mainly because aperture is a main factor affecting the joint permeability, and joints with the same aperture at the same site have similar permeability capacities; (2) The permeability coefficient of artificial jointed sample is slightly higher than that of natural jointed sample, and although they have the same average aperture, the natural jointed sample has greater surface undulation and certain contact area, so that its permeability coefficient is slightly less than that of artificial jointed sample.

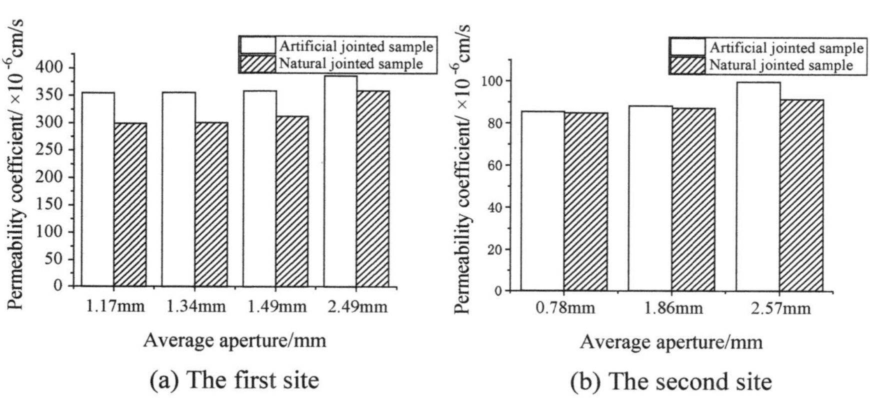

Fig. 7. The permeability coefficient of natural and artificial loess joint

3.4 Influence of Loess Around Joints on Seepage

The comparison of "artificial jointed sample" and "artificial single joint sample" at the second site is shown in Fig. 8.

It can be seen from Fig. 8 that, the permeability capacity of the sample with both loess and joint seeping is more than that of the sample only with joint seeping, for 20×10^{-6} cm/s or so, and much more than that of the complete sample. This means for the sample with both loess and joint seeping, the permeability coefficient is not the sum of permeability coefficients of loess and joint, but joint seepage facilitates the permeability.

From the theory of rock joint seepage, the seepage model of fractured rock can be roughly divided into non-double medium model and double medium model. Wherein the former is mainly used to study the joint permeability, paying no attention to water exchange between rock pore and joint systems, while in the latter, the fractured rock is

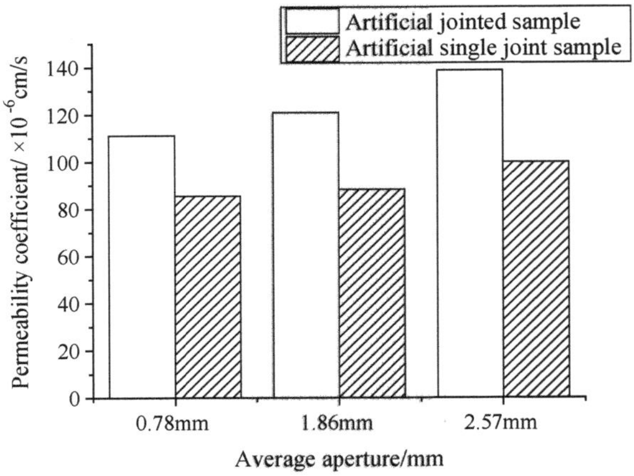

Fig. 8. The permeability coefficient of only artificial joint and when it is surrounding soils

regarded as a unity of fracture system with low porosity but high transmissivity and rock pore system with good porosity but low transmissivity, with consideration of water exchange between these two systems [16]. In the saturated state, the difference between water velocity of joints and that of rock pores etc. causes water pressure difference between the pores and the joints, under which the water exchange occurs between the pores and joints. Low permeability and pore water content of rock results in a small water exchange capacity, so in the study on rock joint seepage, the water exchange between pore system and joint system can be ignored. Whereas high permeability capacity and pore water content of loess causes obvious water exchange between pore and joint systems during seepage of jointed loess, for this cause the surrounding loess facilitates the joint seepage. From Fig. 8, the surrounding loess facilitates the joint seepage during seepage of jointed loess, and the seepage basically meets the assumption for the fracture-pore double medium model [17, 18]. Therefore water exchange shall be considered for joint action of loess and joints, and the seepage model is more similar to double medium model.

4 Conclusion

(1) The sampling and sample preparation methods used in the testing can reduce the samples' disturbance and effectively protect of the original properties of natural loess joints. The method that replaces the standard gasket with fast filter paper is used to control the artificial joint aperture, reduce the effect of the gasket on water-carrying section of joints, effectively protect the joint surfaces, and improve the test precision.

(2) The joint aperture of loess is a main factor affecting the permeability of jointed loess, and its permeability coefficient increases with the increase of such aperture; the natural and artificial joints with the same aperture have similar permeability coefficients, and for the natural joint, the coefficient is less due to surface bulging, contact etc.
(3) The water exchange of some extent between loess and joints during seepage of jointed loess causes the overall permeability capacity greater than sum of permeability of loess and joints.

Acknowledgements. This paper is funded by the National Natural Science Foundation of China (Grant no. 51408464).

References

1. Wang, J.M., Ni, Y.L., Sun, J.Z.: A study on structural joints in loess and its practical applications. J. Eng. Geol. **2**(4), 31–42 (1994). (in Chinese)
2. Ni, Y.L., An, Y.Q., Wang, J.M.: Soil and water loss and loess structure joint in Luochuanyuan. J. Soil Water Conserv. **7**(3), 26–31 (1993). (in Chinese)
3. Wang, Z.G., Kang, G.J., Ma, C.W., et al.: Exploring about formation mechanism of loess vertical joints. Sci. China (series B) **23**(7), 765–770 (1993). (in Chinese)
4. Li, T.L., Wang, H., Fu, Y.K., Liang, Y.: Test simulation on the forming mechanism of loess vertical joints. J. Earth Sci. Environ. **36**(2), 127–134 (2014). (in Chinese)
5. Luo, J., Xiang, W.W., Yun, G., Wang, Y.: Experimental study on formation of loess vertical joints in northern Shaanxi Province. J. Yangtze River Sci. Res. Inst. **27**(3), 38–41 (2010). (in Chinese)
6. Zhong, S.H.: Influence of included tectonic joints on stability of tunnels in the loess. Chin. J. Geotech. Eng. **5**(11), 68–77 (1983). (in Chinese)
7. Tsang, Y.W., Tsang, C.F.: Channel model of flow through fractured media. Water Resour. Res. **23**(3), 467–479 (1987)
8. Xia, C.C., Qian, X., Lin, P., Xiao, W.M.: Experimental investigation of nonlinear flow characteristics of real rock joints under different contact conditions. J. Hydraul. Eng. **143**(3) (2016)
9. Su, B.Y., Zhan, M.L., Zhao, J.: Study on fracture seepage in the imitative nature rock. Chin. J. Geotech. Eng. **17**(5), 19–24 (1995). (in Chinese)
10. Du, Z.D., Zhao, Y.J., Cheng, J.R.: Research on the seepage test of unsaturated loess. Hydrogeol. Eng. Geol. **24**(2), 50–51 (1997). (in Chinese)
11. Luo, Y., Wang, T.H., Liu, X.J., Zhang, H.: Laboratory study on shear strength of loess joint. Arab. J. Sci. Eng. **39**(11), 7549–7554 (2014). (in Chinese)
12. Luo, Y., Wang, T.H., Wang, J.J.: Finite element seepage flow model for unsaturated loess with joints. J. Eng. Geol. **22**(6), 1115–1122 (2014)
13. Zhang, C., Tu, S., Zhang, L., Chen, M.: A study on effect of seepage direction on permeability stress test. Arab. J. Sci. Eng. **41**(11), 1–14 (2016)
14. Wang, T.H., Luo, Y., Zhang, H.: Two-dimensional steady flow rate equation for loess joints. Chin. J. Geotech. Eng. **35**(6), 1115–1120 (2013). (in Chinese)
15. Xia, C.C., Wang, W., Cao, S.D.: Flow characteristics of joints under different contact conditions. Rock Mech. Rock Eng. **29**(7) (2010). (in Chinese)

16. Wang, Y., Su, B., Xu, Z.: Comment on the models of seepage flow in fractured rock masses. Adv. Water Sci. **7**(3), 276–282 (1996). (in Chinese)
17. Barenblatt, G.I., Zheltov, I.P., Kochina, I.N.: Basic concepts in the theory of seepage of homogeneous liquids in fissured rocks [strata]. J. Appl. Math. Mech. **24**(5), 1286–1303 (1960)
18. Warren, J.E., Root, P.J.: The behavior of naturally fractured reservoirs. Soc. Pet. Eng. J. **3**(3), 245–255 (1963)
19. Zhou, C.B., Xiong, W.L.: A generalized cubic law for percolation in rock joints. Rock Soil Mech. **17**(4), 1–7 (1996). (in Chinese)
20. Luo, Y.: Stress and Seepage Properties of Loess Joint. Xi'an University of Architecture and Technology (2014) (in Chinese)

Numerical Investigation of Time-Dependent Deformation in Rock Based on Numerical Manifold Method

Xian-yang Yu, Tao Xu(✉), and Guang-lei Zhou

Center for Rock Instability and Seismicity Research, Northeastern University, Shenyang, People's Republic of China
xutao@mail.neu.edu.cn

Abstract. In this paper, the extended Nishihara model (ENM) is incorporated into numerical manifold method (NMM) to study time-dependent deformation of stressed rock, this model uses a nonlinear viscoelastoplastic component to achieve the tertiary creep of rock by using ENM based on numerical manifold method. The improved NMM based creep model is a combination of creep theory for the time-dependent deformation of rock and traditional NMM. Using time step-initial strain method, the improved NMM program divides viscous strain and large creep time into a series of increment time-step values to calculate creep problems. Parameter sensitivity analysis is performed and the Improved NMM program is validated against experimental data. Meanwhile, the influence of axial stress and confining pressure on time-dependent deformation of rock is also investigated. The fact that numerical simulations agree well with experimental results shows that improved NMM program is suitable for modelling time-dependent creep deformation of rocks.

1 Introduction

Time-dependent deformation properties of rocks are of fundamental importance for understanding the long-term behavior of rock mass in the underground mines, as well as predicting the long-term stability for rock engineering structures. Laboratory tests and in situ rheological observations are the major methods to study time-dependent deformation behaviors of rocks. Considerable creep tests (Scholz 1972; Heap et al. 2009, 2011) have been conducted, and three regimes (primary creep or transient creep, secondary creep or minimum creep, and tertiary creep or unstable creep) are usually observed during creep experiments. Many methods are used to model the time-dependent deformation of rocks. Constitutive laws based on laboratory experiments provide a relation between strain, stress, and strain rate (Voight 1989; Lockner 1998). Wang et al. (2016) simulated the time-dependent behavior of brittle rocks based on damage constitutive law at a mesoscale. Wang and Wong (2015) performed a series of triaxial compression creep tests of the saturated till to investigate the creep behavior at different effective stresses and strains. Zhao et al. (2017a, b) proposed a new viscoelastoplastic creep model to study the non-linear rheological mechanical properties of hard rocks under cyclic incremental loading and unloading. Zhao et al. (2017a, b)

B. Fatahi et al. (eds.), *Sustainable Design and Construction for Geomaterials and Geostructures*, Sustainable Civil Infrastructures,
https://doi.org/10.1007/978-3-319-95753-1_8

used an idealised microstructure mesh model (a random grains separated by idealised grain boundaries) to characterize crack growth at tertiary crack growth stage. Except for traditional experimental and creep model analysis, mathematical method of trust-region reflective least squares algorithm is used to obtain and optimize viscoplastic parameters at creep stage (Le and Fatahi 2016; Le et al. 2016). These models reproduce the creep behavior of different types of rocks under different loading conditions.

Recently, the numerical manifold method (NMM) (Shi 1992) has gained a wide attention and application in rock mechanics and rock engineering due to its efficient treatment to the problems involving continuous and discontinuous deformations in a unified way (Ning et al. 2011; An et al. 2014; Yang et al. 2016). He et al. (2013) put forward a customized contact algorithm in the 3-D NMM to study the jointed rock masses. Using time step-initial strain method, the creep theory is coupled with NMM to simulate the creep of rocks. High-order numerical manifold method is studied to improve the computational efficiency of the numerical manifold method, and solve the complicated geotechnical engineering problems (Liu and Chen 2012a, b).

Though great progress has been made in the application of NMM in rock engineering, there are still some difficult problems to be solved such as time-dependent deformation of rock. Thus, in the present paper, the ENM is incorporated into NMM to study the time-dependent deformation of stressed rock. The improved model is validated against experimental data and parameter sensitivity analysis is performed. Meanwhile, the influences of axial stress and confining pressure on time-dependent creep deformation are also investigated.

2 NMM-Based Nishihara Model

2.1 Numerical Manifold Method

Numerical manifold method is composed of block dynamics, cover system and numerical integration method. The block dynamics solve the mechanical behavior of block systems under loading. The cover system consists of a mathematical cover and a physical cover, and the simplex integration is an important method to calculate integral field, which transforms the area of polygon into many triangles and solve concave polygon problem that FEM cannot deal with, meanwhile, the analytic method is used to find the triangular integration of simplex. The mathematical and physical cover form the cover system of NMM, the former is composed of regular graph, and the latter consists of physical boundary. Compared with the differential manifold, the main function of NMM based on the finite covering technique is defined by the basis of covering and integration by parts. The function is almost discontinuous on the contact interface, however, the main function of differential manifold is highly differentiable, and it is independent on coverage.

In NMM, every physical cover set is a single cover displacement function. Main displacement function is composed of local displacement function, the main displacement function between the two sides of joint is discontinuous, which can adapt to wide variety from continuum to discontinuum medium. Cover functions on the physical cover U_i is $u_i(x)$

$$u(x) = \sum_{i=1}^{n} w_i(x)u_i(x) \tag{1}$$

where $u_i(x)$ can be constant, linear polynomial, high order polynomial and local series, $u_i(x)$ can connect with weight function $w_i(x)$. Main function $u(x)$ is defined with $w_i(x)$ and $u_i(x)$. $w_i(x)$ means weighted average, which is the percent of each covering function $u_i(x)$ for all physical cover U_i containing x. As for each element, nodes have different weight functions with a summary of 1 in total.

$$\sum_{(x,y)\in U_j} w_j(x,y) = 1 \tag{2}$$

local displacement function U_i of physical cover is defined $u_i(x,y)$ and $v_i(x,y)$,

$$U_i = \begin{Bmatrix} u_i(x,y) \\ v_i(x,y) \end{Bmatrix} = PD \tag{3}$$

where P is basic function of the cover function, D is displacement function. If the cover function is a first order function,

$$U_i = \begin{Bmatrix} u_i(x,y) \\ v_i(x,y) \end{Bmatrix} = \begin{pmatrix} 1 & 0 & x & 0 & y & 0 \\ 0 & 1 & 0 & x & 0 & y \end{pmatrix} \begin{Bmatrix} d_{i1} \\ d_{i2} \\ d_{i3} \\ d_{i4} \\ d_{i5} \\ d_{i6} \end{Bmatrix} \tag{4}$$

and the whole displacement function U is calculated by $w_i(x,y)$ and U_i

$$\begin{aligned} U = W_i\, U_i &= \sum_{i=1}^{n} w_i(x,y) \begin{Bmatrix} u_i(x,y) \\ v_i(x,y) \end{Bmatrix} \\ &= \sum_{i=1}^{n} [T_i(x,y)]\{D_i\} \end{aligned} \tag{5}$$

where T is displacement matrix. According to the minimum potential energy principle, total potential equation is

$$KD = F \tag{6}$$

Simplex integration is the exact solution of general shape region in n-dimensional space, and the integration can be any n-dimensional polygon. One-dimensional simplex is line segment, two-dimensional simplex is directed triangle, and three-dimensional simplex is tetrahedron. Main integration difference between simplex and traditional is integration region. The simplex integration uses simplexes as integral region, positive and negative directions are designated as positive and negative volumes, respectively.

2.2 Formulation of Nishihara Creep Constitutive Model

The time-dependent deformation of rock is a common mechanical relationship between the magnitude of stress and strain and time in rock materials, which makes up of creep, stress relaxation and elastic aftereffect. The creep of rock is the most common phenomenon in rock engineering, so it is of great importance to study the creep deformation of rocks. Rheological component method is a common method that makes up of linear elastic spring, linearly viscous dashpot and slider. By combining these components, we can construct a model of linear viscoelasticity or viscoelastoplasticity. The models are all kinds of constitutive equations essentially, which can reflect rock deformation properties. The most representative creep models are the Generalized kelvin model, the Burgers model and the Nishihara model. The Generalized Kelvin model is a three-element model of a linear elastic spring and the other two-element (Kelvin model) in series. The Kelvin model making up of a spring and dash-pot in parallel, is oftern referred to as "solids", since it reflects instantaneously as elastic materials and recover completely upon unloading, but this model can only simulate primary and secondary creep stages. The strain at secondary stage will not change and strain rate is zero. The Burgers model is a four-element model of a Maxwell and a Kelvin model. The difference between Generalized Kelvin and Burgers model is viscous creep strain at the secondary stage, the former holds constant and the latter increases with time. Traditional Nishihara model (Nishihara 1955) making up of Kelvin model and Bingham model in series, is a piecewise function and is unable to describe the accelerating stage of rock creep. The Bingham model just reflects the viscous deformation of the rock under a constant stress and strain rate is a certain definite value. However, the creep deformation of rock in actual condition changes against time. Only simulating the process of deformation, can we simulate the trend of rock failure. So, an extended Nishihara model is implemented in NMM to describe the accelerating creep properties of rock, in which a viscous component is replaced with the linear viscous components. In this way, ENM can be applied to analyze the problem of viscoelasto-ductile. In this part, ENM is considered to study the creep of rock based on NMM. The model of ENM is illustrated in Fig. 1.

The differential piecewise creep constitutive equation is

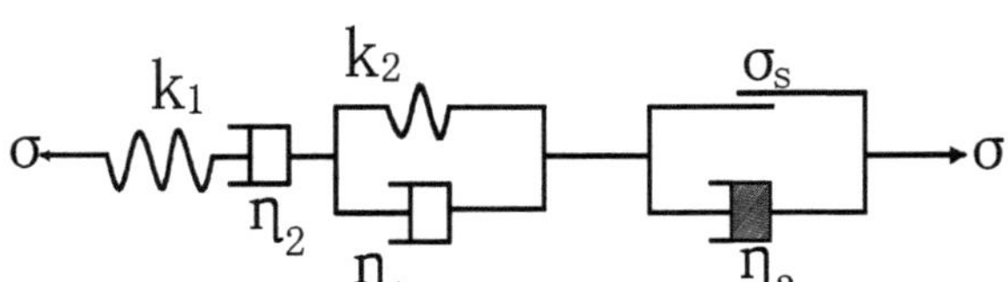

Fig. 1. Extended Nishihara model (ENM)

$$\varepsilon = \begin{cases} \frac{\sigma_0}{k_1} + \frac{\sigma_0}{\eta_2} t + \frac{\sigma_0}{k_2}\left(1 - e^{-\frac{k_2}{\eta_1}t}\right), & \sigma < \sigma_s \\ \frac{\sigma_0}{k_1} + \frac{\sigma_0}{\eta_2} t + \frac{\sigma_0}{k_2}\left(1 - e^{-\frac{k_2}{\eta_1}t}\right) + \frac{(\sigma_0 - \sigma_s)t^n}{n!\eta_3}, & \sigma \geq \sigma_s \end{cases} \quad (7)$$

where k_1 and k_2 are spring stiffness, η_1 and η_2 are the viscosity of material, η_3 is the viscosity of viscous component in the material, ε is total strain, n is time index (Yang et al. 2015), σ_s is the yield stress of plastic component. $(\sigma_0 - \sigma_s) \times t^n/n! \times \eta_3$ in the equation denotes nonlinear component. Nonlinear means a relationship between stress and strain and strain rate in constitutive equation. Traditional creep model is linear model, which can simulate primary and secondary creep stage greatly, but it is unable to present creep failure process, so, nonlinear element can replace linear element to simulate creep nonlinear deformation. If $\sigma < \sigma_s$, ENM will reduce to Burgers model of a four-element model. If $\sigma > \sigma_s$, ENM will reflect rock deformation during accelerating creep stage, the dashpot η_3 in Fig. 1 is the key component of deformation. In order to combine with NMM program, strain differential equation has to turn to the iteration by first order ordinary differential equation.

When $\sigma < \sigma_s$, the viscous strain iteration equation is shown in Eq. (8), which is the bridge of creep theory and NMM.

$$\begin{aligned} \varepsilon_v &= \frac{\sigma_0}{\eta_2} t + \frac{\sigma_0}{k_2}\left(1 - e^{-\frac{k_2}{\eta_1}t}\right) \quad \rightarrow e^{-\frac{k_2}{\eta_1}t} = 1 - \frac{k_2}{\sigma_0}\left(\varepsilon_v - \frac{\sigma_0}{\eta_2} t\right) \\ \varepsilon_v' &= \frac{\sigma_0}{\eta_2} + \frac{\sigma_0}{\eta_1} e^{-\frac{k_2}{\eta_1}t} = \frac{\sigma_0}{\eta_2} + \frac{\sigma_0}{\eta_1} - \frac{k_2}{\eta_1}\varepsilon_v + \frac{k_2\sigma_0}{\eta_1\eta_2} t \\ \varepsilon_v' + \frac{k_2}{\eta_1}\varepsilon_v &= \frac{k_2\sigma_0}{\eta_1\eta_2} t + \frac{\sigma_0}{\eta_2} + \frac{\sigma_0}{\eta_1} \end{aligned} \quad (8)$$

According to the first order linear differential equation, the general solution of ε_v is

$$\varepsilon_v = u(x) e^{-\frac{k_2}{\eta_1}t} \quad (9)$$

Put Eq. (9) into Eq. (8),

$$\begin{aligned} u'(x) &= \frac{k_2\sigma_0}{\eta_1\eta_2} t e^{\frac{k_2}{\eta_1}t} + \left(\frac{\sigma_0}{\eta_2} + \frac{\sigma_0}{\eta_1}\right) e^{\frac{k_2}{\eta_1}t} \\ u(x) &= \left(\frac{\sigma_0}{\eta_2} t + \frac{\sigma_0}{k_2}\right) e^{\frac{k_2}{\eta_1}t} \end{aligned} \quad (10)$$

$$\begin{aligned} \varepsilon_v = u(x) e^{-\frac{k_2}{\eta_1}t} &= \frac{\sigma_0}{\eta_2} t_1 + \frac{\sigma_0}{k_2} \quad \rightarrow \quad \varepsilon_v\left(1 - e^{-\frac{k_2}{\eta_1}t}\right) = \left(\frac{\sigma_0}{\eta_2} t_1 + \frac{\sigma_0}{k_2}\right)\left(1 - e^{-\frac{k_2}{\eta_1}t}\right) \\ \varepsilon_{v(t+\Delta t)} &= \varepsilon_{v(t)} e^{-\frac{k_2}{\eta_1}\Delta t} + \left(\frac{\sigma_0}{\eta_2} t_1 + \frac{\sigma_0}{k_2}\right) A\left(1 - e^{-\frac{k_2}{\eta_1}\Delta t}\right) \end{aligned} \quad (11)$$

where t_1 means total iterative time.

When $\sigma > \sigma_s$, ENM constitutive equation is the combination of Burgers and extended Bingham constitutive equation. For simplicity, we assume the time index n is 2.

$$
\begin{aligned}
&\varepsilon_v = \frac{\sigma_0}{\eta_2}t + \frac{\sigma_0}{k_2}\left(1 - e^{-\frac{k_2}{\eta_1}t}\right) + \frac{(\sigma_0 - \sigma_s)t^2}{2\eta_3} \quad \rightarrow \quad e^{-\frac{k_2}{\eta_1}t} = 1 - \frac{k_2}{\sigma_0}\left[\varepsilon_v - \frac{\sigma_0}{\eta_2}t - \frac{(\sigma_0 - \sigma_s)t^2}{2\eta_3}\right] \\
&\varepsilon_v' = \frac{\sigma_0}{\eta_2} + \frac{\sigma_0}{\eta_1}\left[1 - \frac{k_2}{\sigma_0}\left(\varepsilon_v - \frac{\sigma_0}{\eta_2}t - \frac{(\sigma_0 - \sigma_s)t^2}{2\eta_3}\right)\right] \\
&\varepsilon_v' + \frac{k_2}{\eta_1}\varepsilon_v = \frac{(\sigma_0 - \sigma_s)}{2\eta_1\eta_2}k_2 t^2 + \left[\frac{k_2\sigma_0}{\eta_1\eta_2} + \frac{(\sigma_0 - \sigma_s)}{\eta_3}\right]t + \frac{\sigma_0}{\eta_2} + \frac{\sigma_0}{\eta_1}
\end{aligned}
\tag{12}
$$

Put Eq. (9) into Eq. (12),

$$
\begin{aligned}
u'(x) &= \left[\frac{(\sigma_0 - \sigma_s)}{2\eta_1\eta_2}k_2 t^2 + \left[\frac{k_2\sigma_0}{\eta_1\eta_2} + \frac{(\sigma_0 - \sigma_s)}{\eta_3}\right]t + \frac{\sigma_0}{\eta_2} + \frac{\sigma_0}{\eta_1}\right]e^{\frac{k_2}{\eta_1}t} \\
u(x) &= \left[\frac{(\sigma_0 - \sigma_s)}{2\eta_3}t^2 + \frac{\sigma_0}{\eta_2}t + \frac{\sigma_0}{k_2}\right]e^{\frac{k_2}{\eta_1}t}
\end{aligned}
\tag{13}
$$

$$
\begin{aligned}
&\varepsilon_v = \frac{(\sigma_0 - \sigma_s)}{2\eta_3}t^2 + \frac{\sigma_0}{\eta_2}t + \frac{\sigma_0}{k_2} \quad \rightarrow \quad \varepsilon_v\left(1 - e^{-\frac{k_2}{\eta_1}\Delta t}\right) = \left[\frac{(\sigma_0 - \sigma_s)}{2\eta_3}t^2 + \frac{\sigma_0}{\eta_2}t + \frac{\sigma_0}{k_2}\right]\left(1 - e^{-\frac{k_2}{\eta_1}\Delta t}\right) \\
&\varepsilon_{v(t+\Delta t)} = e^{-\frac{k_2}{\eta_1}\Delta t}\varepsilon_{v(t)} + \left(\frac{\sigma_0}{k_2(\sigma_0 - \sigma_s)} + \frac{\sigma_0}{\eta_2(\sigma_0 - \sigma_s)}t_1 + \frac{t_1^n}{n!\eta_3}\right)A(\sigma_0 - \sigma_s)\left(1 - e^{-\frac{k_2}{\eta_1}\Delta t}\right)
\end{aligned}
\tag{14}
$$

So, the total viscous strain iteration equation is:

$$
\varepsilon_{v(t+\Delta t)} = \begin{cases} e^{-\frac{k_2}{\eta_1}\Delta t}\varepsilon_{v(t)} + \left(\frac{1}{k_2} + \frac{t_1}{\eta_2}\right)A\,\sigma_0\left(1 - e^{-\frac{k_2}{\eta_1}\Delta t}\right), & \sigma < \sigma_s \\ e^{-\frac{k_2}{\eta_1}\Delta t}\varepsilon_{v(t)} + \left(\frac{\sigma_0}{k_2(\sigma_0 - \sigma_s)} + \frac{\sigma_0}{\eta_2(\sigma_0 - \sigma_s)}t_1 + \frac{t_1^n}{n!\eta_3}\right)A(\sigma_0 - \sigma_s)\left(1 - e^{-\frac{k_2}{\eta_1}\Delta t}\right), & \sigma \geq \sigma_s \end{cases}
\tag{15}
$$

where Δt is the time step of creep simulation. A is constant matrix connected with Poisson's ratio υ, A in the plane strain problem is

$$
A = \begin{bmatrix} 1 & \frac{u}{u-1} & 0 \\ \frac{u}{u-1} & 1 & 0 \\ 0 & 0 & \frac{2}{1-\mu} \end{bmatrix}
\tag{16}
$$

If $\sigma < \sigma_s$, unloading the stress at time t_1, the elastic deformation of linear elastic spring recovers immediately, axial creep strain will take a long time to recover. According to the principle of stress superposition, when the time $t > t_1$, the stress $(-\sigma_0)$ is coupled with σ_0, the unloading equation is

$$\varepsilon_{v(t+\Delta t)} = e^{-\frac{k_2}{\eta_1}\Delta t}\, \varepsilon_{v(t)} + A\,\sigma_0 \frac{t_1}{\eta_2}\left(1 - e^{-\frac{k_2}{\eta_1}\Delta t}\right) \tag{17}$$

Viscous strain will generate the additional viscous load F' according to the strain energy equation,

$$\begin{aligned} \{D_e\}^T[F'] &= \iint_{\Omega} \frac{1}{2}[\varepsilon]^T[\sigma]dxdy \\ &= \iint_{\Omega} \frac{1}{2}\{D_e\}^T[B_e]^T[E]\,\varepsilon'\,dxdy \\ \rightarrow \quad [F'] &= \iint_{\Omega} [B_e]^T[E]\,\varepsilon'\,dxdy \end{aligned} \tag{18}$$

where B_e is strain matrix, E is elastic stiffness matrix, Ω is integral area. By putting the viscous load into total potential energy equation (Eq. 6), we will get D_e, and then we will get displacement U by Eq. (4).

2.3 Implementation of ENM into NMM

Time-dependent strain can be divided into a series of incremental values in each time step, and parameters are regarded as constants in each incremental time step, that is to say, a series of linear problems are used to approximate the solution of problem, this approach is named as time step-viscous strain method. For time-dependent creep problem, we add viscous strain to each time step and solve creep problems finally, then we get the relationship between time and both stress and strain, as well as the deformation of rock.

We take two steps to incorporate time-dependent creep into NMM. Firstly, the program will calculate the initial elastic deformation of rock specimen under constant stress, the system will gradually balance according to block dynamics, and calculate elastic strain. The time of elastic strain is negligible comparing with creep time, so it can be regarded as instantaneous. Secondly, according to viscous strain iteration equation (Eq. 15) and strain energy equation (Eq. 18), new additional viscous load will break the balance of system, the new viscous displacement is generated until the system balances again. The stress acting on the specimen will induce a deformation, which increase to a specified limit (i.e. the strain rate approaches to a constant) with time.

The improved NMM program is used to simulate the time-dependent deformation of rock specimen under constant stress loading and unloading condition. In order to obtain the input parameters of rock samples, the uniaxial compressive experiment was conducted numerically. The sandstone specimen of 50 mm in diameter and 100 mm in length is divided into 236 elements by physical cover system of NMM. The physical and mechanical parameters are listed in Table 1 (Liu et al. 2012). The specimen is under the single-stage axial stress load of 40 MPa, the bottom of the specimen is fixed, and the both sides of the specimen are free. Volumetric creep is the sum of axial strain and lateral strain of specimen under triaxial compression condition. The volumetric strain experiences a non-linear change process. Shear creep is the study of the creep

Table 1. Physico-mechanical parameters of numerical model

Items	Specimen
Spring stiffness K_1, MPa	3775
Spring stiffness K_2, MPa	6000
Viscosity coefficient η_1, MPa h	4164.58
Viscosity coefficient η_2, MPa h	1200
Viscosity coefficient η_3, MPa h	10,000
Time index n	3
Poisson's ratio υ	0.29
Time step Δt, h	0.083

deformation of specimens under the action of constant shear stress in direct shear test. The creep in this paper is mainly the axial creep under constant axial stress of uniaxial compressive condition. Figure 2 shows the simulation model under uniaxial compression test. The measuring point A (25,100) at the middle upper end of specimen is set to record the deformation of specimen under loading. Normal direction is same as cartesian coordinate system.

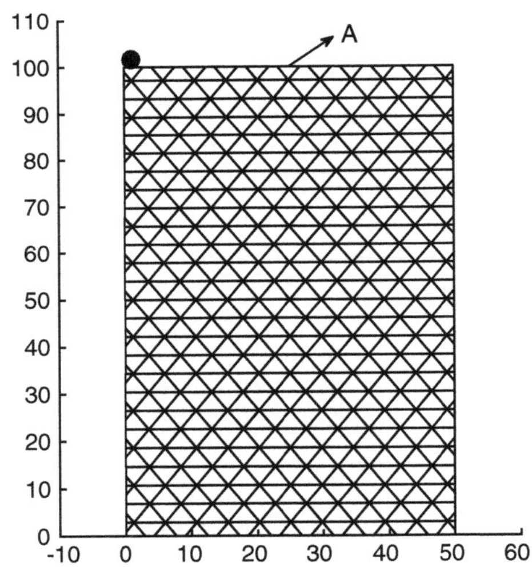

Fig. 2. Numerical model in uniaxial compression

Traditional Nishihara model cannot capture accelerating creep stage, so, the linear Newton component in Fig. 1 is replaced by a new nonlinear viscoelastoplastic component. According to the viscous strain iteration equation (Eq. 15), we applied 40 MPa axial stress and 36 MPa yield stress. We get the creep strain curve of ENM in Fig. 3, which shows the changing process of creep strain, strain rate and strain acceleration against time, in which the strain will go through the primary, secondary and tertiary

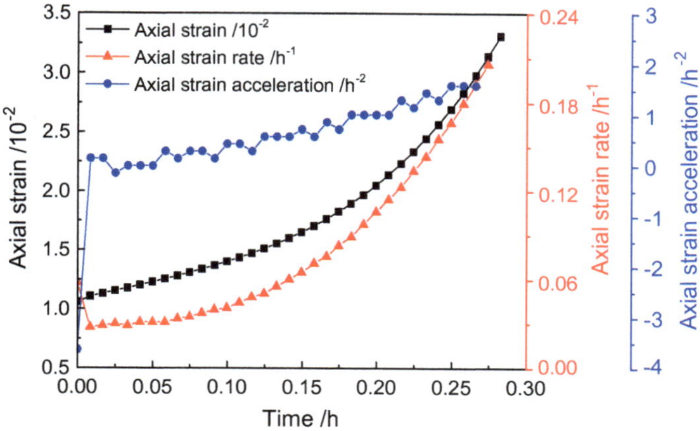

Fig. 3. Axial strain, strain rate and strain acceleration curve

creep stages. At primary creep, with the decrease of strain rate and strain acceleration, the axial strain increased slowly, at secondary creep, strain increases with time ae a constant strain value, and at tertiary creep, the strain increases faster with the increasing of strain rate and strain acceleration.

Plasticity is the property that the deformation of an object after being unloaded cannot be fully recovered, this property has been considered in this paper. According to the unloading equation (Eq. 17) and the creep program of NMM, we can get the unloading curve of the specimen in Fig. 4. Elastic deformation OA was produced by

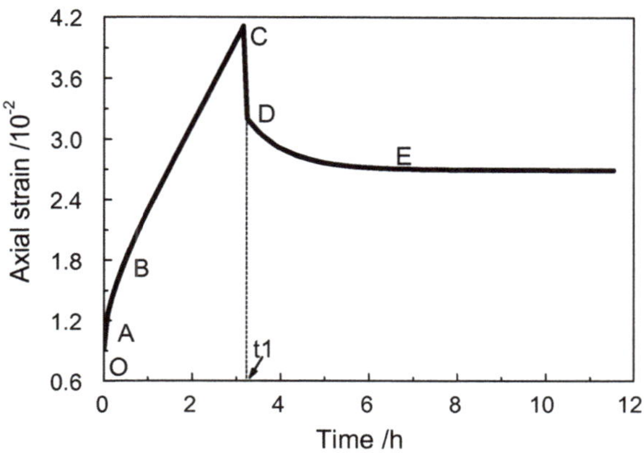

Fig. 4. Stress unloading curve of extended Nishihara model under uniaxial compression

spring k_1, which was generated immediately when stress applied to the specimen. Viscous strain AB was transient creep stage, the longer the time, the larger the strain, and the slower the strain rate until approaching to zero. Then the deformation would go into a minimum strain rate stage. If unloading the stress at time t_1, elastic strain would recover immediately at CD, viscous strain would slowly return to zero over time, DE was strain recovery stage, the strain generated in minimum creep would not recover.

Many parameters in ENM control the curve, different parameters have different effects on experimental results. Parameter sensitivity analysis is performed to study the influence of parameters on calculation result. The parameters of elastic modulus, viscosity coefficient, Poisson's ratio and time index are shown in Table 2. The values of parameters are −40, −20, 0, 20, 40% of initial values, respectively. Axial stress is 40 MPa, yield stress is 36 MPa, the effect of the parameters on experimental creep results are shown in Fig. 5.

Table 2. Parametric sensitivity analysis scheme

Parameter	A	B	C	D	E
	−40%	−20%	0	20%	40%
Spring stiffness K_1, MPa	2265	3020	3775	4530	5285
Spring stiffness K_2, MPa	3600	4800	6000	7200	8400
Viscosity coefficient η_1, MPa h	2500	3331	4164	4497	5830
Viscosity coefficient η_2, MPa h	720	960	1200	1440	1680
Viscosity coefficient η_3, MPa h	6000	8000	10,000	12,000	14,000
Time index n	1.8	2.4	3	3.6	4.2
Poisson's ratio υ	0.174	0.232	0.29	0.348	0.406

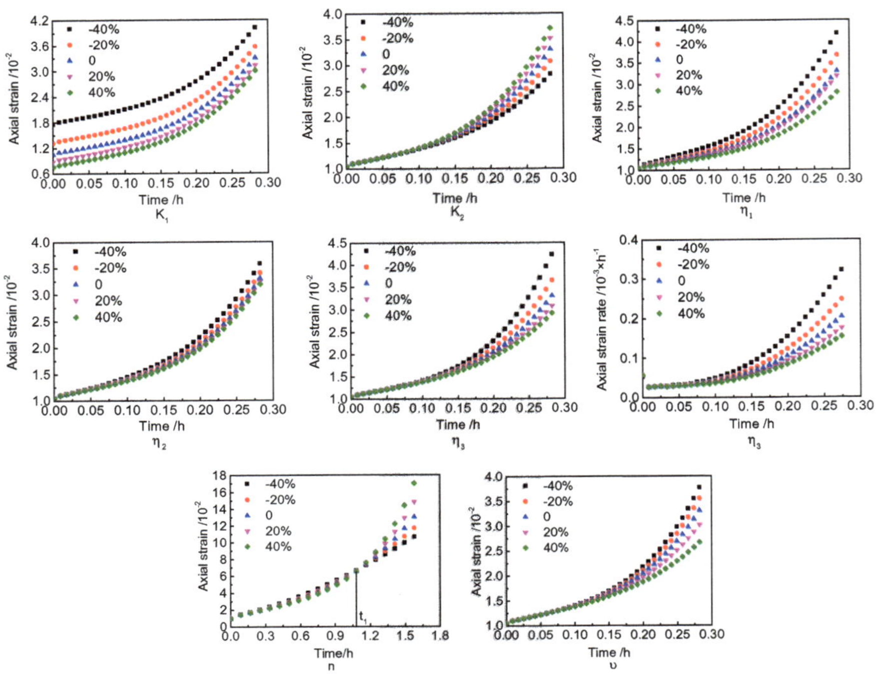

Fig. 5. Strain-time curves with different parameters

In Fig. 5, the elastic strain of rock specimen increases with the decreasing of spring stiffness k_1. It is indicated that k_1 is negative with the axial strain, and has little influence to strain rate and strain acceleration, while, which has more effect on elastic strain. Spring stiffness k_2 is a complicated parameter, accelerating strain rate increases with the increase of k_2 and the creep strain also increases, it has no influence on elastic, decelerating and minimum creep strain, but it has a significant impact on tertiary creep stage. Both viscosity η_1 and η_2 have the same effect on creep strain, the elastic strain doesn't change, the creep strain will decrease at the same time but not at the same amplitude with the increasing of viscosity, the influence of η_1 on strain is more remarkable than that of η_2. As for viscosity coefficient η_3, we can find that there are more remarkable effect on accelerating creep stage, the smaller the η_3, the larger of tertiary creep strain and the strain rate. Time index n is a factor controlling the strain acceleration, the effect of time index on tertiary strain is positive if n is greater than 1. The effect of Poisson's ratio υ is just like η_3, the smaller Poisson's ratio υ, the greater the tertiary creep strain.

3 Numerical Simulation of Creep Deformation

3.1 Model Set-up

The improved NMM program is validated against the experimental results of sandstone tested by a servo-controlled rock testing system. The geometrical dimension of red sandstone specimen is 50 mm in diameter and 100 mm in length. Figure 6 is the

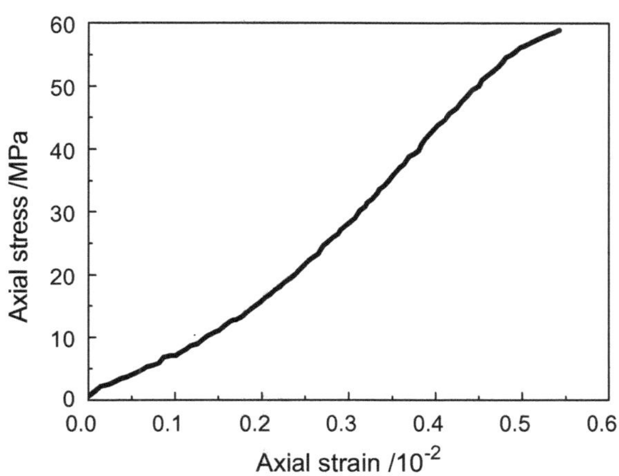

Fig. 6. Stress-strain curve in uniaxial compression

complete stress-strain curve of red sandstone in uniaxial compression test (Zhao et al. 2013). The uniaxial compressive strength, elastic modulus and Poisson's ratio of the sandstone specimen are 58.37 MPa, 9.37 GPa and 0.14, respectively. The numerical model of specimen is the same as the model in Fig. 2. The multi-step axial stresses of 28.03, 33.63, 39.24, 44.84 and 50.45 MPa are applied on the top of the specimen. For triaxial compressive creep tests, confining pressure is applied to the present value and hold constant. After axial stress loading, the confining pressure will be used to test the characteristics of rock. Creep test are performed to study the influence of different confining pressures on the creep behavior of the sandstone specimen.

3.2 Modelling of Creep Under Uniaxial Stress

Multi-step creep experimental curve is shown in Fig. 7. The load at each stress-step is maintained constant for several hours until the specimen fails. At the first 2.65 h, the stress remains at 28.03 MPa, instantaneous displacement is 0.20 mm, creep displace-

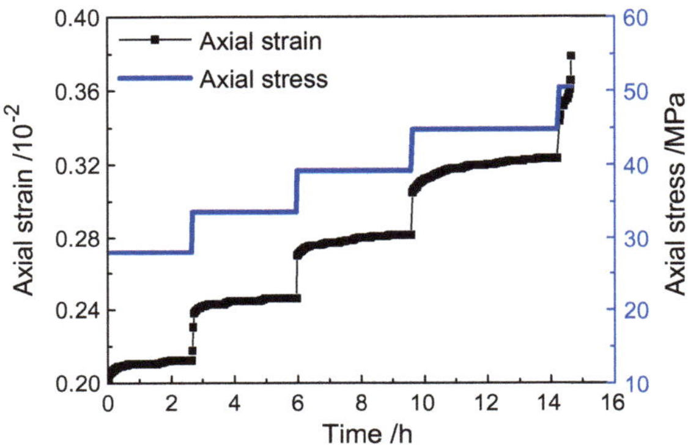

Fig. 7. Multi-step creep test curve

ment is 0.013 mm. When the stress increases to 33.63 MPa, creep deformation in rock specimen increases gradually, instantaneous axial displacement is 0.0022 mm, creep displacement is 0.011 mm, this period lasts 3.26 h. When the stress increases to 39.24 MPa, instantaneous displacement is 0.024 mm, creep displacement is 0.012 mm and creep time will continue 3.68 h. The next stage will last 4.62 h, stress is 44.84 MPa, instantaneous displacement is 0.028 mm, creep displacement is 0.014 mm. Under the stress level of 50.45 MPa, the creep curve at every stage makes up of the primary creep and secondary creep phase. When the stress is up to 50.45 MPa, accelerating stage, can be found and the specimen finally fails. Based on numerical model and ENM, numerical simulations will be performed on red sandstone at laboratory test. Axial stresses of 28.03, 33.63, 39.24, 44.84 and 50.45 MPa are applied on model specimen. The influence of the model parameters on the axial strain is shown in Fig. 5, some adjustments to the parameters (Zhao et al. 2013) are made to simulate the creep of rock. The input mechanical parameters for the numerical model based on the least squares method are listed in Table 3.

The simulation results are shown in Fig. 8. When $\sigma < 50.45$ MPa, the simulated curve is in good agreement with the test curve at each stage. With the increase of the

Table 3. Physico-mechanical parameters of numerical model

Items	Specimen				
Axial stress σ, MPa	28.03	33.63	39.24	44.84	50.45
Spring stiffness K_1, GPa	13.8	14.3	15.6	15.4	15.6
Spring stiffness K_2, GPa	421	540	217	363	560.5
Viscosity coefficient η_1, GPa h	54.2	118.8	85.1	214.1	214.1
Viscosity coefficient η_2, GPa h	2224.6	4472	3000	3000	20,000
Viscosity coefficient η_3, GPa h	0				29,000
Time index n	0				14.9
Poisson's ratio υ	0.14				
Time step Δt, h	0.05				

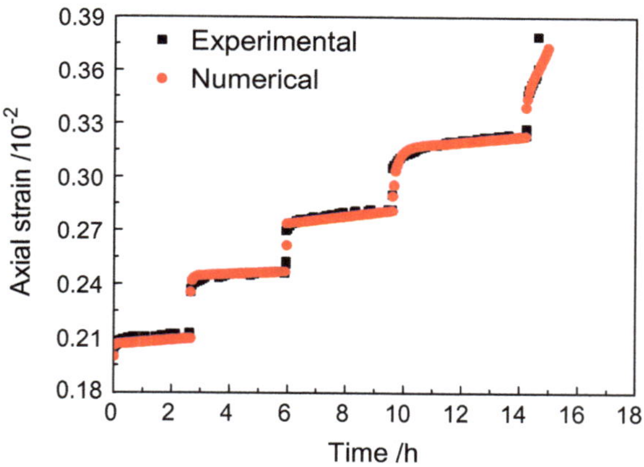

Fig. 8. Comparisons between numerical and experimental curves for multi-step creep

stress, creep deformation becomes more and more obvious and the numerical creep curve agrees well with experimental curve. When $\sigma > 50.45$ MPa, the specimen finally fails, and creep curve exhibits typical three creep stage, and axial strain increases sharply at tertiary creep stage. Figure 9 is the effect of axial stress on minimum axial strain rate. We can clearly see that axial stress has a great influence on the minimum axial creep strain rate, the small increase of axial stress leads to a great increase of the minimum axial creep strain rate, a very small stress perturbation can have a great effect on the minimum creep strain rate. Figure 10 is creep trimodal strain curve under the axial stress of 50.45 MPa, the specimen undergoes three stages of deformation, i.e., primary creep, secondary creep, and tertiary creep, and finally fails. Figure 11 is the strain rate curve under axial stress of 50.45 MPa, it is a classic bathtub curve. We can find that primary creep phase is characterized by an initially high strain rate that decreases with time to reach a quasilinear secondary phase that is often interpreted as minimum creep. After an extended period of time, a tertiary creep phase is reached, characterized by the accelerating strain. This eventually results in macroscopic failure

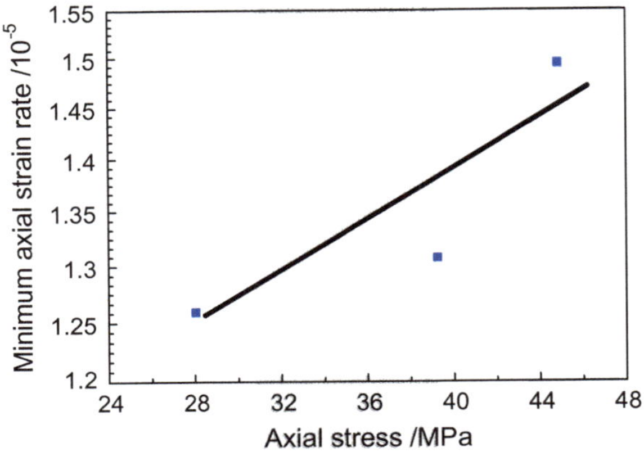

Fig. 9. Effect of axial stress on minimum axial strain rate

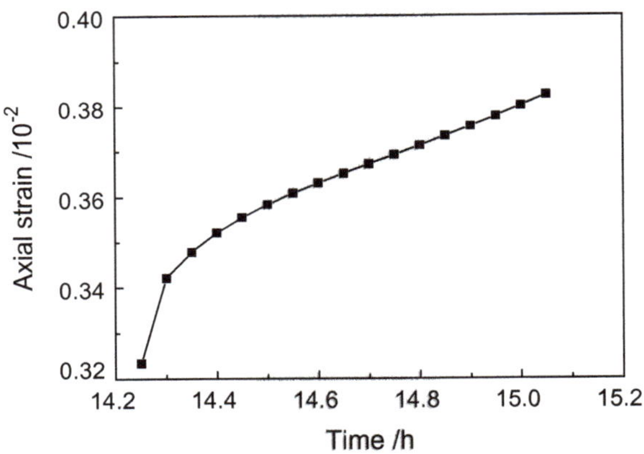

Fig. 10. Creep curve under 50.45 MPa

of the samples by propagation of a shear fault. The displacement of the specimen under axial stress of 44.84 MPa is balanced in Fig. 12. The displacement of the specimen decreases from the top to the bottom of the specimen. The maximum displacement is 0.32 mm, it is similar to the experimental strain curve of the rock sample under axial stress of 44.84 MPa.

3.3 Modelling of Creep Under Confining Pressures

To study the influence of confining pressure on rock creep, we use stress-stepping method to investigate the influence of confining pressure at 5, 10 and 25 MPa. The axial stresses of 28.03, 33.63, 39.24 and 44.84 MPa are applied on the specimen. Axial creep strain curves under different confining pressures are plotted in Fig. 13. It can be

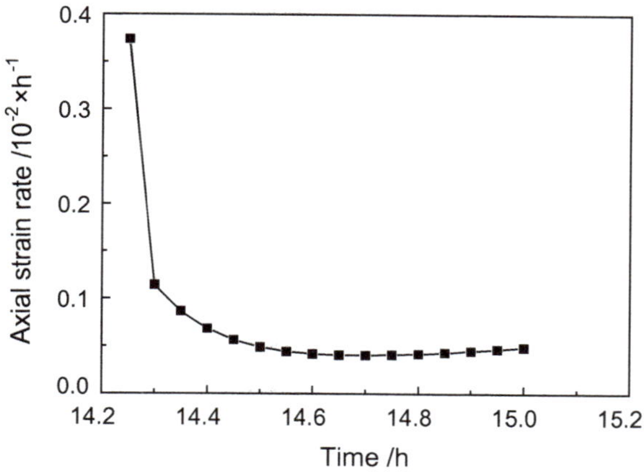

Fig. 11. Creep strain rate curve under 50.45 MPa

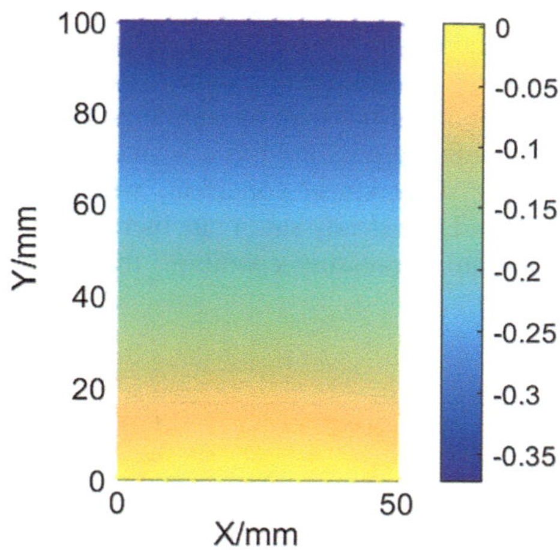

Fig. 12. Stress cloud image of rock specimen under uniaxial compression

seen that the variation of creep strain in different stress stages and confining pressures. The red circle line stands for the confining pressure of 5 MPa, the blue up triangle line represents confining pressure of 10 MPa, the pink down triangle line represents confining pressure of 25 MPa, and the black square line stands for controlled test with the confining pressure of 0 MPa. As the stress increases, axial creep strain increases gradually. The greater the stress, the grater the strain. We can see from Fig. 19 that as the applied axial stress increases, axial creep strain increase gradually, the greater the applied axial stress, the greater the axial creep strain. While the strain is decreased with the increasing confining pressure, the greater the confining pressure, the smaller the axial strain value. Comparing with strain values in four stress levels under different

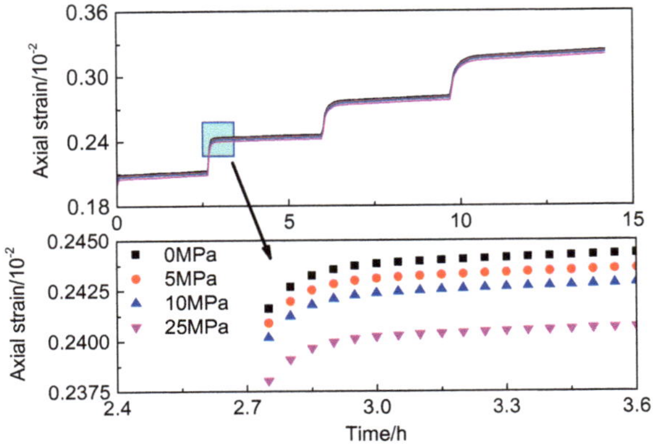

Fig. 13. Influence of confining pressure on axial creep strain behavior of sandstone specimen at 5, 10 and 25 MPa

constant confining pressures, we can find that with the increase of confining pressure, the strain has a certain reduction, but the confining pressure does not change the creep deformation trend, the greater the confining pressure, the greater the strain reduction. Figure 14 is the effect of confining pressure on minimum axial strain rate. We can see clearly from Fig. 14 that confining pressure has a weakening effect on minimum axial strain rate under the same axial stress condition, which inhibits the increase of axial strain. On the other hand, the steady strain rate increases with the increase of axial stress under the same confining pressure condition, the greater the axial stress, the higher the steady strain rate.

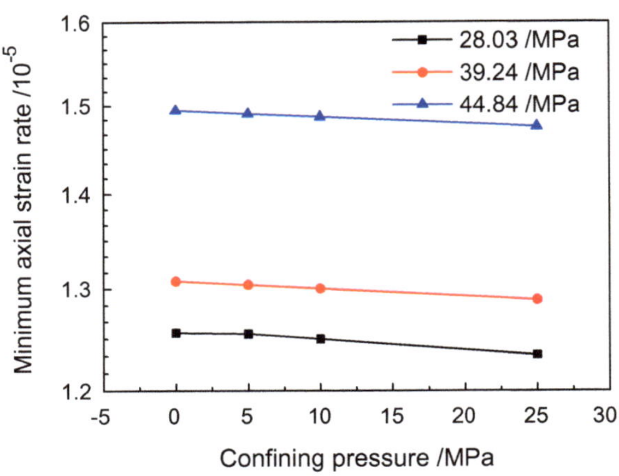

Fig. 14. Effect of confining pressure on minimum axial strain rate

4 Discussions

We have present an ENM for creep to replicate time-dependent deformation under loading conditions and different confining pressures. In order to reflect the accelerating creep stage, the linear Newton component was replaced by a new viscoelastoplastic component, to achieve the tertiary creep. Traditional creep differential equation cannot couple with numerical manifold method, and the creep time is much longer than the time step of numerical manifold method. To solve this problem, there are two methods (time step-viscous initial stress and linear non-homogeneous differential equation) to deal with them. The former is used to divide a series of incremental values in each time step, and the latter is used to change creep differential equation to iteration equation, which can couple with NMM. After the construction of constitutive model, we performed parametric sensitivity analysis to study the correctness of model and characteristics of creep. By this way, we find that spring stiffness k_1 plays a major role in elastic deformation, k_2 has a great influence on tertiary creep stage, with the increase of k_2, the strain gradually decreases. Both viscosity η_1 and η_2 affect the creep strain of first and secondary creep stages, which is the same as k_2, but the influence of η_1 was greater than that of η_2. As viscoelastoplastic component, η_3 mainly affects the tertiary creep stage, with the increase of η_3, the strain and strain rate gradually decreases. Time index n is a factor controlling the accelerating strain rate, if the iteration time was larger than 1 (based on current unit), the larger the factor n, the larger the accelerating rate. Poisson's ratio has little effect on primary and secondary creep stage, with the increase of υ, the tertiary creep strain gradually decreases. After the sensitivity analysis of parameter, we perform a comparative test between laboratory tests and numerical simulation under axial pressure and confining pressure by the improved NMM. According to the results, we can see that the simulation results are in good agreement with the test, the ENM can greatly simulate the trend of creep deformation. The present creep model allows us to investigate the creep deformation with different axial stresses and confining pressures, and we will get the full strain deformation curve to study the creep behavior of rock specimen.

5 Conclusions

In the present paper, the Nishihara creep constitutive equations are incorporated into numerical manifold method by rheological differential equations on MATLAB platform to study the creep properties of rocks, in which the linear Newton component is replaced by a viscoelastoplastic component (in Fig. 1). By this way, the tertiary creep is achieved in the program. In the improved model, the viscous strain is devided into a series of increment time-step values due to time inconsistency between the creep theory and NMM, and the relative parameters are regarded as constants in each time-step increment, this method can solve the time problem between creep and NMM. Parameter sensitivity analysis is performed and the improved NMM is validated against experimental data. Parameter sensitivity analysis reveals the influence of different parameters on the creep deformation. Meanwhile, the influences of axial stress and confining pressure on time-dependent creep deformation are also investigated. The fact that numerical simulations agree well with experimental results shows that the improved NMM based Nishihara model is suitable for modelling of time-dependent creep deformation of rocks.

Acknowledgements. The support provided by Natural Science Foundation of China (Grant Nos. 51474051 and 41672301), National Basic Research Program (973) of China (Grant No. 2014CB047100), Fundamental Research Funds for the Central Universities of China (N150102002).

References

An, X.M., Ning, Y.J., Ma, G.W., et al.: Modeling progressive failures in rock slopes with non-persistent joints using the numerical manifold method. Int. J. Numer. Anal. Meth. Geomech. (2014). https://doi.org/10.1002/nag.2226

He, L., An, X.M., Zhao, X.B., et al.: Investigation on strength and stability of jointed rock mass using three-dimensional numerical manifold method. Int. J. Numer. Anal. Meth. Geomech. (2013). https://doi.org/10.1002/nag.2147

Heap, M.J., Baud, P., Meredith, P.G., et al.: Time-dependent brittle creep in darley dale sandstone. J. Geophys. Res. Atmos. (2009). https://doi.org/10.1029/2008jb006212

Heap, M.J., Baud, P., Meredith, P.G., et al.: Brittle creep in basalt and its application to time-dependent volcano deformation. Earth Planet. Sci. Lett. (2011). https://doi.org/10.1016/j.epsl.2011.04.035

Le, T.M., Fatahi, B.: Trust-region reflective optimisation to obtain soil visco-plastic properties. Eng. Comput. (2016). https://doi.org/10.1108/EC-11-2014-0236

Le, T.M., Fatahi, B., Khabbaz, H., et al.: Numerical optimisation applying trust-region reflective least squares algorithm with constraints to optimise the non-linear creep parameters of soft soil. Appl. Math. Model. (2016). https://doi.org/10.1016/j.apm.2016.08.034

Liu, J., Chen, Q.: A numerical manifold method for simulating creep of rocks. Rock Soil Mech. (2012a). https://doi.org/10.3969/j.issn.1000-7598.2012.04.036

Liu, J., Chen, Q.: Study of high-order numerical manifold method in viscoelastic creep of rock mass. Rock Soil Mech. (2012b). https://doi.org/10.3969/j.issn.1000-7598.2012.07.038

Lockner, D.A.: A generalized law for brittle deformation of Westerly granite. J. Geophys. Res. Solid Earth (1998). https://doi.org/10.1029/97JB03211

Ning, Y.J., An, X.M., Ma, G.W.: Footwall slope stability analysis with the numerical manifold method. Int. J. Rock Mech. Min. Sci. (2011). https://doi.org/10.1016/j.ijrmms.2011.06.011

Nishihara, M.: A method for predicting creep characteristics. Doshisha Kogaku Kaishi (1955)

Scholz, C.H.: Static fatigue of quartz. J. Geophys. Res. (1972). https://doi.org/10.1029/jb077i011p02104

Shi, G.H.: Discontinuous deformation analysis: a new numerical model for the statics and dynamics of deformable block structures. Eng. Comput. (1992). https://doi.org/10.1108/eb023855

Voight, B.: A relation to describe rate-dependent material failure. Science (1989). https://doi.org/10.1126/science.243.4888.200

Wang, Q.Y., Zhu, W.C., Xu T., et al.: Numerical simulation of rock creep behavior with a damage-based constitutive law. Int. J. Geomech. (2016). https://doi.org/10.1061/(asce)gm.1943-5622.0000707

Wang, Z.C., Wong, R.C.K.: Strain-dependent and stress-dependent creep model for a till subject to triaxial compression. Int. J. Geomech. (2015). https://doi.org/10.1061/(ASCE)GM.1943-5622.0000583

Yang, S.Q., Xu, P., Xu, T.: Nonlinear visco-elastic and accelerating creep model for coal under conventional triaxial compression. Geomech. Geophys. Geo-Energy Geo-Resourc. (2015). https://doi.org/10.1007/s40948-015-0014-y

Yang, Y.T., Tang, X.H., Zheng, H., et al.: Three-dimensional fracture propagation with numerical manifold method. Eng. Anal. Boundary Elem. (2016). https://doi.org/10.1016/j.enganabound.2016.08.008

Zhao, B.Y., Liu, D.Y., Zheng, Y.R.: Uniaxial compressive creep test of red sandstone and its constitutive model. J. Min. Saf. Eng. (2013)

Zhao, L., Xu, L.Y., Nikbin, K.: Predicting failure modes in creep and creep-fatigue crack growth using a random grain/grain boundary idealised microstructure meshing system. Mater. Sci. Eng. Struct. Mater. Prop. Microstruct. Process. (2017a). https://doi.org/10.1016/j.msea.2017.08.035

Zhao, Y.L., Wang, Y.X., Wang, W.J., et al.: Modeling of non-linear rheological behavior of hard rock using triaxial rheological experiment. Int. J. Rock Mech. Min. Sci. (2017b). https://doi.org/10.1016/j.ijrmms.2017.01.004

Numerical Seepage Model for Joint Type Loess

Yang Luo(✉), Wen-chieh Cheng, and T.-H. Wang

Institute of Tunnel and Underground Structure Engineering, School of Civil Engineering, Xi'an University of Architecture and Technology, Xi'an 710055, China
{y.luo.geo,w-c.cheng,wangtx}@xauat.edu.cn

Abstract. Performing tunnel excavation in the joint type loess is very challenging as the associated seepage behavior cannot be predicted accurately, severe geo-hazards like sinkhole, water ingress, and ground surface settlement are thus easily to be triggered in such region. This study presents the governing equations that account the effect of the loess joint on the seepage behavior. A numerical seepage simulation model for joint type loess is then established. This study also carried out the numerical simulation to characterise the development pattern of the wetting front in joint/non-joint type loess. The results obtained from this study provide insights into future tunnel construction or deep excavation in such joint type loess region.

1 Introduction

Loess has been extensively distributed in arid areas in China and a series of joints may be developed in loess because of the effect of drying-shrinkage. Since the joints significantly control the seepage behavior and play a leading role in the stability of loess caverns (Xi-An et al. 2009; Liang et al. 2011), loess landslide (Wang 2009), tunneling or deep excavation (Ni and Cheng 2011; Cheng and Ni 2009) in such loess region, the investigation is very challenging. In spite that there are many analytical models available for simulating the seepage behavior of rock joint (Xiao et al. 2012; Chen and Zhang 1994; Chen et al. 2008), analytical model developed for simulating the seepage behavior of the loess joint is rare. This paper thus presents the seepage governing equations of the loess joint. A numerical seepage simulation model for joint type loess is also established. The simulated results of wetting front for joint type loess and non-joint loess are then compared, thereby verifying the correctness of the developed numerical model.

2 Simplification of Seepage Governing Equations for Unsaturated Loess

Seepage in the unsaturated loess is a non-stationary flow and has been considered as a two-dimensional plane problem. The water phase partial differential equation can be obtained by considering the continuity for the water phase. The net flux of water

B. Fatahi et al. (eds.), *Sustainable Design and Construction for Geomaterials and Geostructures*, Sustainable Civil Infrastructures,
https://doi.org/10.1007/978-3-319-95753-1_9

through an element of unsaturated soil (Fig. 1) can be computed from the volume rates of water entering and leaving the element within a period of time:

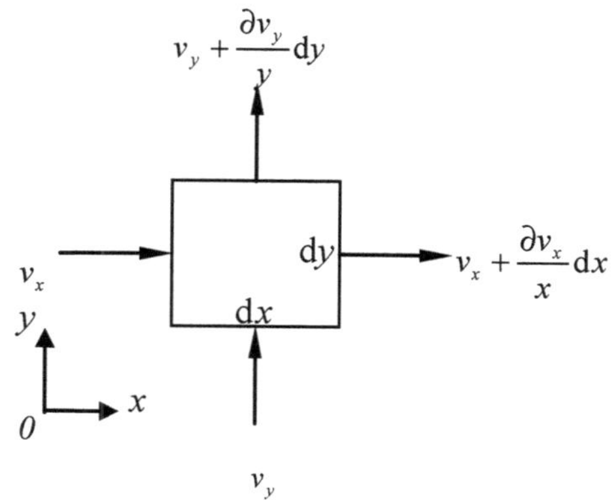

Fig. 1. Two-dimensional non-stationary flow for particle element M

$$\frac{\partial v_x}{\partial x} + \frac{\partial v_y}{\partial y} = \frac{\partial (V_w/V)}{\partial t} \tag{1}$$

where V_w represent the volume of water in the element; V represent initial overall volume of the element. According to the soil mechanics, the porosity of particle element (n) should be V_v/V, and the degree of saturation (S) should be V_w/V_v then V_w/V at the right-hand side of Eq. (1) equal to $S\,n$. Meanwhile, the volumetric water content (θ) is expressed as $\theta = S{\cdot}n$. It changes with time during water penetration, largely affects the degree of saturation and the suction water head.

What's more, as seepage in the unsaturated soil also satisfies the Darcy law, so that $v_x = k_x\,(\partial h/\partial x)$ and $v_y = k_y\,(\partial h/\partial y)$. Where k_x & k_y represent the coefficient of permeability in the x and y directions, respectively; h corresponds to the pressure head.

With those discussions above, Eq. (1) can be derived as the following equation.

$$\frac{\partial}{\partial x}(k_x \frac{\partial h}{\partial x}) + \frac{\partial}{\partial y}(k_y \frac{\partial h}{\partial y}) = \frac{\partial (S \cdot n)}{\partial t} = \frac{\partial \theta}{\partial t} \tag{2}$$

In soil-water dynamics, the change in the volumetric water content due to the change in the unit matric suction is called as "the specific water capacity" and denoted as $C(\theta)$, where $C(\theta) = \partial\theta/\partial h$. The right side of Eq. (2) can be rewritten as Eq. (3). Additionally, the coefficient of permeability $k(\theta)$ is defined as Eq. (4). Where $D(\theta)$ represents the diffusivity of the unsaturated soil.

$$\frac{\partial \theta}{\partial t} = \frac{\partial \theta}{\partial h} \cdot \frac{\partial h}{\partial t} = C(\theta) \cdot \frac{\partial h}{\partial t} \tag{3}$$

$$k(\theta) = D(\theta) \cdot C(\theta). \tag{4}$$

Substituting Eqs. (3) & (4) into Eq. (2), it can be simplified and expressed as Eq. (5) for the unsaturated loess.

$$\frac{\partial}{\partial x}[D(\theta)\frac{\partial h}{\partial x}] + \frac{\partial}{\partial y}[D(\theta)\frac{\partial h}{\partial y}] = \frac{\partial h}{\partial t} \tag{5}$$

3 Analytical Analysis of Governing Equation for Loess Joint

For Numerical calculation, the joint element should work with loess element together. This makes the two kinds of elements will share some element node, as well as the head of the node. Here we can make a assumption that the V-G model (Genuchten 1980) can also be used to describe the relationship between the degree of saturation and the suction water head for the joint, as shown in Eq. (6).

$$\frac{\theta - \theta_r}{\theta_s - \theta_r} = \frac{1}{[1 + |\alpha \cdot h|^n]^m} \tag{6}$$

where θ_s represents the saturated water content in %; θ_r represents the residue water content in %; α, n and m are test parameters. Ideally the porosity n of loess joint should be 1. Since there is no medium in joint, the associated θ_r will be 0 and the θ_s will be 1.

Then the right-hand side of Eq. (2) can also be expressed as follows:

$$\frac{\partial \theta}{\partial t} = \frac{\partial \theta}{\partial h} \cdot \frac{\partial h}{\partial t} = \frac{mn|\alpha \cdot h|^n}{[1 + |\alpha \cdot h|^n]^{m+1}} \cdot \frac{\partial h}{\partial t} \tag{7}$$

Since previous studies ignore the seepage perpendicular to the joint strike direction, determining the coefficient of permeability for the loess joint should consider the direction both perpendicular to the joint and along, as illustrated in Fig. 2.

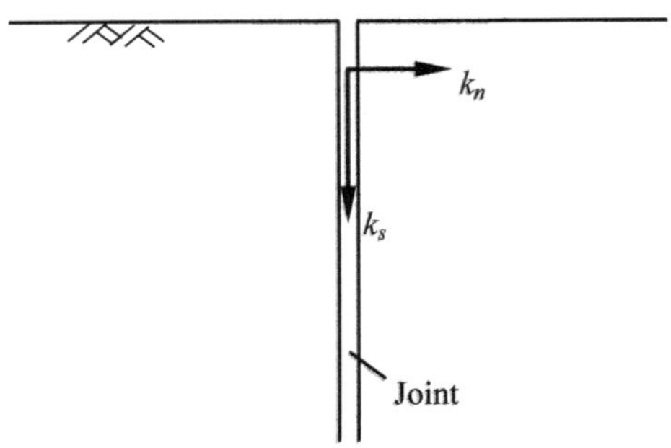

Fig. 2. Coefficients of permeability for the joint type loess

When water migrates from loess to joint, it pass through one material of porosity n to another material of porosity $n = 1$. We use Eq. (8) to describe the permeability coefficient perpendicular to the joint strike direction (k_n).

$$k_n = \frac{k_\sigma}{n} \tag{8}$$

where k_σ is permeability coefficient on the boundary surface at unsaturated soil side. It is a function of volumetric water content of loess.

Besides, assuming the joint surface is smooth, flat, and straight with large opening a in droughts and the width of this opening however is far less than the depth of joint l (Fig. 3) (Tie-hang et al. 2013), the fluid motion inside the joint should satisfy the N-S equation and substituting the boundary conditions into the N-S equation leads to the transversal average flow rate inside the joint along the joint strike direction, as shown in Eq. (9).

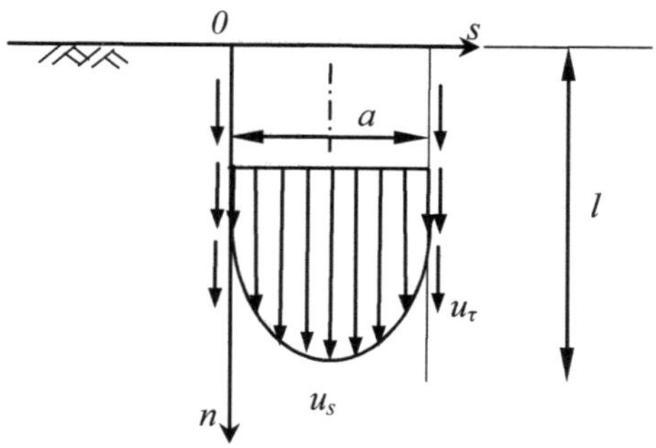

Fig. 3. Schematic illustration of the fluid motion in the joint type loess

$$u_s = u_\tau + \frac{\gamma_w}{\mu} \cdot \frac{a^2}{12} J \tag{9}$$

where u_s represents the transversal average flow rate along the joint strike direction; u_τ represents the flow rate of joint boundary along the surface tangential direction; μ is the dynamic viscosity coefficient of water; J is the hydraulic gradient.

To describe how the width of joint can be narrowed or widened with the increasing or decreasing water content of loess, joint opening a can be described in Eq. (10):

$$a = \frac{w_p - w}{w_p - w_0} a_0 \tag{10}$$

where w represents mass water content of loess, and take shrinkage limit water content w_s in the case of $w < w_s$, with hydraulic aperture of joints reaching maximum; take plastic limit water content w_p in the case of $w > w_p$, with hydraulic aperture of joints then is zero (influence from joints on seepage can be neglected); a_0 indicates geometrical width of joints; w_0 indicates initial water content of loess corresponding to a_0.

Thus, with Eqs. (9) and (10), the coefficient of permeability for a joint along the joint strike can be calculated through the Darcy law as Eq. (11) shows.

$$k_s = \frac{u_s}{J} = k_\tau + \frac{\gamma_w}{12\mu} \cdot \left(\frac{w_p - w}{w_p - w_0} a_0\right)^2 \tag{11}$$

where k_s represents the coefficient of permeability along the joint strike direction; k_τ represents the permeability of joint boundary along the surface tangential direction.

With Eqs. (7), (8) and (11), Eq. (2) can be further expressed as follow

$$\begin{aligned} \nabla \cdot (\mathbf{k} \cdot \nabla h_p) &= \frac{mn|\alpha \cdot h|^n}{[1 + |\alpha \cdot h|^n]^{m+1}} \cdot \frac{\partial h}{\partial t} \\ [\mathbf{k}] &= \begin{bmatrix} & \frac{k_\sigma}{n} & \\ k_\tau + \frac{\gamma_w}{12\mu} \cdot \left(\frac{w_p - w}{w_p - w_0} a_0\right)^2 & & \end{bmatrix} \end{aligned} \tag{12}$$

4 FEM of Joint Type Loess for Seepage Analysis

Equation (6) is to describe the seepage of loess, while Eq. (12) is to describe the seepage of joint. This two governing equation together can be used to simulation seepage problems in joint type loess.

Finite element equation for seepage of loess joints can be established by using Galerkin weighted residual method.

$$[D]\{h\} + [E]\frac{\partial \{h\}}{\partial t} = [F] \tag{13}$$

where, **[D]** is called stiffness matrix, related with element permeability coefficient; **[E]** is called capacity matrix, related with water volume coefficient of unsaturated soil; **[F]** is called flow boundary conditions, a unit column vector reflecting infiltration or evaporation amount; $\{\boldsymbol{h}\}$ represents unit head column vector; t indicates time.

Stiffness matrix **[D]** is expressed as follows:

$$[D] = \int_\Omega [B]^\mathrm{T}[k][B]d\Omega \tag{14}$$

where, **[B]** is derivative matrix of element shape function against coordinate, n indicates number of element node. $[\boldsymbol{k}]$ indicates permeability coefficient tenser in element.

$$[B] = [B_1\ B_2\ \cdots\ B_n];\ [B_i] = \begin{bmatrix} \frac{\partial N_i}{\partial x} & \frac{\partial N_i}{\partial y} \end{bmatrix}^\mathrm{T} \tag{15}$$

As isotropic loess, [***k***] is expressed as Eq. (16). For joint medium, [***k***] can be calculated with Eq. (12).

$$[k] = \begin{bmatrix} D(\theta) & \\ & D(\theta) \end{bmatrix} \tag{16}$$

D(*θ*) is called diffusivity and is the function of water content in soil, which can be obtained through test and calculated with the Equation below if dry density of loess is considered (Wang et al. 2007)

$$D(\theta) = \exp\,(\mathrm{a} \cdot \theta^2 + \mathrm{b} \cdot \theta + \mathrm{c}) \tag{17}$$

wherein, a, b and c are testing constants related to dry density,

$$\begin{cases} \mathrm{a} = 84.5\rho_d - 57.1 \\ \mathrm{b} = -56.0\rho_d + 61.8 \\ \mathrm{c} = 5.7\rho_d - 12.0 \end{cases} \tag{18}$$

Capacity matrix [**E**] is expressed as follows.

$$[E] = \lambda \int\limits_{\Omega} [N]^{\mathrm{T}}[N] d\Omega \tag{19}$$

According to Eq. (5), λ should be 1 for unsaturated loess medium. For joint medium, λ should be calculated from Eq. (7).

[**F**] is a matrix reflecting flow boundary conditions. If flow rate (*v*) is specified in normal direction of boundary, then

$$[F] = \int\limits_{\Gamma_2} [N]^{\mathrm{T}} \bar{v} d\Gamma \tag{20}$$

Positive sign represents infiltration, and node is regarded as "source"; negative sign represents evaporation water loss, and node is regarded as "convergence".

In [**D**], [**E**] and [**F**], [*N*] refers to the shape function. When the quadrilateral isoparametric element is adopted, the shape function on each node and the displacement mode at any point in the element respectively are

$$N_i = \frac{1}{4}(1 + \zeta_i\zeta)(1 + \eta_i\eta) \quad (i = 1, 2, 3, 4) \tag{21}$$

where, ζ_i, η_i are local coordinate values of point *i*.

If backward difference method is used to disperse time domain, finite element equation is as follows.

$$([D] + \frac{[E]}{\Delta t})\{h\}_{t+\Delta t} = \frac{[E]}{\Delta t}\{h\}_t + [F] \tag{22}$$

where, $\{h\}_t$ indicates node head array at the time of t, $\{h\}_{t+\Delta t}$ indicates node head array at the time of $t + \Delta t$; Δt is time step.

5 A Case Analysis: Wetting Front for Joint/Non-joint Type Loess

Here is a case for instruction. Field infiltration test is designed to study the wetting front of water migration in joint and non-joint type loess, as shown in Fig. 4. This case choose typical loess field with vertical joint. It needs to story water on its surface. The water infiltrates vertically along joint direction from loess surface. Wetting front is observed in joint loess after field test. Meanwhile, we choose non-joint loess field to conduct the same test previously mentioned.

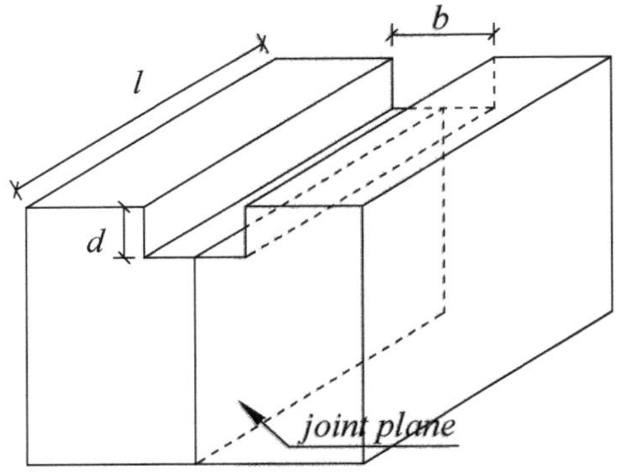

Fig. 4. Model of test pit

The test was terminated after 6 h infiltration. The test result is shown in Fig. 5. It is obviously shown in Fig. 5 that the maximum saturation depth in joint type loess is greater than that in non-joint type loess in vertical direction. The maximum saturation depth value in joint testing site appeared nearby the joint. Wetting front was in obvious sinking shape along joint. In comparison, the saturation depth value in non-joint test site also appeared nearby the vertical axis, but its trend of sinking along vertical axis is not obvious and wetting front is smooth. On the other hand, in horizon direction the maximum saturation width in joint loess is lightly greater than that in non-joint loess.

In the numerical simulation, a total of 500 quadrilateral linear elements were utilised for discretising the simulated domain with a single 1-m long vertical joint, as shown in Fig. 6. As can be seen, this vertical joint was consisted of 20 joint elements of 1.0 cm by 5.0 cm, which is much longer than the width of this vertical joint. Soils element in contact with both sides of this joint were 4.5 cm by 5.0 cm in size, and the

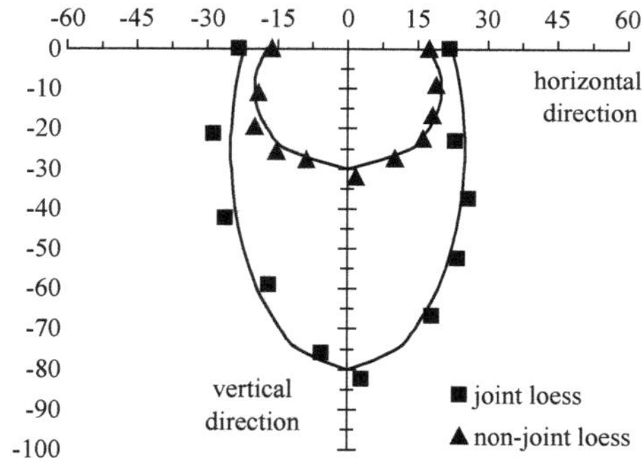

Fig. 5. Wetting front of joint/non-joint type loess

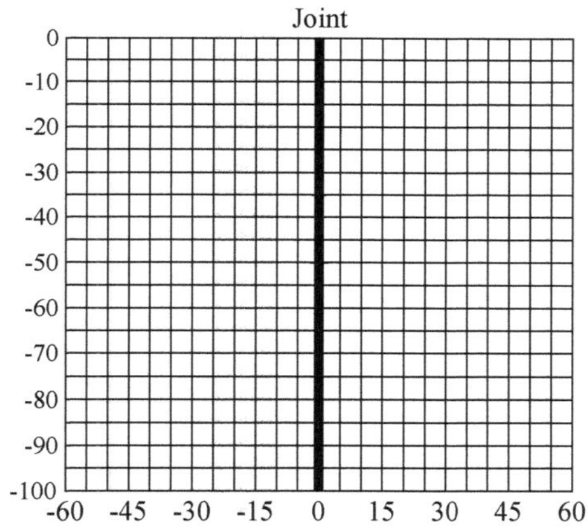

Fig. 6. Finite element mesh of the joint type loess

others were 5.0 cm by 5.0 cm. Assuming the infiltration was around this vertical joint only, the saturated area was set as a range of 10.0 cm against either side of this vertical joint, in which the pressure head of the element node was zero, while the level head depended on the y coordinate. To consider the anisotropic nature of the joint type loess, the ratio of the coefficient of permeability in the x direction to that in the y direction was assumed to be 1:4 (Liang et al. 2012). The coefficient of permeability for this joint was calculated through Eq. (12), considering $n = 0.483$, $\gamma_w = 9.8$ kN/m^3, $\mu = 1.002 \times 10^{-3}$ Pa s. The simulation adopted 1-h time step and the numerical iteration was terminated as the water head was converged with the head difference equal to or less than 1%.

Figure 7 presents the development pattern of the wetting front in joint/non-joint type loess. It can be seen from Fig. 7 that in the x direction, the development of wetting front was more or less the same in joint/non-joint type loess. In the y direction, the wetting front reached more than 50 cm for the simulation with the embedded vertical joint, which is greater than 23 cm from the simulation in non-joint type loess.

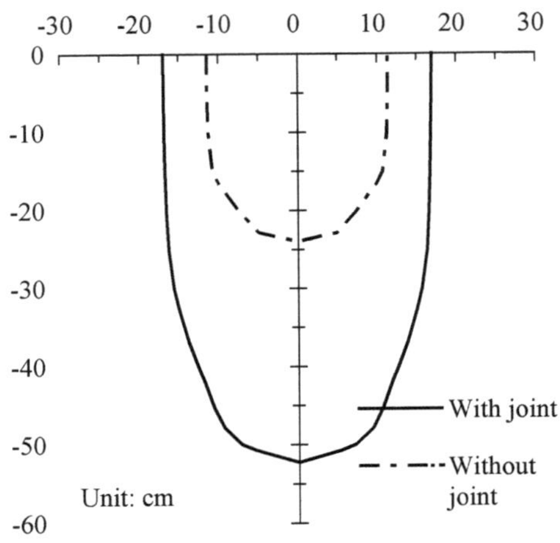

Fig. 7. Numerical calculation of wetting front in joint/non-joint type loess

From the comparison of field test (Fig. 5) and numerical calculation (Fig. 7), the simulation result fit the test result well. This numerical seepage model can be used to research the behavior of joint type loess.

6 Conclusions

(1) This study presents the governing equations for describing the seepage behavior of the joint type loess.
(2) In order to do the numerical seepage calculation of joint type loess, this work simplifies the seepage governing equation for loess, and determines the equation for loess joint.
(3) The concave-shaped wetting front is numerical calculated with a certain time. The development pattern of the wetting front in joint/non-joint type loess is carried out. A contrast of seepage test result and numerical calculation result is to verify the correctness of the analytical analysis and the numerical method.

References

Chen, P., Zhang, Y.: Coupling analysis of seepage/stress for jointed rock. Chin. J. Rock Mechan. Eng. **13**(4), 299–308 (1994)

Chen, Y.F., Zhou, C.B., Sheng, Y.Q.: Generalized cubic law for seepage flow through rock joints considering post-peak mechanical response. Rock Soil Mech. **29**(7), 1825–1831 (2008)

Cheng, W.C., Ni, J.C.: Feasibility study of applying SOFO optical fiber sensor to segment of shield tunnel. Tunn. Undergr. Space Technol. **24**(3), 331–349 (2009)

Genuchten, M.T.V.: A closed-form equation for predicting the hydraulic conductivity of unsaturated soils. Soil Sci. Soc. Am. J. **44**(44), 892–898 (1980)

Liang, Q.G., Jie, L.I., De-Wu, L.I., Zhou, Y.Y.: Some problems on surrounding rock classification of loess tunnels. Chin. J. Geotech. Eng. **33**(Supp. 1), 170–176 (2011)

Liang, Y., Xing, X.L., Li, T.L., Xu, P., Liu, S.L.: Study of the anisotropic permeability and mechanism of Q3 loess. Yantu Lixue/rock Soil Mech. **33**(5), 1313–1318 (2012)

Ni, J.C., Cheng, W.C.: Shield machine disassembly in grouted soils outside the ventilation shaft: a case history in taipei rapid transit system (TRTS). Tunn. Undergr. Space Technol. **26**(2), 435–443 (2011)

Tie-hang, W., Yang, L., Hui, Z.: Two-dimensional steady flow rate equation for loess joints. Chin. J. Geotechn. Eng. **35**(6), 1115–1120 (2013)

Wang, J.: Stability assessment of Tashan loess landslide at Qianyang County of Baoji region in Shaanxi province. J. Eng. Geol. **17**(5), 619–627 (2009)

Wang, T.H., Yue, C.K., Lu, J., Liu, L.P.: Numerical analysis on moisture migration in loess subgrade under rainfall. Xi'an University of Architecture and Technology. (Natural Science Edition) (2007)

Xi-An, L.I., Peng, J.B., Run-Yong, M.A., Qiao, X.Y., Kang, J.H., Liang, L.I.: On the hazards and evolution processes of loess tunnels. J. Xian Univ. Sci. Technol. (2009)

Xiao, W.M., Xia, C.C., Wang, W., Bian, Y.W.: Study of void model for fluid flow through rough joints-taking Reynolds equation as governing equation. Yantu Lixue/rock Soil Mech. **33**(12), 3680–3688 (2012)

Effect of Waveform Mircopile on Foundation Underpinning During Building Remodeling with Vertical Extension

Cheng Can Wang[1], Young Eun Jang[1], Seok Jung Kim[2], and Jin Tae Han[2(✉)]

[1] Department of Geo-Space and Geotechnics, University of Science & Technology, 283, Goyangdae-Ro, Ilsanseo-Gu, Goyang-Si, Gyeonggi-Do ST 10233, Korea
{wangchengcan, yejang}@kict.re.kr
[2] Department of Geotechnical Engineering, Korea Institute of Civil Engineering and Building Technology, 283, Goyangdae-Ro, Ilsanseo-Gu, Goyang-Si, Gyeonggi-Do ST 10233, Korea
{seokjungkim, jimmyhan}@kict.re.kr

Abstract. During the vertical extension of an existing building, the foundation would be subjected to additional loads vertically. In order to enhance bearing capacity and reduce settlement of the existing foundation, a new type of micropile named waveform micropile was applied to underpin the foundation in this study. Finite element method (FEM) was used to evaluate the micropile's load-settlement behavior and the underpinning performance of the waveform micropile. The load distribution ratio (LDR) of existing piles and the reinforcing micropile was estimated before and after applying additional loads. The underpinning performance of a conventional micropile was also included for comparison with the waveform micopile. Numerical results load-settlement response for single pile showed bearing capacity of waveform micropiles was higher than that of conventional micropile because of the shape of the waveform micropile. Moreover, when the additional loads applied on the reinforced foundation, the waveform micropile showed better reinforcing performance compared to the conventional micropile.

1 Introduction

Foundation reinforcement is essential to enhance the bearing capacity and reduce the settlement of an existing foundation for a building to be remodeled with vertical extension. A number of technologies are available to underpin the foundations (Thorburn 1993). Due to the limited spacing between existing piles and applicability for in situ construction, micropile technology is widely used in foundation reinforcement.

A micropile is a small-diameter, drilled, and grouted pile with a central steel bar. Generally, micropiles are between 100 and 300 mm in diameter, 20–30 m in length, and 300–1000 kN in compressive or tensile service load (FHWA 2005) The installation of micropiles causes minimal disturbance to adjacent structures, soils, and the

B. Fatahi et al. (eds.), *Sustainable Design and Construction for Geomaterials and Geostructures*, Sustainable Civil Infrastructures,
https://doi.org/10.1007/978-3-319-95753-1_10

environment. Also, it can be installed in restrictive conditions at any angle. The technology of micropililing was introduced by Lizzi in the early 1950s. Since then, it has been widely used to reinforce existing foundations in static and dynamic zones and support slopes since the 1980s (Bruce 1996). Load-displacement responses and load transfer mechanism of micropiles under different kinds of loading have been investigated previously (Kyung et al. 2016; Han and Ye 2006a). Micropiles as reinforcing elements and their reinforcing performances in the static and dynamic zones have also been evaluated by many researchers (Isam et al. 2012; Han and Ye 2006b; Sadek and Isam 2004; Babu et al. 2004; Esmaeili et al. 2012).

Up to now, mircopile technology has been continuously evolved. The first generation of micropile introduced in the 1950s and it was called "root pile" or "palo radice". It was a small-diameter and grouted pile that can carry load of less than 100 kN (Lizzi 1982). The second generation of micropile developed in the 1970s named mini pile, pin pile or needle pile is a pressure-grouted pile that can carry a load of more than 1500 kN. A new generation of micropiles named Titan Injection Bore (IBO) micropile using a continuously threaded hollow steel bar that allows drilling and grouting to proceed simultaneously without casing was presented by Ernst Ischebeck (Contech System 2011). The ultimate bond strength of hollow bar micropiles has been demonstrated to be significantly greater than that of conventional micropiles in various soil conditions (Gomez et al. 2007; Bruce 2009). In Korea, a new type of micropile named waveform micropile was developed by Jang and Han (2014). The main idea of a waveform micopile is to develop the waveform shape by jet grouting construction method to enhance its skin resistance along the shaft of the pile. The bearing capacity of the waveform micropile has been verified to be higher than that of a conventional micropile by full-scale field tests, centrifuge tests, and numerical analysis (Jang and Han 2015, 2016). However, the application of waveform micropile as a reinforcing element has not been sufficiently investigated previously.

Therefore, the objectives of this study were: (1) evaluation of the axial performance of waveform micropile; (2) investigation of the reinforcing effect of waveform micropile in building remodeling by estimation of LDR and (3) evaluation of the behavior of waveform micropile compared to the conventional micropile. To accomplish these goals, three dimensional finite element analysis was used.

2 Waveform Micropile

Waveform micropile is a new type of micropile with shear keys. It uses jet grouting construction method to strengthen the bond strength along the pile's shaft. The construction process is shown in Fig. 1. The construction method was demonstrated more economical than that of conventional micropile (Jang and Han 2014). Figure 2 shows the schematics of conventional micropile and the waveform micopile. Waveform micropile has two diameters. The diameter of shear keys is 500 mm, which is 1.7 times larger than the diameter of the pile's shaft. The bearing capacity of the waveform micropile has been verified to be about 50% greater than that of conventional micropile according to full-scale tests, centrifuge tests, and numerical analysis (Jang and Han 2014, 2015, 2016).

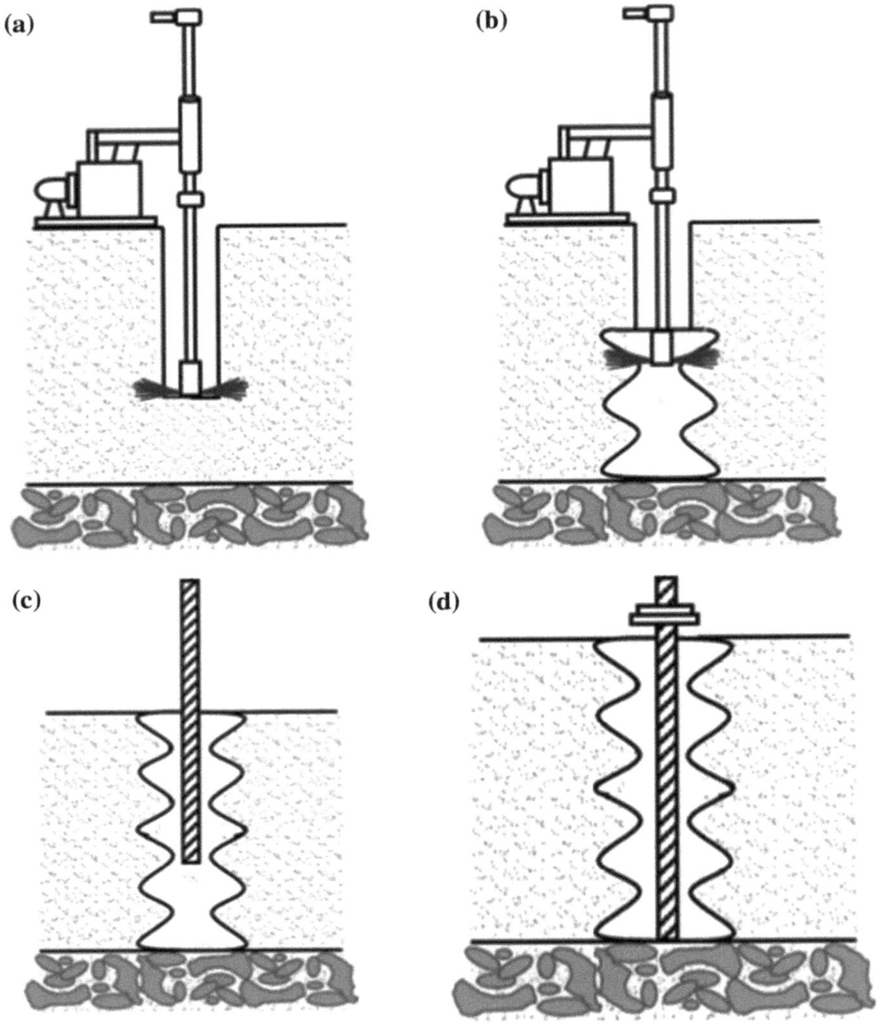

Fig. 1. Construction process of waveform micropile boring **a** Boring; **b** Jet grouting; **c** Reinforcing; **d** Completion. Jang and Han (2016)

3 Method of Analysis

3.1 Numerical Model for Pile Foundation

The numerical study was carried out with 3D finite element code PLAXIS 3D. A $4 \times 4 \times 1$ m raft with four existing piles (EP) and one reinforcing pile (RP) were considered in this study. General PC piles which were widely used as foundation building materials were modeled as existing piles. Two kinds of micropiles, conventional micropile (CMP) and waveform micropile (WMP), were used as reinforcing pile.

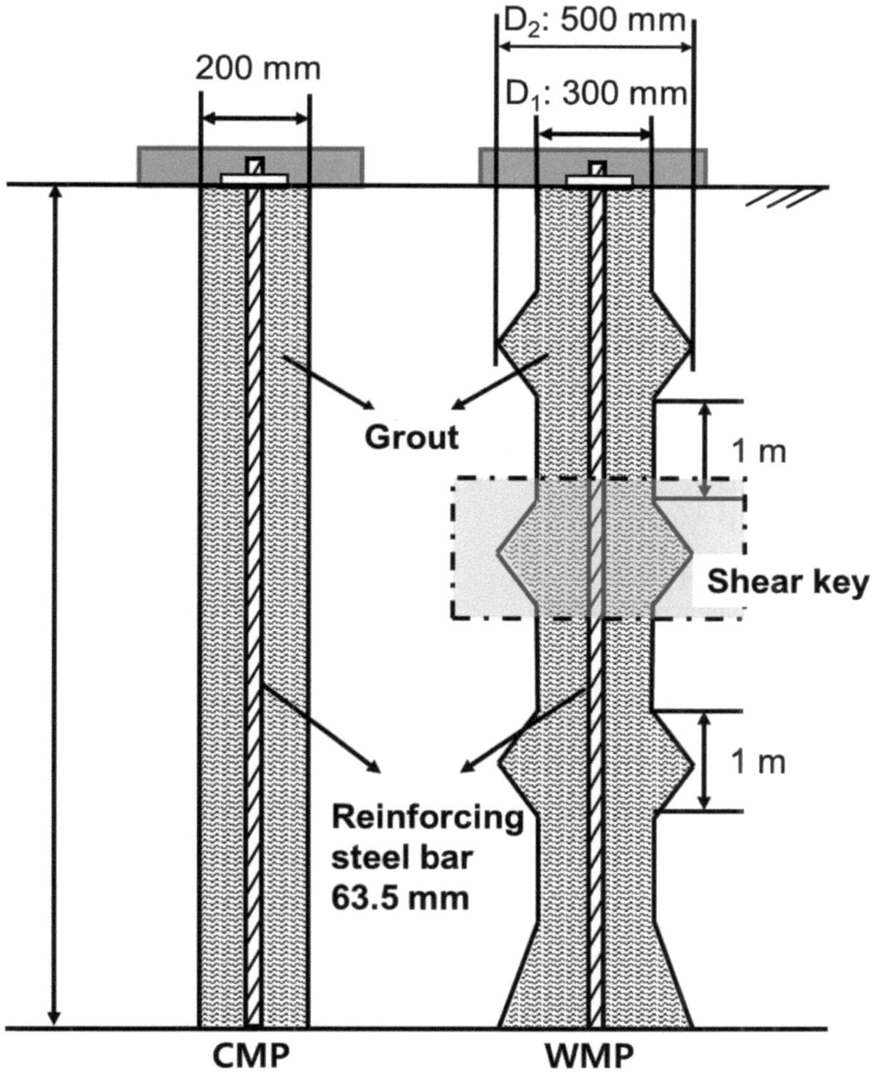

Fig. 2. Schematics of conventional and waveform micropile

To avoid any influence of the boundary effect, the mesh was extended in both horizontal direction to a width of 10 m and vertical direction to a height of 20 m. Only two soil layers were applied in this analysis: 0–8 m of sand layer and rock. Both piles were installed until reaching the rock stratum. Figure 3 exhibits the geometry of the soils and pile foundation. Soil behavior was determined using Mohr-Coulomb model. PC piles were modeled as linear-elastic solid elements. Micropiles consisted of grout material and a central steel bar. Grout was also modeled as linear-elastic solid model, while the steel bar was modeled as beam elements. Tables 1 and 2 show properties of piles and soils, respectively. The soil-pile interface was described by R_{inter}, the interface strength reduction factor (Plaxis 2005). R_{inter} was defined as:

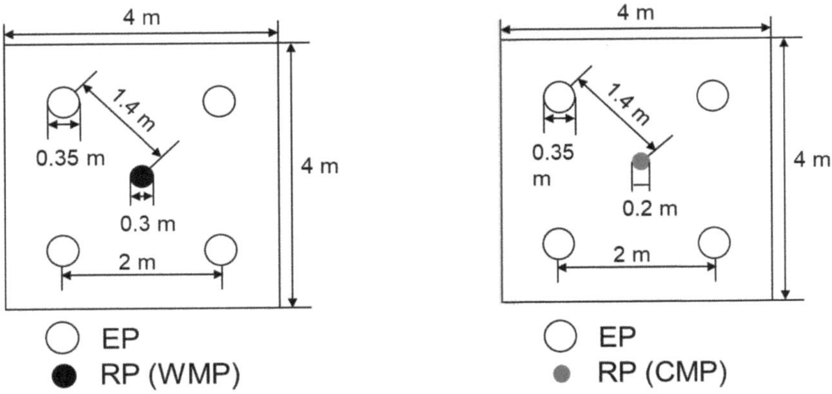

(a) Top view of pile foundation reinforced by WMP

(b) Top view of pile foundation reinforced by CMP

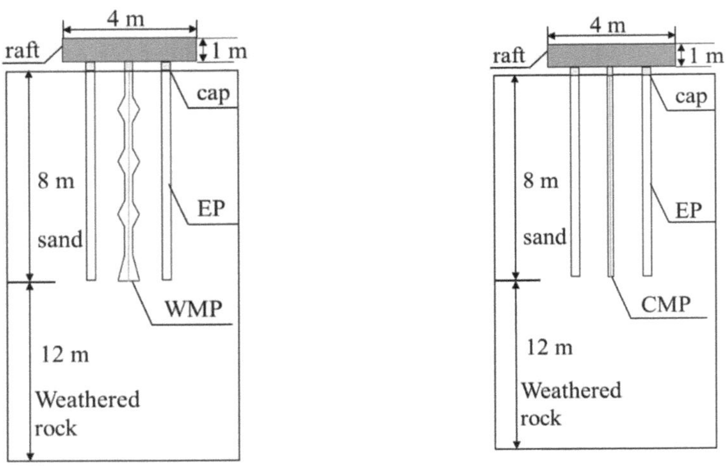

(c) Front view of pile foundation reinforced by WMP

(d) Front view of pile foundation reinforced by CMP

Fig. 3. Geometry of pile foundation consdering conventiona micropile and waveform micropile

Table 1. Properties of soils

Description	Sand	Weathered rock
Depth (m)	0–8	8–20
Unit weight (kN/m^3)	19	21
Material model	MC	MC
Interface strength factor, R_{inter}	0.67	0.67
Frictional angle, Φ (°)	34	39
Cohesion, c (kN/m^2)	10	30
Dilatacy angle, ψ (°)	4	9
Young's modulus, E (KPa)	3.5E4	3.0E5
Poisson Ratio, v	0.3	0.28

Table 2. Properties of piles and the raft

Description	EP (PC pile)	RP 1 (CMP)	RP 2 (WMP)		Raft
Material	Concrete	Concrete	Concrete	Steel	Concrete
Length (m)	8	8	8	8	3 × 3 × 1
Diameter (mm)	350	200	D_1: 300 D_2: 500	63.5	
Unit weight,	23.5	23.5	23.5	78.5	23.5
Material model	Linear elastic	Linear elastic	Linear elastic	Beam	Linear elastic
Young's modulus, (KPa)	2.4E7	4.3E7	2.4E7		2.4E7
Poisson ratio, v	0.167	0.167	0.167	0.167	0.167

EP existing pile, *RP* reinforcing pile, *CMP* conventional micropile, *WMP* waveform micropile

$$c_i = R_{inter} c_{soil} \tag{1}$$

$$\tan\Phi_i = R_{inter} \tan\Phi_{soil} \tag{2}$$

where,

c_i and Φ_i are cohesion and frictional angles of the interface; c_{soil} and Φ_{soil} are cohesion and frictional angles of the soil; $R_{inter} = 0.67$, was a representative value in Plaxis 3D.

4 Results and Discussion

4.1 Load-Settlement Behavior of Piles

Figure 4 presents the load-settlement curves which show the effect of pile's type on bearing capacity. Diameter of conventional micopile, waveform micropile, and PC pile were 200, 300 and 350 mm, respectively.

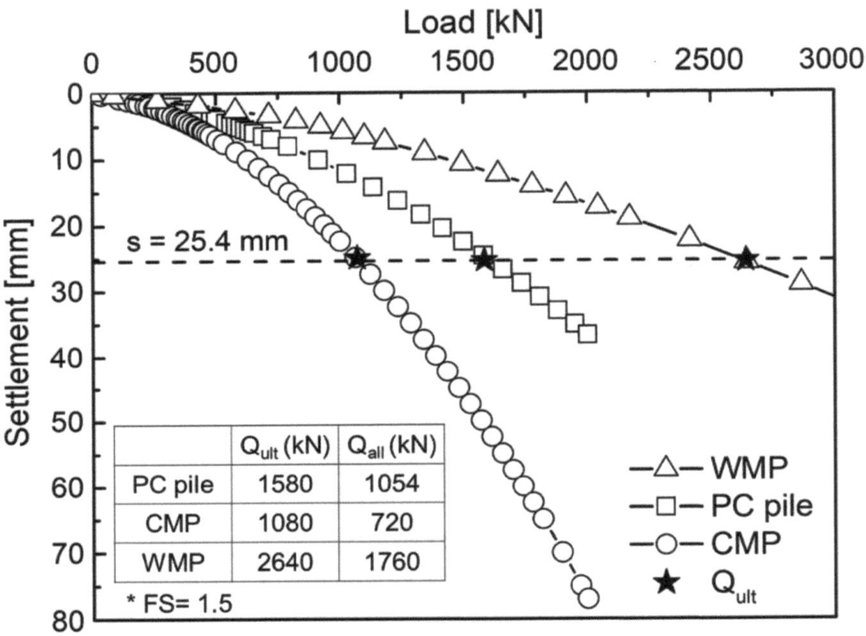

Fig. 4. Load settlement behavior for EP, CMP and WMP

The ultimate bearing capacity of each pile was estimated corresponding to pile head settlement of 25.4 mm (Terzaghi and Peck 1967; Touma and Reese 1974). Ultimate bearing capacities of PC pile, conventional mircopile, and waveform micropile were 1580, 1080, and 2640 kN, respectively. As FS of 1.5 was applied, allowable bearing capacities of each pile were 1054, 720, and 1760 kN, respectively. It can be seen that pile's type has a significant influence on the bearing capacity for a pile. Although micropile was reinforced with a central steel bar, bearing capacity of PC pile was greater than the conventional micopile due to the large diameter of PC pile. Waveform micropile with shear keys along the shaft showed the highest bearing capacity. Its ultimate bearing capacity is 1.67 times more compared to PC pile and 2.4 times more compared to the conventional micropile.

4.2 Simulation Procedure Considering the Remodeling Construction of a Building

Structural load of an existing building which applied on the foundation consists of frame load, finishing load and live load. When extending buildings vertically, the finishing material load and live load that occupy about 40% of the total load in the design will be removed expect of the frame load (KICT 2013). An existing building was remodeled with vertical extension using the following steps: (1) removing the finishing material load and live load; (2) drilling a hole and installing the reinforcing pile; (3) applying additional floors; and (4) recovering the finishing material load and live load. In this analysis, the process of building remodeling construction was described as: (1) loading stage (construction of an existing building); (2) unloading stage (removal of finishing material load and live load); (3) installation of a micropile; and (4) reloading stage (building remodeling with vertical extension). Table 3 outlines

Table 3. FEM simulation cases of procedure of a building remodeling construction with vertical extension

No.	Construction procedure in FEM simulation					
	Initial stage	Foundation construction	Loading (kN)	Unloading (kN)	Underpinning	Vertical extension[a] (kN)
Case 1	Soil preparation	Installation of PC pile and plate	1500	1500 → 900	Installation of CMP	1500 → 2250
Case 2					Installation of WMP	

[a]Vertical extension phase: Reloading from 1500 to 2250 kN in 5 steps with an increment of 150 kN

two cases (existing foundation reinforced by conventional micropile and waveform micropile) considering the procedure of remodeling construction stages of a building in this study. In the loading stage, the applied load was determined to be 1500 kN based on allowable bearing capacity of the existing pile. After installation of existing foundations, a load of 1500 kN (100%) was applied on the foundation for simulation of the existing building. In the unloading stage, 40% of the load (600 kN) was removed. Then a reinforcing pile was installed to underpin the existing foundation. At the reloading stage, a load from 1500 to 2250 kN was applied in an increment of 150 kN as the vertical extension process.

4.3 Load Transfer Behavior for Reinforcing Piles During Reloading Stages

Load transfer curves of reinforcing piles are plotted in Fig. 5. At each reloading stage, the carried load on the head of waveform micropile was greater compared to that on the conventional micropile. The skin frictional resistance along the shear key of the

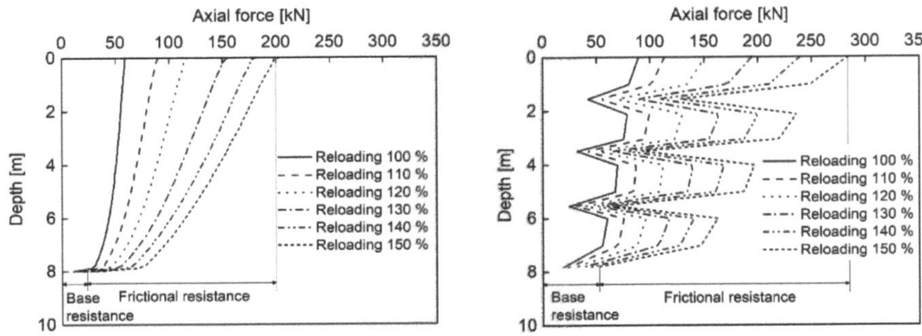

(a) Axial force of CMP (b) Axial force of WMP

Fig. 5. Load transfer curve of the reinforcing piles during the reloading phases

waveform micropile was increased greatly than that of the conventional micropile. Although the tip resistance of micropiles is ignored in FHWA (2005), the tip resistance of the waveform micropile accounted for about 15% of the total carried load due to tip expanded shape of the pile.

4.4 Load Distribution Behavior of Reinforced Foundation Considering Loading Stages

Due to different bearing capacities among the PC pile, conventional micropile, and waveform micropile as shown in Fig. 3, the reinforcing effect of the conventional micropile and waveform micropile was compared through the carried load of the pile when applying load. Load distribution ratio (LDR) was used to describe the percentage of the carried load of a pile by the total applied load. The carried load of a pile depends on various factors, such as pile's stiffness, type, soil condition, applied load and so on. In this study, load distribution ratio (LDR) was used to describe the percentage of the carried load of a pile by the total applied load. The definition of LDR for existing pile and reinforcing pile is shown as follows.

$$LDR = \frac{Carried\ load\ of\ each\ pile}{Total\ applied\ load} \tag{3}$$

The carried load of a pile was measured on the head of the pile. Figure 6 illustrates the LDR of the existing pile and reinforcing pile when the existing foundation underpinned by conventional micropile. During the loading and unloading phases, the applied loads were carried by existing piles. Each PC pile carried 25% of the total load. After installation of conventional micropile and reloading to the reinforced foundation, additional loads were shared by PC piles and the micropile. After loading was completed, the LDR of the existing pile was decreased from 25 to 22.7%. On the contrary, the micropile started to carry load up to 9.3% with the increase of the applied loads.

Figure 7 exhibits the influence of pile shape on foundation underpinning. Load carrying capacities of existing piles and the reinforcing waveform micropile after

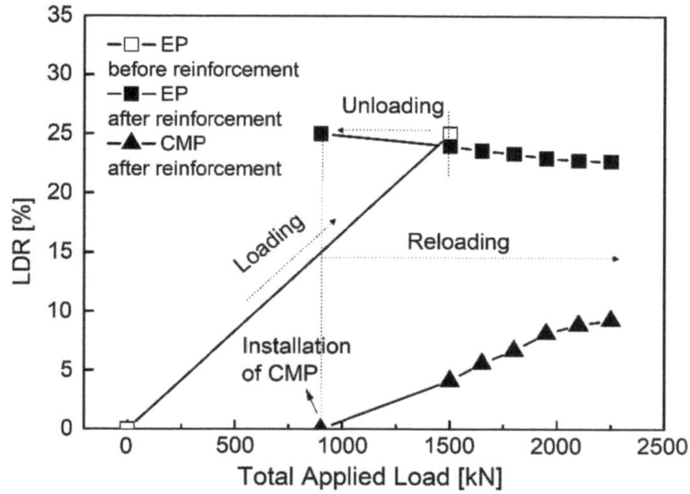

Fig. 6. Load carrying capacity of EP and CMP versus total load applied on the plate

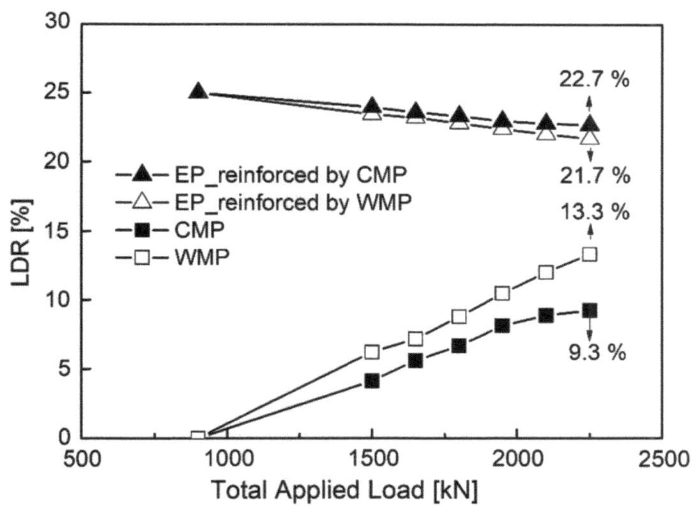

Fig. 7. Effect of micropile types on foundation reinforcement

underpinning were compared to those reinforced with the conventional micropile. After underpinning of the foundation, with the increase of the total load, load distribution ratio of the existing pile was decreased from 25% gradually. While the reinforcing pile commenced to carry loads, the slope of the curve for reinforcing piles showed waveform micropile carried more loads than conventional micropile at each loading stage. When the applied load increases to the finishing loading stage, load distribution ratios of conventional micropile and waveform micropile reached 9.3 and 13.3%, respectively. The underpinning effect of waveform micropile was about 1.4 times higher than that of conventional micropile for load distribution ratio (LDR). Therefore, waveform micropile has significant influence on foundation reinforcement.

5 Conclusions

In this study, FEM numerical analysis was carried out to investigate the waveform micropile's reinforcing effect on foundation underpinning for building remodeling with vertical extension. Results were compared to those of foundation underpinning with conventional micropile. After installing reinforcing piles and applying loads to the reinforced plate, the reinforcing pile started to carry the extra loads. The waveform mircopile carried approximately 13.3% of the total loads, which was 1.4 times higher than the conventional micropile. Therefore, the application of waveform micropile would be an economical and effective method on the foundation underpinning construction.

Acknowledgements. This research was supported by a grant from "Developing & Verifying the Cost Saving and Highly Efficient Remodeling Technologies of Vertical Extension for the Aged Multi-Family Housing" which is funded by the Korea Agency for Infrastructure Technology Advancement.

References

Babu, G.S., Murthy, B.S., Murthy, D.S.N., Nataraj, M.S.: Bearing capacity improvement using micropiles: A case study. In: GeoSupport 2004: Drilled Shafts, Micropiling, Deep Mixing, Remedial Methods, and Specialty Foundation Systems. pp. 692–699 (2004)

Bruce, D.A.: Micropile for Structural Support and In Situ Reinforcement (1996)

Bruce, J., Gurpersaud, N.: Grout Flushed micropiles foundations for a new rail overpass. In: Proceedings of the 9th International Workshop on Micropiles, pp. 10–13 (2009)

Contech System: A New Dimensions of Grout Engineering from Ischebeck Titan. Vancouver, BC (2011)

Esmaeili, M., Nik, M.G., Khayyer, F.: Experimental and numerical study of micropiles to reinforce high railway embankment. Int. J. Geomech. **13**, 729–744 (2012)

FHWA: Micropile Design and Construction: Reference Manual, Federal Highway Administration (FHWA). U.S. Department of Transportation, Washington, D.C. Publication No. FHWA-NHI-05-039 (2005)

Gomez, J.E., Rodriguez, C.J., Robinson, H.D., Mikitka, J.: Hollow core bar micropiles-installation, testing, and interpolation of design parameter of 260 micropiles. In: Proceedings of the 8th International Workshop on Mircopiles, Toronto, pp. 22–26 (2007)

Han, J., Ye, S.L.: A field study on the behavior of micropiles in clay under compression or tension. Can. Geotech. J. **43**(1), 19–29 (2006a)

Han, J., Ye, S.L.: A field study on the behavior of a foundation underpinned by micropiles. Can. Geotech. J. **43**, 30–42 (2006b)

Isam, S., Hassan, A., Mhamed, S.: 3D elastoplastic analysis of the seismic performance of inclined micropiles. Comput. Geotech. **39**, 1–7 (2012)

Jang, Y.E., Han, J.T.: A study on the constructability and load bearing characteristics of the waveform micropile by field tests (2016)
Jang, Y.E., Han, J.T.: Development on the Micropile for Applying to Artificial Ground above Railroad Site (2014)
Jang, Y.E., Han, J.T.: Study of load capacity of waveform micropile by centrifuge test. In: The Twenty-fifth (2015) International Ocean and Polar Engineering Conference. International Society of Offshore and Polar Engineers, pp. 700–706 (2015)
KICT: Development of Pre-loading Method for Reinforcement Piles of Apartment Remodeling (1). KICT2013-260, 23-61 (2013). (in Korean)
Kyung, D., Kim, G., Lee, J.: Vertical load-carrying behavior and design models for micropiles considering foundation configuration conditions. Can. Geotech. J. **54**(2), 234–247 (2016)
Lizzi, F.: The pali radice (root piles). In: Symposium on Soil and Rock improvemet Techiques including Geotextiles, Reinforced Earth and Modern Piling Methods, Bangkok, Paper D-3 (1982)
Plaxis, B.V.: PLAXIS User's manual (2005)
Sadek, M., Isam, S.: Three-dimensional finite element analysis of the seismic behavior of inclined micropiles. Soil Dyn. Earthq. Eng. **24**, 473–485 (2004)
Terzaghi, K., Peck, R.B.: Soil Mechanics in Engineering Practice, 2nd edn. Wiley, New York (1967)
Thorburn, S., Littlejohn, G.S.: Underpinning and Retention. Springer, Berlin (2013)

Influence of Adjacent Substructures on Deformation Behavior and Lateral Earth Pressure Development of Flexible Retaining Walls at Non-limit State

Olivier Demeijer[1], Ming-Guang Li[1(✉)], Jin-Jian Chen[1,2], and Yong-Mao Hou[3]

[1] Department of Civil Engineering, Shanghai Jiao Tong University, Shanghai 200240, China
lmg20066028@sjtu.edu.cn

[2] State Key Laboratory of Ocean Engineering, Shanghai Jiao Tong University, Shanghai 200240, China

[3] Shanghai Tunnel Engineering Co., Ltd., Shanghai 200232, China

Abstract. Rapid development of urban areas leads to a higher density of substructures. Because of the limited space in these areas, new excavations cannot be carried out with excavated slopes and vertical excavation solutions with retaining walls are needed. Much research has investigated the response of substructures adjacent to a deep excavation but there is a lack of research focusing on the impact of substructures on the deformation behavior of nearby retaining walls. In this study, the well-documented case-history of the Tapei National Enterprise Center (TNEC) excavation was used to create a numerical model using finite difference software. The Mohr-Coulomb soil model was used as constitutive model. After the verification with field data, three types of substructures where added 5 m behind the flexible retaining wall. The substructures are two piles; a wall and a three-layered substructure. The deformation of the retaining wall and developed non-limit lateral earth pressure was analyzed. Results showed that the confined soil width has a significant impact on both the deformation as well as on the earth pressure development for all models. Furthermore, it showed that the model with three-layered substructure had the most impact on the deformation and earth pressure, the model with piles in the backfill the least. This can be attributed to the stronger development of soil arching.

Keywords: Flexible retaining wall · Adjacent substructures · Lateral deformation · Lateral earth pressure · Non-limit state · Soil arching

1 Introduction

Rapid development of urban areas leads to a high density of substructures. Because of the limited space in urban areas, new excavations cannot be carried out with excavated slope so a vertical excavation solution with retaining walls is needed. Prior to the excavation process, the retaining wall is constructed to resist the earth pressure that will

B. Fatahi et al. (eds.), *Sustainable Design and Construction for Geomaterials and Geostructures*, Sustainable Civil Infrastructures,
https://doi.org/10.1007/978-3-319-95753-1_11

occur during the excavation. Although the retaining wall is constructed to control soil movements, interactions between walls and substructures will still occur if the substructure is in the close vicinity to the excavation.

Much research has focused on the influence of excavations on adjacent substructures (Chang et al. 2001; Liyanapathirana and Nishanthan 2016; Sharma et al. 2001; Zhang et al. 2013) but less studies have investigated the impact of adjacent substructures on retaining walls. Qiu and Grabe (2012) reported about the earth pressure occurring at a quay wall with piles in the retained soil. They found that the presence of piles reduced the earth pressure working on the quay wall. Ding et al. (2015) investigated a case of two adjacent foundation pits constructed with vertical retaining walls and found that the earth pressures on the walls were being reduced. Liao et al. (2015) investigated the influence of a nearby metro station on the deformation behavior of a retaining wall. Their results indicated that because of the barrier effect of the metro station, the lateral deformation of the retaining wall was lower than when there was no metro station nearby. The reduction of earth pressure exerted on retaining walls can be contributed to the soil arching effect, which was first described by Terzaghi (1943) as the transfer of pressure from a yielding mass of soil onto adjoining stationary parts. In a case of retaining wall with nearby substructures, the soil width between wall and structures is of great importance. This because previous studies indicated that the lateral earth pressure on walls decreases with decrease in soil width due to the soil arching effect (Fan and Fang 2010; Frydman and Keissar 1987; O'Neal and Hagerty 2011; Peck 1969; Take and Valsangkar 2001). This was recently confirmed by other research (Chen et al. 2017; Li et al. 2016).

Due to the shortage of research about the impact of substructures on the deformation behavior of adjacent retaining walls, it is necessary to carry out a realistic study of a well-documented case-history. A better understanding of the interaction behavior of retaining walls is needed since this still remains unclear. As a result, findings can improve the cost-efficiency of wall designs and lead to improved safety control of excavations. In this investigation, numerical models will be created and used to simulate the deformation behavior of a non-rigid diaphragm wall at non-limit state. Three types of numerical models with different substructures placed in the active side of retained soil will be compared with each other; a model with two piles; model with wall, and a model with a three-layered substructure. The specific substructures are added close to the retaining wall so its influences will be clear to notice. Comparisons will be carried out for lateral deflection profile of the retaining wall, lateral earth pressure distribution profile and total lateral earth pressure development.

2 Problem Description and Numerical Analysis Method

2.1 Case-History Description

In the current study, the well-documented case history of the Taipei National Enterprise Center (TNEC) excavation was used as basis for the numerical models (Ou et al. 1998). The TNEC is an 18-story building and contains five basement levels. It covers a total area of 3500 m^2. Figure 1 shows the profile of excavation. The construction of the

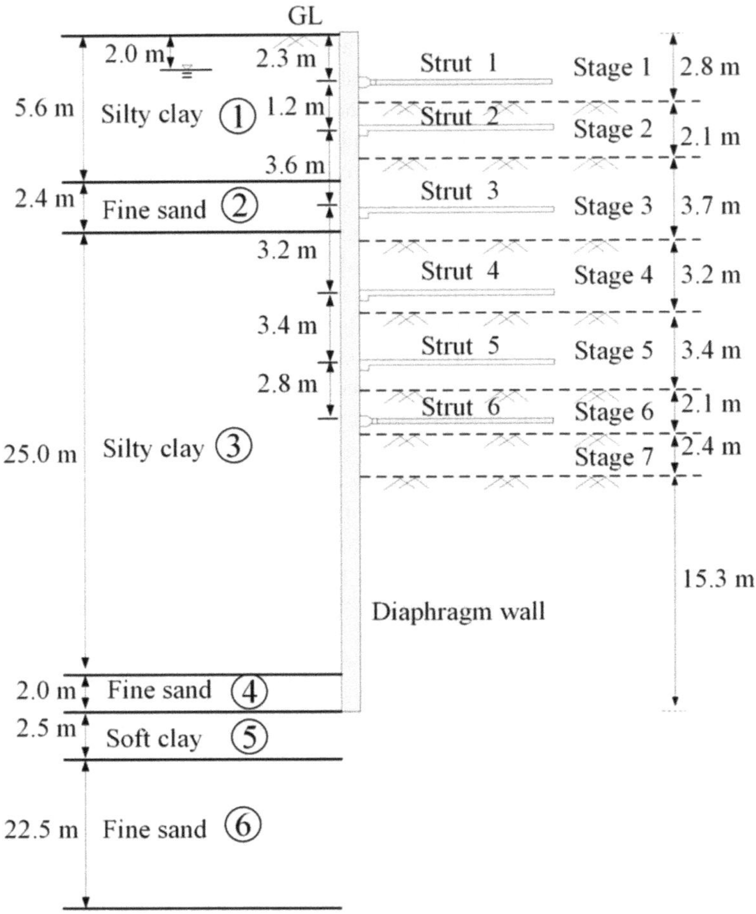

Fig. 1. Excavation profile

excavation was completed using the top-down construction method. As earth retaining structures, diaphragm walls of 0.90 m thick and 35 m deep were used. As supporting structures, five 150 mm thick concrete floor slabs and two temporary steel struts were constructed. The total excavation depth of 19.7 m was completed in seven excavation stages. The properties of the structures are shown in Table 1.

Table 1. Structure properties

Structures	Stage	A (m^2)	Thickness (m)	Spacing (m)	E (GPa)	υ
Concrete slabs	2–5	–	0.15	–	21	0.15
Steel Struts	1	0.012	–	8	210	0.2
	6	0.0219	–	3.4	210	0.2
Diaphragm wall	–	–	0.9	–	21	0.15

2.2 Numerical Model

The case history was modeled using a finite difference model created with the commercial software program FLAC^{3D}. The verification model, further referred to as the green field model, is created based on previous research of Lim et al. (2017). The model measures 121.5 m in length (x), 8 m in width (y) and 60 m in depth (z), the excavation measures 21.5 m × 8 m × 19.7 m (xyz). The grid points of the bottom boundary were restricted in their movements in all directions (ux = uy = uz = 0). The grid points at the three other boundaries were restricted in their movements in the direction perpendicular to their surface boundaries. The grid points at the top boundary were free to move. Furthermore, because of the symmetry of the case, symmetric boundary conditions were adopted for the boundary nodes of struts and slabs. Hong and Ng (2013) showed that this type of model is suitable to model a comparable case. Prior to the construction of the diaphragm wall and struts, substructures were added to the green field. Three different models were created with three different, common found substructures in retained soil: a pile foundation, a wall and a multi-layered substructure, each presenting a different kind of rigidity and shape, with the pile foundation as least rigid and multi-layered substructure as most rigid. The structure surfaces were all given the same roughness, i.e. friction angles of 20°. These three substructures have frequently been reported in the literature due to its increased appearance in congested urbanized areas (Bilotta and Russo 2010; Finno et al. 1991; Li et al. 2017; Luo and Wei 2012; Tan et al. 2015; Wei 2010; Yao et al. 2012; Zhai et al. 2010). This led to more and more cases where these structures interfered with new deep excavations. Also, some comparable investigated cases can be found in the literature to state simulation results (Ding et al. 2015; Liao et al. 2015; Qiu and Grabe 2012). Prior to the excavation simulations, all initial soil displacements were set to 0. The model with pile foundation, as shown in Fig. 2a and further referred to as model with pile, consists of the retaining wall with two piles placed in the retained soil. Each pile measures 40 m in length and 0.8 m in diameter. The spacing is 4 m, the S/d equals 5, with S the spacing and d the diameter of the piles, which is a common used proportion in practical engineering. The Young's modulus of the piles is 20,000 MPa. The model with wall, as shown in Fig. 2b, contains a concrete wall placed in the retained soil of the retaining wall. The concrete wall has identical dimensions and properties as the retaining wall. The model with multi-layered substructure, as shown in Fig. 2c, will be further referred to as model with M-L substructure. The substructure contains two concrete walls, identical to the retaining wall, three concrete slabs, with the same properties as the slabs used in the excavation, and five piles, identical to the piles used in the model with pile. In total the substructure has three levels, with the lowest concrete slab located at a depth of 13.7 m and measures 60 m in length. The pile, wall and M-L substructures are placed 5 m behind the retaining wall. This distance was used because previous research indicated that the impact of adjacent substructures on the behavior of retaining walls is more significant when the distance in between is lower (Chen et al. 2017; Li et al. 2016).

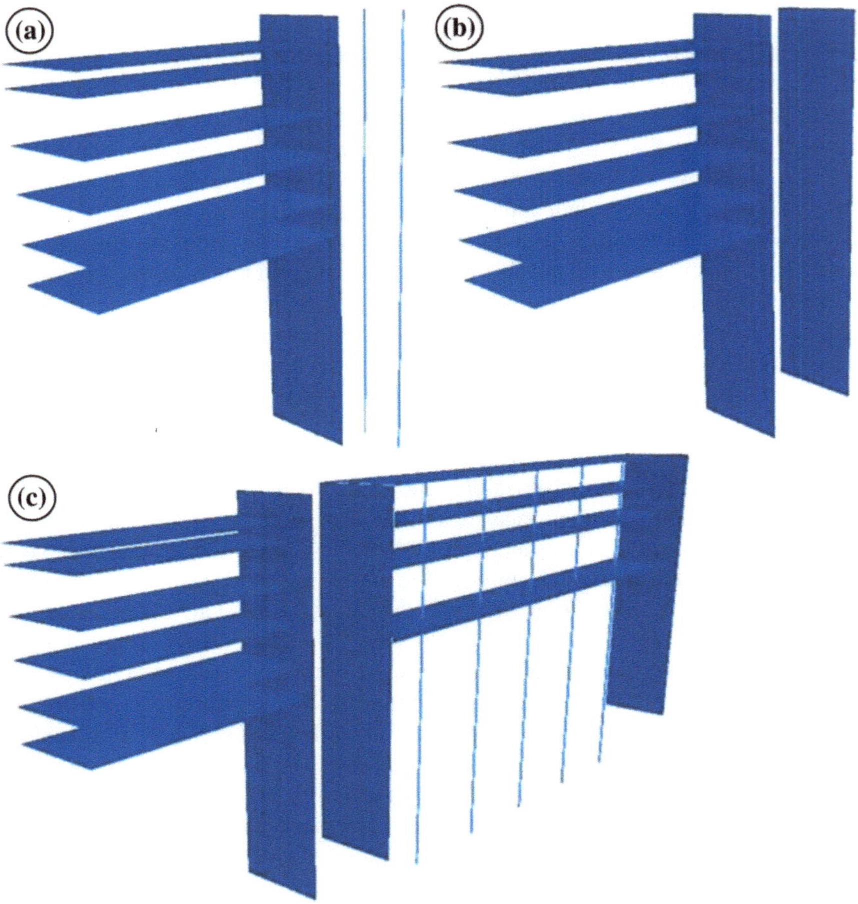

Fig. 2. Models with substructures: **a** model with pile, **b** model with wall, and **c** model with M-L substructure

2.3 Constitutive Model

The used constitutive model is the classic Mohr-Coulomb model and is a frequently used model in commercial software programs. It's an elastic perfectly-plastic model and a combination of Hooke's law and Coulomb's failure criterion. Four parameters have to be determined, Young's modulus E and Poisson's ratio υ (from Hooke's law), and friction angle φ and cohesion intercept c (from Coulomb's failure criterion) (Lim et al. 2010). Because in the TNEC project the soil consists of alternating clay and sand layers, both the undrained as drained conditions, respectively, must be considered. If considering the undrained conditions, the undrained Poisson's ratio υ_u is equal to 0.5 (0.495 used). Furthermore, because consolidation must be considered, the undrained friction angle φ_u of the soil is equal to 0. Consequently, the undrained cohesion intercept c_u is equal to the undrained shear strength S_u (Day 2010; Lim et al. 2010). According to Lim (2010) the assumption: $E_u/500 = c_u$ gives acceptable results for deflection of latter excavation stages. For sand layers, the drained conditions are being used. Poisson's ratio υ_d is set to 0.3, cohesion intercept c_d equals 0. Per Ou and Lai (1994) the Young's modulus E for a deep excavation can be estimated by the following formulas:

$$E_0 = 2(1 + v)G_0 \tag{1}$$

$$G_0 = \rho V_s^2 \tag{2}$$

$$V_s^2 = 93.11(N + 1)^{0.33} \tag{3}$$

$$E = 2\beta E_0 \tag{4}$$

with reduction factor $\beta = 0.5$ based on previous findings (Lim et al. 2010).
The used properties in the models are shown in Table 2.

Table 2. Soil properties

Soil layer	Depth (m)	Drained/Undrained	γ (kN/m^3) γ_t (kN/m^3)	S_u (kPa)	E_u (kPa) E′ (kPa)	ϕ' (°)	υ_{ur} υ'
1-a	0–2	Undrained	18.25	4.6	3851	0	0.495
1-b	2–4	Undrained	18.25	11.4	9203	0	0.495
1-c	4–5.6	Undrained	18.25	15.3	11,963	0	0.495
2	5.6–8	Drained	18.93	0	68,351	30	0.3
3-a	8–12	Undrained	18.15	28.5	15,947	0	0.495
3-b	12–16	Undrained	18.15	37.5	22,176	0	0.495
3-c	16–20	Undrained	18.15	46.5	27,777	0	0.495
3-d	20–24	Undrained	18.15	55.5	32,022	0	0.495
3-e	24–27	Undrained	18.15	63.4	35,336	0	0.495
3-f	27–30	Undrained	18.15	70.1	39,571	0	0.495
3-g	30–33	Undrained	18.15	105.4	42,328	0	0.495
4	33–35	Drained	19.62	0	265,473	33	0.3
5	35–37.5	Undrained	19.13	154.4	49,361	0	0.495
6	37.5–60	Drained	19.62	0	300,247	35	0.3

3 Results

3.1 Model Verification

The used green field model is based on the model of Lim et al. (2017). They optimized the soil properties and obtained good results by using the classic Mohr-Coulomb soil model. Figure 3 shows the comparison of results for the green field model with field

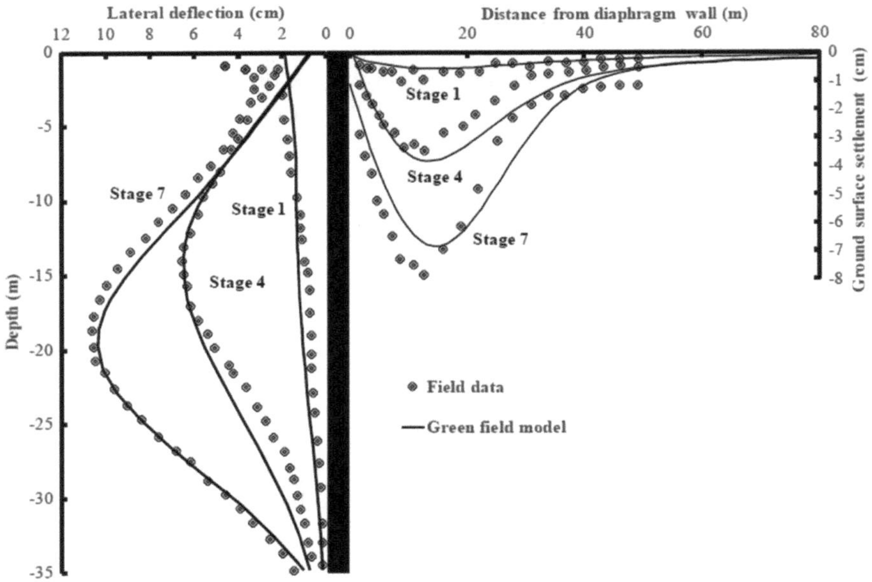

Fig. 3. Comparison field data with green field model for wall deflection and ground surface settlement

data for diaphragm wall deflection profile and ground settlement profile. All the presented data are taken in the middle of the wall (y = 4 m). Results at excavation stage 1, 4 and 7 are shown. After analyzing the results, it is concluded that the last excavation stage shows the best agreement with field data for the deflection profiles. The magnitude and position of maximum deflection correspond well with field data for both stage 4 and 7. The results for ground surface settlement also show a good agreement with field data. At stage 4, the green field model corresponds well with the field data until 10 m, the distance where the maximum settlement occurs. At this stage, the green field model slightly overestimates the maximum settlement and continues to overestimate the settlements until 50 m. At final excavation stage, the green field model shows an underestimation of settlements until 16 m and after 16 m an overestimation. These results are comparable to other previous research of the TNEC case, using numerical software. The disparities in results compared to field data are inherent to numerical simulations since studies that used the same and other soil constitutive models for the TNEC case showed similar results (Kung et al. 2009; Lim et al. 2010; Lim and Ou 2017; Ou et al. 2000). This can be attributed to the estimations made in all different kinds of soil constitutive models. Therefore, the presented results are acceptable.

3.2 Influence of Substructures

In the following paragraphs, comparisons are being made between the green field model and the three models with a substructure placed in the retained soil. The retaining wall—substructure distance is 5 m and remains constant. Results are presented for lateral deflection profiles, lateral earth pressure distribution, and total earth pressure development per excavation stage. All the presented data are taken in the middle of the model (y = 4 m).

3.3 Lateral Deflection Profile of Retaining Wall

Figure 4 presents the lateral deflection profiles of the models with substructures and green field model. It is noticed that the retaining wall with M-L substructure in the retained soil shows the highest overall reduction of lateral deflection, the contributions of pile and wall placed in retained soil is significant lower. The position of maximum deflections for model with pile, model with wall and M-L substructure are similar, the position of the green field model is slightly lower. To quantify the impact of different

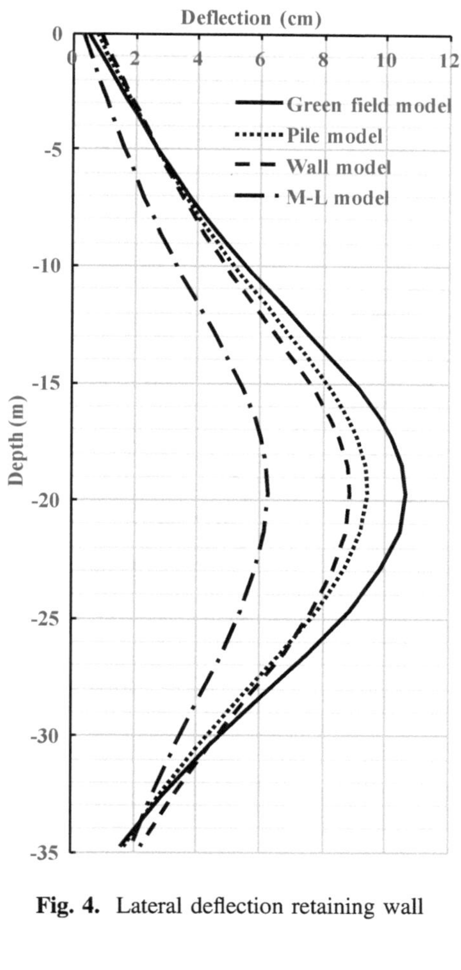

Fig. 4. Lateral deflection retaining wall

substructures on the maximum deflection of the retaining wall, which is an important design parameter, a reduction ratio R is introduced and is defined as:

$$R = 100\% - \frac{maximum\ wall\ deflection_{model\ with\ substructure}}{maximum\ wall\ deflection_{green\ field\ model}} \quad (5)$$

The reduction ratios are 12, 17 and 41 for the case with pile, wall and M-L substructure, respectively.

3.4 Lateral Earth Pressure Distribution on Retaining Wall

Results of lateral earth pressure distribution working on the retaining wall are presented in Fig. 5. The results are less obvious than the ones of the deflection profiles, but the overall tendency is comparable. The model with M-L substructure shows the highest reduction in earth pressure, then model with wall and finally model with pile. The irregular shaped pressure distribution is a result of the used boundary conditions. More specific, the flexibility and non-fixity of the used structures, both diaphragm wall as substructures, allowing structure rotation and non-linear deformation.

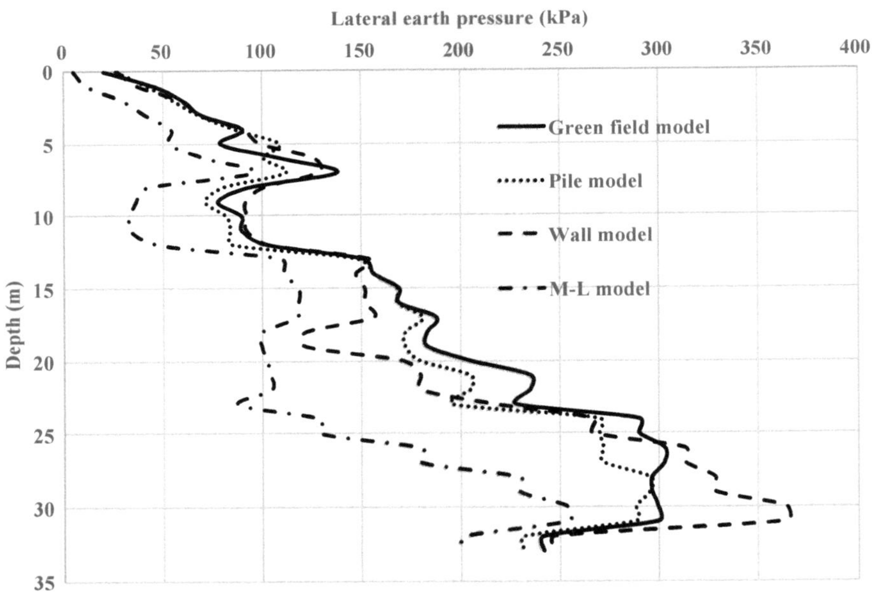

Fig. 5. Lateral earth pressure distribution exerting on retaining wall

3.5 Total Lateral Earth Pressure Development

Results of total lateral earth pressure working on the retaining wall for the models with substructures and green field model are presented in Fig. 6. Every dot on the lines

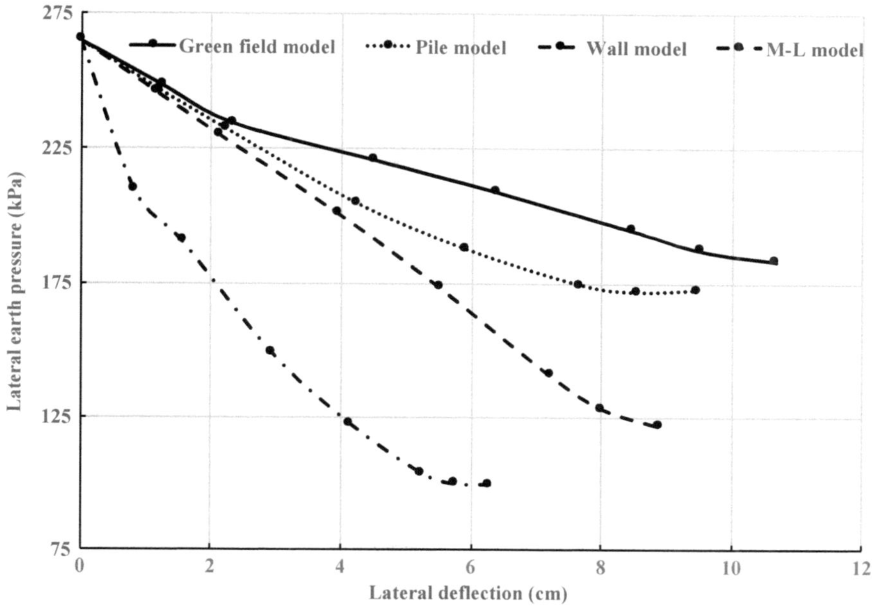

Fig. 6. Lateral earth pressure development per excavation stage over lateral deflection retaining wall

present the total pressure after an excavation stage, 7 in total. The total pressure before excavation is equal for all models, i.e. 265 kPa. The results confirm previous results, the model with M-L substructure shows the highest reduction in both earth pressure as lateral deflection during the entire excavation process. Furthermore, it is noticed that the development curves are not parallel to each other. The differences in earth pressure tend to increase with increase of excavation depth, with exception of the final excavation stage for model with M-L substructure, indicating that the influence of type of substructure in retained soil is more significant in deeper excavations.

3.6 Working Phenomena

Figure 7 shows the comparison between a case with an excavation without an adjacent substructure, where the soil width can be considered as infinite (a). And a case where a substructure is in the close vicinity of the excavation, where the soil becomes confined (b). The vertical contour arches are shown for both cases. Vertical frictional resistance forces, occurring at the contact surface of the structures and yielding soil, decrease the vertical soil stress significantly (soil arching). Therefore, the vertical stress close to structures is lower than further away. In case (a), the diaphragm wall reduces the vertical stress close to its surface. Further away from the wall, the reduction of vertical stress is much lower and the stress increases rapidly and remains constant after a certain distance from the wall. In case (b), the substructure also decreases the vertical stress close to its surface. The small soil width between wall and substructure causes that the

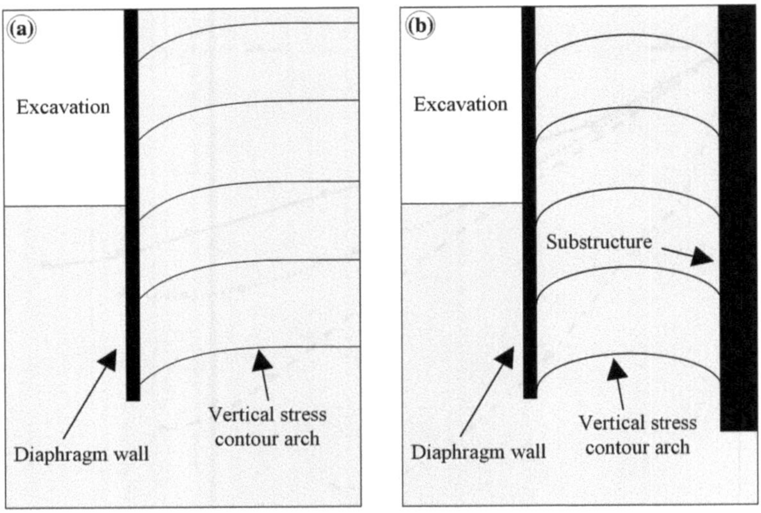

Fig. 7. Vertical stress contour arches in case of diaphragm wall with infinite retained soil width (**a**) and in case with confined soil width by substructure (**b**)

combined vertical stress reduction of wall and substructure results in large vertical stress reductions. The stress contours are now arch shaped, with the maximum vertical stress occurring somewhere in the middle between the wall and structure. It can be stated that the presence of an adjacent substructure increases the development of the soil arching effect, resulting in much lower vertical stresses occurring in the confined soil, compared to a case without adjacent substructure.

Furthermore, the numerical results show that a substructure with a higher rigidity in retained soil behind a retaining wall causes lower lateral deflections and earth pressures than a substructure with a lower rigidity. The combined stiffness of piles, concrete slabs and two retaining walls result in a very rigid M-L substructure. Therefore, the structure can be considered as almost stationary, which maximizes the frictional resistance forces. The relatively small difference in results between the model with pile and wall are somehow remarkable since the wall is significantly more rigid and the interface surface with soil is larger. A possible explanation is that for the model with pile the soil will be squeezed around and in between the piles which lowers the lateral soil movement and eventually the lateral earth pressure. The presence of soil squeezing between piles and its reduction on soil movement was confirmed by Chen and Martin (2002). It is likely that when the retaining wall—substructure distance increases, the influence of the pile and related soil squeezing effect will reduce rapidly. It can be concluded that the presented results confirm previous findings by Qiu and Grabe (2012), Ding et al. (2015) and Liao et al. (2015), for a pile, wall and M-L substructure in retained soil, respectively.

4 Conclusions

In this paper, the influence of different types of substructures in retained soil, adjacent to a diaphragm wall for the TNEC case-history, was investigated by using numerical models and field data to simulate the deformation behavior of the diaphragm wall and its exerting earth pressure. Based on the presented work, the following observations and conclusion can be made for a case with small confined soil width:

- Substructures with higher rigidity cause an increased development of the soil arching effect, resulting in an increased reduction of lateral earth pressure exerting on walls and lateral deflection of walls.
- The difference in total earth pressure between models with substructures increases with excavation depth, indicating that for deep excavations a higher rigidity of adjacent substructure can be an important asset for construction works.
- The interface surface between piles and surrounding soil is remarkable lower than that of a complete wall, the difference in reduction of wall deflections and earth pressure is not significant. This can possibly be attributed to the soil squeezing effect in a case with piles.

Acknowledgements. The financial support from the National Natural Science Foundation of China (NSFC Grant No. 41602283), and Science and Technology Commission of Shanghai Municipality (Grant No. 16QB1403400) is gratefully acknowledged.

References

Bilotta, E., Russo, G.: Use of a line of piles to prevent damages induced by tunnel excavation. J. Geotech. Geoenviron. Eng. **137**(3), 254–262 (2010)

Chang, C.-T., Sun, C.-W., Duann, S.W., Hwang, R.N.: Response of a Taipei Rapid Transit System (TRTS) tunnel to adjacent excavation. Tunn. Undergr. Space Technol. **16**(3), 151–158 (2001)

Chen, C.-Y., Martin, G.R.: Soil–structure interaction for landslide stabilizing piles. Comput. Geotech. **29**(5), 363–386 (2002)

Chen, J.-J., Li, M.-G., Wang, J.-H.: Active earth pressure against rigid retaining walls subjected to confined cohesionless soil. Int. J. Geomech. **17**(6), 6016041 (2017)

Day, R.W.: Foundation engineering handbook: design and construction with the 2009 international building code. McGraw-Hill Companies (2010)

Ding, Z., Wang, J., Zhou, Y., Wei, X., Zhang, S.: Monitoring analysis of the interrelationship of synchronous construction in adjacent foundation pit. China Civil Eng. J. **48**, 124–130 (2015)

Fan, C.-C., Fang, Y.-S.: Numerical solution of active earth pressures on rigid retaining walls built near rock faces. Comput. Geotech. **37**(7), 1023–1029 (2010)

Finno, R.J., Harahap, I.S., Sabatini, P.J.: Analysis of braced excavations with coupled finite element formulations. Comput. Geotech. **12**(2), 91–114 (1991)

Frydman, S., Keissar, I.: Earth pressure on retaining walls near rock faces. J. Geotech. Eng. **113** (6), 586–599 (1987)

Hong, Y., Ng, C.W.W.: Base stability of multi-propped excavations in soft clay subjected to hydraulic uplift. Can. Geotech. J. **50**(2), 153–164 (2013)

Janssen, H.A.: Versuche über getreidedruck in silozellen. Zeitschr. d. Vereines deutscher Ingenieure **39**(35), 1045–1049 (1895)

Kung, G.T.-C., Ou, C.-Y., Juang, C.H.: Modeling small-strain behavior of Taipei clays for finite element analysis of braced excavations. Comput. Geotech. **36**(1–2), 304–319 (2009). (Elsevier Ltd.)

Li, M.-G., Chen, J.-J., Wang, J.: Lateral earth pressure on rigid retaining wall subjected to confined granular backfill: numerical and theoretical analysis. Granular Matter (2016)

Li, M.-G., Wang, J.-H., Chen, J.-J., Zhang, Z.-J.: Responses of a newly built metro line connected to deep excavations in soft clay. J. Perform. Constructed Facil. Am. Soc. Civil Eng. **31**(6), 4017096 (2017)

Liao, S.-M., Wei, S.-F., Shen, S.-L.: Structural responses of existing metro stations to adjacent deep excavations in Suzhou, China. J. Perform. Constructed Facil. **0**(0), 4015089 (2015)

Lim, A., Ou, C.-Y.: Stress paths in deep excavations under undrained conditions and its influence on deformation analysis. Tunn. Undergr. Space Technol. **63**, 118–132 (2017). (Elsevier Ltd.)

Lim, A., Ou, C.-Y., Hsieh, P.-G.: Evaluation of clay constitutive models for analysis of deep excavation under undrained conditions. J. GeoEng. **5**(1), 9–20 (2010)

Liyanapathirana, D.S., Nishanthan, R.: Influence of deep excavation induced ground movements on adjacent piles. Tunn. Undergr. Space Technol. **52**, 168–181 (2016). (Elsevier Ltd.)

Luo, C.-H., Wei, J.-H.: Method for decreasing initial displacement of surrounding environment of deep excavations in soft soils. Yantu Gongcheng Xuebao/Chin. J. Geotech. Eng. **34** (Suppl.), 49–53 (2012)

O'Neal, T.S., Hagerty, D.J.: Earth pressures in confined cohesionless backfill against tall rigid walls—a case history. Can. Geotech. J. **48**(8), 1188–1197 (2011)

Ou, C.-Y., Lai, C.-H.: Finite-element analysis of deep excavation in layered sandy and clayey soil deposits. Can. Geotech. J. **31**(2), 204–214 (1994)

Ou, C.-Y., Liao, J.-T., Lin, H.-D.: Performance of diaphragm wall constructed using top-down method. J. Geotech. Geoenviron. Eng. **124**(9), 798–808 (1998)

Ou, C.Y., Shiau, B.Y., Wang, I.W.: Three-dimensional deformation behavior of the Taipei National Enterprise Center (TNEC) excavation case history. Can. Geotech. J. **37**(2), 438–448 (2000)

Peck, R.B.: Deep excavation and tunneling in soft ground. In: Proceedings of the 7th International Conference on Soil Mechanics and Foundation Engineering, Mexico City (1969)

Qiu, G., Grabe, J.: Active earth pressure shielding in quay wall constructions: numerical modeling. Acta Geotech. **7**(4), 343–355 (2012)

Sharma, J.S., Hefny, A.M., Zhao, J., Chan, C.-W.: Effect of large excavation on deformation of adjacent MRT tunnels. Tunn. Undergr. Space Technol. **16**(2), 93–98 (2001)

Spangler, M.G., Handy, R.L.: Soil Engineering. Harper and Row, New York (1984)

Take, W.A., Valsangkar, A.J.: Earth pressures on unyielding retaining walls of narrow backfill width. Can. Geotech. J. **38**(6), 1220–1230 (2001)

Tan, Y., Wei, W., Zhou, X., Diao, Y.P.: Lessons learned from construction of Shanghai Metro Stations: importance of quick excavation, prompt propping, timely casting, and segmented construction. J. Perform. Constructed Facil. **29**(4) (2015)

Terzaghi, K.: Arching in ideal soils. In: Theoretical Soil Mechanics, pp. 66–76. Wiley, Hoboken (1943)

Wei, F.: Application of isolation piles to deformation control of deep foundation pits close to buildings with shallow foundation. Chin. J. Geotech. Eng. **32**(Supp.1) (2010)

Yao, A., Yang, X., Dong, L.: Numerical analysis of the influence of isolation piles in metro tunnel construction of adjacent buildings. Proc. Earth Planet. Sci. **5**, 150–154 (2012)

Zhai, J.Q., Jia, J., Xie, X.L.: Practice of partition wall in building protection projects near deep excavation. Chin. J. Undergr. Space Eng. **6** (2010)

Zhang, Z.-G., Huang, M.-S., Wang, W.-D.: Evaluation of deformation response for adjacent tunnels due to soil unloading in excavation engineering. Tunn. Undergr. Space Technol. **38**, 244–253 (2013)

Assessment of Soil Plasticity Effects on Seismic Response of Mid-Rise Buildings Resting on End-Bearing Pile Foundations

Ruoshi Xu(✉) and Behzad Fatahi

School of Civil and Environmental Engineering, University of Technology Sydney (UTS), Sydney, Australia
{Ruoshi.Xu, Behzad.Fatahi}@uts.edu.au

Abstract. In this study, the seismic response of mid-rise buildings supported by end-bearing pile foundations under the influence of six different soil Plasticity Indices (PI) is investigated. FLAC3D is utilised to develop the soil-pile-structure system and perform fully nonlinear analysis in time domain adopting the direct method. The results in terms of shear forces developed in the superstructure, lateral building displacement, maximum inter-storey drift, lateral pile displacement and shear forces and bending moments developed along piles are presented. The results indicate the seismic response of the soil-pile-structure system is sensitive to the variation of PI. By considering the effects of PI, the seismic performance of the superstructure may shift from life safe zone to near collapse level. Moreover, the shear force imposed on the piles increases with PI, particularly in the top section of the pile near the ground surface due to the inertial interaction. Also, as PI increases, the bending moment developed in the piles increases and no need to mention the lateral pile displacement. Due to the load bearing mechanism, the maximum bending moment occurs at the pile toe socketed in the bedrock. Thus, appropriate soil Plasticity Index should be considered to obtain the safe and reasonable seismic design of mid-rise buildings resting on end-bearing pile foundations.

1 Introduction

End-bearing pile foundation is a common solution to resolve issues induced by unfavourable geotechnical conditions such as soft soils. For routine engineering design, the effect of Seismic-Soil-Pile-Structure Interaction (SSPSI) is ignored or greatly simplified due to the complexity of the effect and unavailability of standard and validated analysis techniques. In recent three decades, the SSPSI has drawn increasing attention in the earthquake engineering society. Many research works concluded that the SSPSI would greatly influence the seismic response of the superstructures, particularly, for moment resisting buildings in soft soils (e.g. Carbonari et al. 2011; Luo et al. 2016; Nguyen et al. 2017).

The dynamic behaviour of soft soil is nonlinear and hysteretic. In the numerical simulation, the behaviour can be represented in the forms of backbone curves expressing the correlation between the shear modulus and damping of the soil as a function of cyclic shear strain. In fact, the backbone curves play a vital role in the

B. Fatahi et al. (eds.), *Sustainable Design and Construction for Geomaterials and Geostructures*, Sustainable Civil Infrastructures,
https://doi.org/10.1007/978-3-319-95753-1_12

SSPSI analysis. Some research works have been done in determining the curves for cohesive and cohesionless soils (e.g. Hardin and Drnevich 1972; Sun et al. 1988). Particularly, in cohesive soils, soil plasticity is one of the essential aspects affecting the backbone curves (Vucetic and Dobry 1991). Indeed, the soil plasticity presents the ability of the soil to deform irreversibly without cracking or crumbling; and it can be measured by Plasticity Index (PI) which indicates the size of the range of water contents where the soil behaves plastically. The liquid limit (LL) and plastic limit (PL) are the upper and lower limits of the water contents of the soil, respectively. Thus, PI can be determined employing Eq. (1)

$$PI = LL - PL \tag{1}$$

Due to the fact that many numerical studies were carried out without considering the effect of Plasticity Index on the backbone curves of cohesive soils (e.g. Galal and Naimi 2008; Tabatabaiefar and Fatahi 2014; Hokmabadi and Fatahi 2016), this study investigates the influence of the soil plasticity due to the variation of PI (from PI = 0 to PI = 200) on the seismic response of mid-rise buildings supported by end-bearing pile foundations.

2 Numerical Soil-Pile-Structure Model

According to Kramer (1996), to study the seismic response of a soil-structure system, it is necessary to treat the structure and soil with the same rigours. Thus, FLAC3D V5.0 (Itasca 2011), a finite difference software, is employed in this study. In fact, several researchers (e.g. Comodromos and Papadopoulou 2012; Kwon et al. 2013) have used this software package to study the problem of SSPSI and reported reasonable results.

In this study, the direct method is adopted. Since this method considers the structure, foundation and soil in one single step, the fully coupled system response can be obtained. Accurate and reasonable outcomes have been achieved by many research works adopting the direct method to simulate the problem of SSPSI (e.g. Chu and Truman 2004; Hokmabadi et al. 2014). Furthermore, the fully nonlinear analysis is performed in the present study, and the advantages of the analysis are underscored by Beaty and Byme (2001) and Byrne et al. (2006). Moreover, according to the recommendation made by Chu (2005), the analysis of the soil-pile-structure system is carried out in the time domain.

2.1 Modelling of the Superstructure and Foundation

In the present study, a reinforced concrete moment resisting building containing fifteen storeys and three equal bays is designed to represent the conventional mid-rise buildings. As shown in Fig. 1, the building is 45 m in height and 12 m in width in both horizontal directions. The adopted mass density and the specified compressive strength of the concrete are 2400 kg/m^3 and 40 MPa, respectively, whereas the mass density and yield strength of the steel reinforcement are 7800 kg/m^3 and 500 MPa, respectively. The structural analysis and design are conducted using SAP2000 V14

(CSI 2010) according to relevant Australian building codes (AS1170.1 2002; AS1170.4 2007; AS3600 2009), and according to ACI318 (2008), cracked section factors are considered by modifying the stiffness of the structural members. Table 1 presents the designed structural sections. The mass and natural frequency of the building are 3667.5 tonnes and 0.78 Hz, respectively. Beam structural elements available in FLAC3D software are used to model designed columns and beams, while shell elements are used to model floor slabs. Inelastic behaviour of the structural elements is considered by introducing plastic hinges. Also, 5% structural damping ratio is assigned to the building model.

As shown in Fig. 1, the end-bearing pile foundation consists of a 16 × 16 × 1 m reinforced concrete slab and a group of 4 × 4 reinforced concrete piles, which is

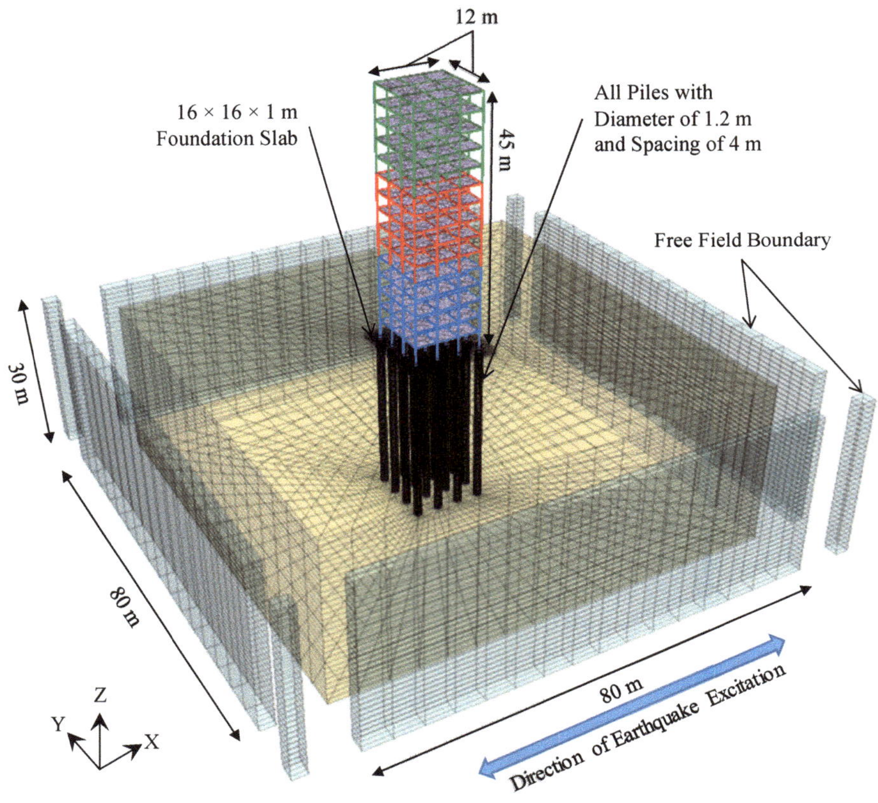

Fig. 1. The layout of adopted numerical soil-pile-structure model

designed following AS2159 (2009). Each pile is 1.2 m in diameter, and 30 m in length and the pile spacing is 4 m centre to centre and aligns with each column of the superstructure so that the load of the superstructure can be readily transmitted into the piles. In fact, the adopted pile spacing has a good agreement with other research works

Table 1. Designed structural members of the building

Structural member	Column			Beam			Slab
Distribution levels	1–5	6–10	11–15	1–5	6–10	11–15	1–15
EA (MPa m^2)	20,992	18,450	16,072	13,858	11,808	9922	8200 (PMR)
EI (MPa m^4)	1120	865	656	488	354	250	43 (PMR)
Crack section factors [According to ACI318 (2008)]	0.7			0.35			0.25

(e.g. Shelke and Patra 2008; Nguyen et al. 2016). Moreover, the pile toes are socketed in the bedrock. Figure 1 presents the layout of the soil-pile-structure system utilised in this study. Solid elements are adopted to model the foundation. Furthermore, elastic-perfectly plastic behaviour with yielding criteria is considered to control any inelastic behaviour in the foundation.

2.2 Modelling of the Soil Deposit

The mass density and Poisson's ratio of the adopted soil are 1470 kg/m^3 and 0.35, respectively. The shear wave velocity of the soil is 150 m/s, and the effective cohesion and friction angle are 3 kPa and 29°. According to Australian seismic code (AS1170.4 2007), the evaluation of local site effect should be based on the top 30 m of the soil profile, as the main part of amplification and attenuation occurs within this range. Thus, as shown in Fig. 1, a 30 m deep soil deposit is modelled assuming the bedrock level at the bottom of the deposit. Solid elements are used to model the deposit, and Mohr-Coulomb constitutive model is used to simulate the plastic behaviour and natural damping of the soil during the earthquake.

According to Vucetic and Dobry (1991), Plasticity Index (PI) is the main factor controlling the variations of the shear modulus degradation curves of cohesive soils. To simulate the modulus degradation and generate corresponding soil damping during the earthquake, hysteretic damping algorithm incorporated in FLAC3D software is utilised. Indeed, the shear modulus degradation curves shown in Fig. 2 imply nonlinear backbone (stress-strain) curves. In this study, the shear stresses developed in the soil are assumed to depend only on the shear strains so that the tangent modulus (M_t) can be derived from the degradation curves employing Eq. (2):

$$M_t = M_s + \gamma \frac{d(M_s)}{d\gamma} \tag{2}$$

where, M_s is the secant modulus and γ is the shear strain. Therefore, the strain rate calculation is modified implementing Eq. (2) and the mean strain rate tensor is calculated before the implementation of any constitutive model. At this stage, the hysteretic logic is invoked, returning a modulus multiplier, which is passed to any called constitutive model. The model then uses the multiplier to adjust the apparent value of

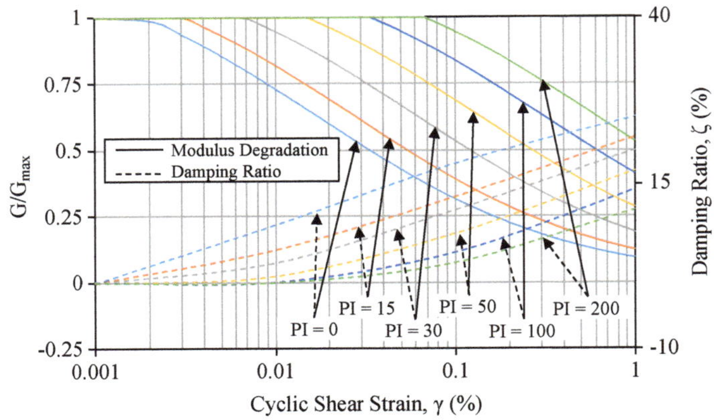

Fig. 2. Shear modulus degradation curves and corresponding damping ratio curves adopted in this study based on SIG-III model

the tangent shear modulus. In particular, a sigmoidal model, SIG-III, is used to define the secant modulus (M_s) using Eq. (3):

$$M_s = \frac{a}{1 + exp\left(-\frac{\log(\gamma) - x_0}{b}\right)} \tag{3}$$

where a, b, and x_0 are the curve fitting parameters. The sigmoidal curve is monotonic and asymptotic in a given range, thus, the function is capable of representing the modulus degradation curves as shown in Fig. 2. Table 2 lists the curve fitting parameters of SIG-III model to represent the modulus reduction curves reported by Vucetic and Dobry (1991). As shown in Fig. 2, it is evident that the shear modulus degradation factor (G/G_{max}) increases with PI while the corresponding damping ratio deceases with the increase in PI.

Table 2. SIG-III model parameter for various Plasticity Indices

Plasticity Index (PI)	SIG-III model parameters		
	a	b	x_0
PI = 0	1.25	−0.7	−1.75
PI = 15	1.25	−0.7	−1.55
PI = 30	1.25	−0.7	−0.7 to 1.2
PI = 50	1.25	−0.7	−0.85
PI = 100	1.25	−0.7	−0.7 to 0.5
PI = 200	1.25	−0.7	−0.2

2.3 Modelling of the Interface and Boundary

The interface is assigned between the pile foundation and the soil deposit to consider the different mechanical characteristics of contacting elements and to capture any possible gapping and sliding of the piles during the earthquake excitation. Each interface element adopted in this study consists of normal and shear spring-slider systems considering Mohr-Coulomb failure criterion. The properties of the adopted interface elements are summarised in Table 3.

Table 3. Properties of the adopted interface elements

Interface properties	Shear stiffness (kPa/m)	Normal stiffness (kPa/m)	Shear strength (kPa)	Friction angle (°)	Tensile strength (kPa)	Dilation angle (°)
Value	5.67e6	5.67e6	3	29	0	0

Two different boundary conditions are applied for static and dynamic analysis in this study. For the static analysis where only gravity loads are analysed, the bottom of the soil deposit is fully fixed in all directions, while only vertical movement is allowed at four sides of the deposit. For the dynamic analysis, free field boundary is applied to the soil sides replacing the preliminary boundaries. Thus, plane waves propagating upward undergo no distortion at the boundaries as the free-field grid supplies conditions identical to those in an infinite model. A near-field seismic acceleration, the 1994 Northridge Earthquake (Fig. 3), is utilised to perform dynamic analysis in the time domain. The duration of the earthquake is 28 s and the dominant frequency and peak ground acceleration are 0.9 Hz and 0.65 g, respectively. For all cases conducted in this study, seismic input motion applies at the bedrock level (i.e. the bottom of the soil deposit).

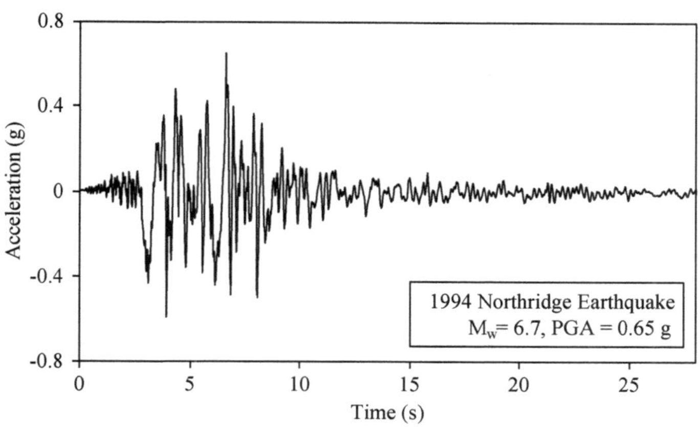

Fig. 3. Acceleration record of the 1994 Northridge Earthquake

3 Results and Discussions

The results of the seismic analysis for the mid-rise building supported by the end-bearing pile foundation in soft soils having different plasticity indices are derived time from history records. The results in terms of response spectrum, shear forces developed in the superstructure, the maximum rocking angle of the foundation slab, lateral building displacement, maximum transient inter-storey drift and the seismic response of foundation regarding lateral pile displacement and shear forces and bending moments developed along piles are reported and compared.

3.1 Response Spectrum

Figure 4 depicts the response spectra derived from the motions of the foundation slabs under the influence of the 1994 Northridge Earthquake in conjunction with six different PI. It is evident that the spectral acceleration increases with PI in the range of period from 0 to 1.5 s, between which the natural period of the structure sits. This observation indicates that in the case analysed, more seismic energy could be transmitted to the superstructure sitting in the stiffer soil deposit (i.e. higher PI referring to Fig. 2).

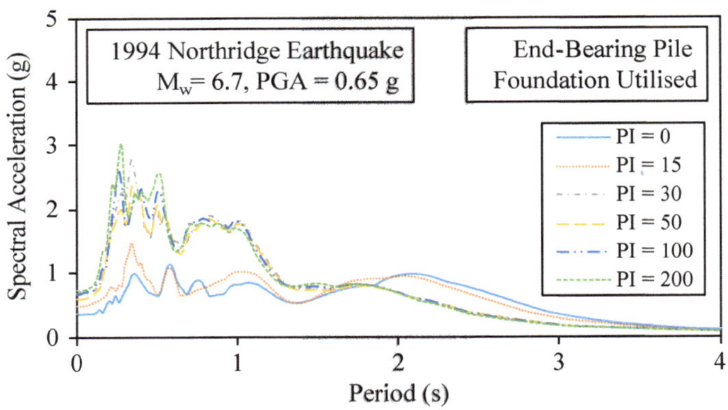

Fig. 4. Response spectrum derived from the motion of foundation slab with 5% damping ratio under the applied earthquake with different PI

3.2 Shear Forces Developed in the Superstructure

Figure 5 shows the envelope of the shear forces experienced by the superstructure under the influence of the applied earthquake considering the variation of Plasticity Index (PI). According to Fig. 5, it is evident that the base shear increases with PI. This is due to the fact that, referring to Fig. 2, the soil stiffness increase with PI while the

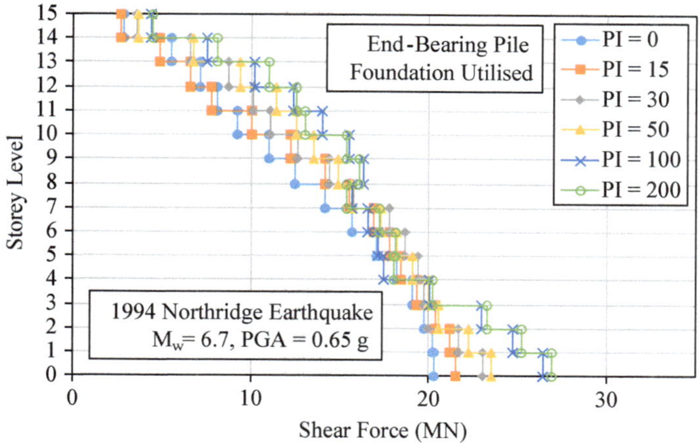

Fig. 5. Shear force developed along the building height under the influence of the applied earthquake considering the variation of PI

corresponding soil damping decreases as PI increases so that less seismic energy is dissipated due to soil deformation and consequently more energy is transmitted to the superstructure and induces more building base shear. Moreover, as observed from Fig. 5, the variation of PI evidently influences the shear force distribution along the building. For example, the shear force recorded at level 4 is increased from 19.1 to 20.2 MN (i.e. increased by 5.9%) when PI increases from 0 to 200, while the corresponding force recorded at level 9 is increased by 29.3% (i.e. from 12.5 to 16.1 MN). Although the other soil properties remain the same in the analysis, soil plasticity changes the dynamic characteristics (natural frequency and damping) of the system due to the shear modulus degradation. As a result, the input excitation attracts different portions of the structural higher mode responses contributing to the fluctuations of the shear force distribution along the superstructure. Therefore, it is necessary to consider the effect of soil plasticity in the real design.

3.3 Displacement and Inter-storey Drift

Figure 6 shows the maximum lateral building displacement recorded when the top level reached the maximum displacement under the excitation of the applied earthquake. As shown in Fig. 6, the maximum displacement increases with Plasticity Index. Comparing the cases of PI = 0 and PI = 200, the maximum building displacement is increased from 460 to 550 mm (i.e. increased by 15%). In fact, the structural displacement consists of two aspects which are the structural distortion due to shear forces developed in the building and the rocking of the foundation slab.

A foundation slab rocks when the generated inertial force in the building induce an increase in compression on one side and a decrease in compression or tension on the other side of the foundation slab, which results in foundation rotation and consequently increases the lateral building displacement. As the piles are socketed in the bedrock, the

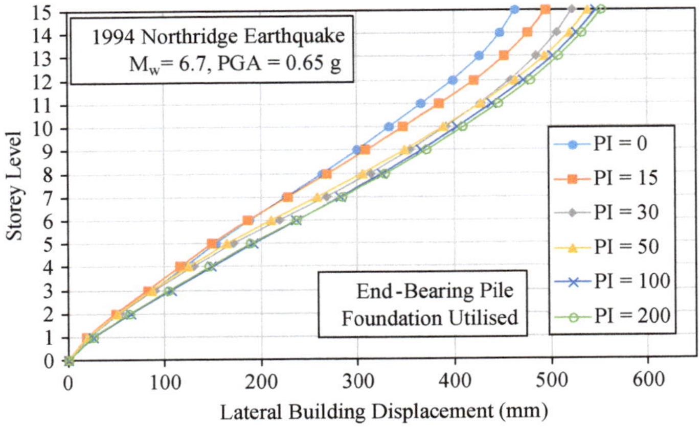

Fig. 6. Maximum lateral building displacement in conjunction with the variation of PI

foundation rocking is only due to the pile bending under the influence of the earthquake in this study. Figure 7 shows an example of the numerical prediction of the rocking of the foundation slab under the influence of the 1994 Northridge Earthquake when PI = 30. According to Table 4, the increase in PI results in an increase in the maximum rocking angle experienced by the building. For instance, when PI = 0, the maximum rocking angle is 0.09°, while the corresponding value is 0.17° when PI = 200 (i.e. increased by 47%).

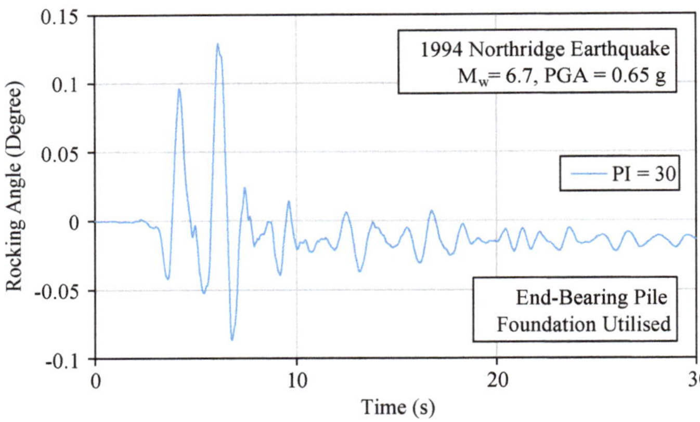

Fig. 7. An example of the numerical prediction of the rocking of the foundation slab under the influence of the 1994 Northridge Earthquake

Table 4. Maximum rocking angle experienced by the foundation slab in conjunction with the variation of PI

Plasticity Index (PI)	PI = 0	PI = 15	PI = 30	PI = 50	PI = 100	PI = 200
Maximum rocking angle (°)	0.09	0.11	0.12	0.13	0.16	0.17

It is obvious that in this study the structural distortion has more pronounceable influence than the rocking on the formation of the building displacement as the end-bearing pile foundation is employed. Figure 8 shows the maximum inter-storey drift experienced by the building considering the variation of PI subjected to the applied earthquake. The inter-storey drift is calculated according to Australian seismic code (AS1170.4 2007). Since the maximum inter-storey drift is directly influenced by the lateral building displacement, the drift tends to increase with PI. For example, the building experienced 1.25 and 1.63% of the maximum inter-storey drift when PI are 0 and 200, respectively. In other words, by considering different PI, it may result up to

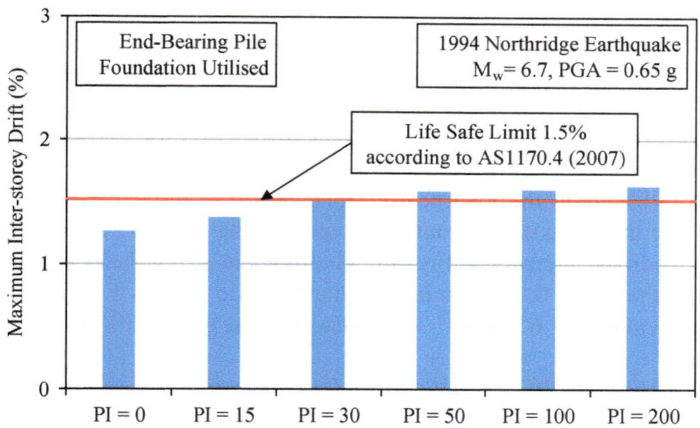

Fig. 8. Maximum inter-storey drift of the building under the influence of the 1994 Northridge Earthquake in conjunction with the variation of PI

30% difference in the recorded inter-storey drift and consequently, it may vary the performance level of the structure from life safe zone to near collapse level referring to AS1170.4 (2007).

3.4 Seismic Response of End-Bearing Pile Foundation

Figure 9 presents the lateral pile displacement along pile depth under the earthquake loading when the maximum lateral deflection occurs at the pile head in conjunction with different Plasticity Index. According to Fig. 9, the lateral pile displacement increases with PI. For instance, when PI = 0, the maximum lateral pile displacement is 145 mm while the corresponding value is 437 mm in the case of PI = 200, which is 66% increase.

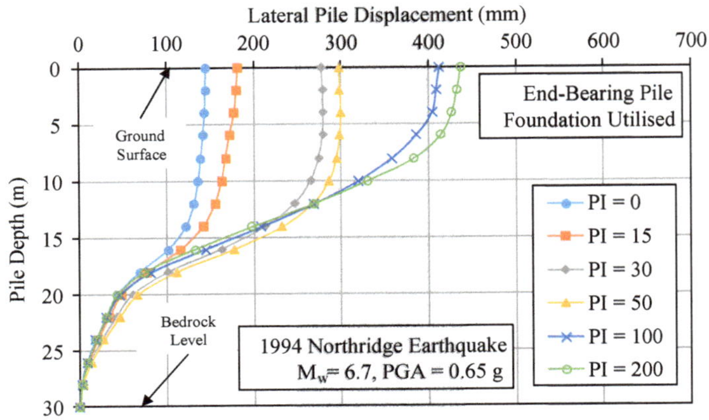

Fig. 9. Maximum lateral pile displacement experienced by the foundation due to the 1994 Northridge Earthquake with the variation of PI

Figure 10 shows the envelope of the shear forces imposed on piles for different PI cases subjected to the seismic loads. It is evident that shear forces on the piles increase with Plasticity Index. For example, at pile head (the connection between the pile and the foundation slab), the shear forces are 1.42 and 2.47 MN for the cases of PI = 0 and 200, respectively. Figure 11 depicts the bending moment envelope along the piles influenced by PI. Due to the load bearing mechanism of the end bearing foundation, piles experience their maximum bending moment at the toe, where they are connected to the strong bedrock referring to Fig. 11. The moment at pile head increases with PI. For example, when PI = 0, the bending moment at pile head is 1.09 MN m while the corresponding value is 1.65 MN m (i.e. 34% increase) when PI = 200. Referring to

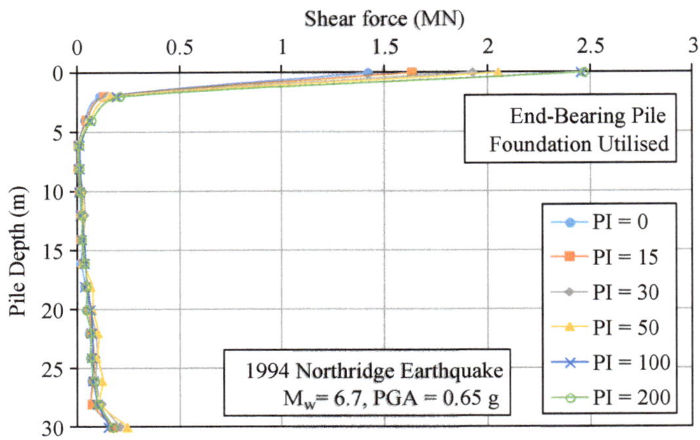

Fig. 10. Shear force imposed on pile foundation under the influence of the applied earthquake in conjunction with the variation of PI

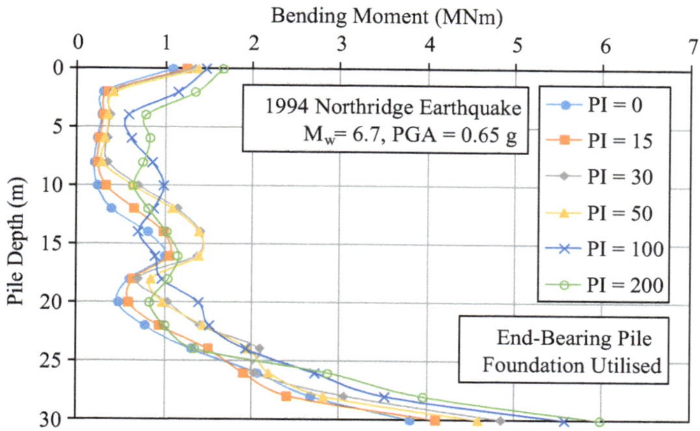

Fig. 11. Bending moment experienced by the pile foundation under the 1994 Northridge Earthquake in conjunction with the variation of PI

Figs. 9 and 11, the maximum mobilised bending moment in the pile increased as the deflection in the head increased.

The loads applied to the soil-pile-structure system are a combination of inertial and kinematic loading. The inertial loading stemmed from the superstructure generates forces and moments in the top section of the pile near the ground surface. Usually, the effects of inertial loading decay rapidly with the depth as the source of the inertial lateral loading is the superstructure (see Figs. 10 and 11). Whereas, the kinematic loading is induced by the soil movement surrounding the piles caused by the seismic excitation and referring to Figs. 10 and 11, the considerable shear forces and bending moments at the pile toe level are the consequence of the kinematic loading. Although results of this study clearly show that soil Plasticity Index (PI) can influence seismic response of buildings, it is recommended to expand this research to capture structures with different heights, soils with different average shear wave velocities and different earthquake excitations to quantify the extend PI can influence the predictions for these conditions. Also, referring to Ho and Fatahi (2015), dynamic loading can induce matric suction and excess pore air pressure variations in soil, and therefore considering unsaturated soil conditions is recommended for further research.

4 Conclusions

The present study investigates the influence of the soil Plasticity Index on the seismic response of mid-rise buildings resting on end-bearing pile foundations. A three-dimensional soil-pile-structure model is developed using FLAC3D. The fully nonlinear analysis is conducted in time domain adopting the direct method which simulates the structure, foundation and soil simultaneously. Different plasticity indices of the soil (i.e. PI = 0–200) are considered employing SIG-III model to simulate corresponding modulus degradation curves and damping ratio curves.

The numerical results show that as Plasticity Index increases (thus, G/G_{max} increases and damping ratio decreases), the building base shear amplifies. Correspondingly, the dynamic response of the structure regarding the maximum rocking angle of the foundation slab, lateral building displacement and the maximum inter-storey drift increases with PI due to the type of the foundation (end-bearing pile foundation) used in this study and eventually, considering the effects of soil PI may change the structural performance level. Thus, it is necessary to consider soil Plasticity Index in the real design. Furthermore, the results also indicate that the soil PI can have significant impact on the seismic response of the foundation. In this study, shear forces imposed on the piles increases with PI, typically in the top section of the piles near the ground surface due to inertial interaction. Also, bending moment developed in the piles increases as PI increases. Due to the load mechanism of the foundation (end-bearing pile foundation), the maximum moment occurs at the pile toe. Therefore, it can be concluded that the variation of PI could considerably alter the response of the system and it should be treat with extreme care in order to obtain reliable results.

References

ACI318: Building Code Requirements for Structural Concrete and Commentary (ACI 318-08). American Concrete Institute Washington DC, USA (2008)

AS1170.1: Structural Design Actions; Part 1: Permanent, Imposed and Other Actions. Standards Australia, Sydney, Australia (2002)

AS1170.4: Structural Design Actions; Part 4: Earthquake Actions in Australia. Standards Australia, Sydney, Australia (2007)

AS2159: Piling Design and Installation. Standards Australia, Sydney, Australia (2009)

AS3600: Concrete Structures. Standards Australia, Sydney, Australia (2009)

Beaty, M.H., Byme, P.M.: Observations on the San Fernando dams. In: International Conferences on Recent Advances in Geotechnical Earthquake Engineering and Soil Dynamic (2001)

Byrne, P.M., Naesgaard, E., Seid-Karbasi, M.: Analysis and design of earth structures to resist seismic soil liquefaction. In: 59th Canadian Geotechnical Conference & 7th Joint CGS/IAH-CNC Groundwater Specialty Conference, Vancouver, Canada (2006)

Carbonari, S., Dezi, F., Leoni, G.: Linear soil–structure interaction of coupled wall–frame structures on pile foundations. Soil Dyn. Earthq. Eng. **31**, 1296–1309 (2011)

Chu, D.: Three-dimensional nonlinear dynamic analysis of soil-pile-structure interaction (2005)

Chu, D., Truman, K.Z.: Effects of pile foundation configurations in seismic soil-pile-structure interaction. In: 13th World Conference on Earthquake Engineering, Canadian Association for Earthquake Engineering, Vancouver, BC, Canada, 1551 (2004)

Comodromos, E., Papadopoulou, M.: Response evaluation of horizontally loaded pile groups in clayey soils. Géotechnique **62**, 329 (2012)

CSI: SAP2000 V14. 1 Analysis Manual: Inelastic Time History Analysis. CSI (Computers and Structures Inc.), Berkley, USA (2010)

Galal, K., Naimi, M.: Effect of soil conditions on the response of reinforced concrete tall structures to near-fault earthquakes. Struct. Des. Tall Spec. Build. **17**, 541–562 (2008)

Hardin, B.O., Drnevich, V.P.: Shear modulus and damping in soils: measurement and parameter effects. J. Soil Mech. Found. Div. **98** (1972)

Ho, L., Fatahi, B.: Analytical solution for the two-dimensional plane strain consolidation of an unsaturated soil stratum subjected to time-dependent loading. Comput. Geotech. **67**, 1–16 (2015)

Hokmabadi, A.S., Fatahi, B.: Influence of foundation type on seismic performance of buildings considering soil–structure interaction. Int. J. Struct. Stab. Dyn. **16**, 1550043 (2016)

Hokmabadi, A.S., Fatahi, B., Samali, B.: Assessment of soil–pile–structure interaction influencing seismic response of mid-rise buildings sitting on floating pile foundations. Comput. Geotech. **55**, 172–186 (2014)

Itasca, F.D.: V5. 0, Fast Lagrangian Analysis of Continua in 3 Dimensions, User's Guide. Itasca Consulting Group, Minneapolis, Minnesota (2011)

Kramer, S.L.: Geotechnical Earthquake Engineering. Pearson Education India, Uttar Pradesh, India (1996)

Kwon, S., Kim, M., Kim, S., Choi, J.: 3D Dynamic Numerical Modeling for Soil-Pile-Structure Interaction in Centrifuge Tests" Proc of the 18th International Conference on Soil Mechanics and Geotechnical Engineering, Paris (2013)

Luo, C., Yang, X., Zhan, C., Jin, X., Ding, Z.: Nonlinear 3D finite element analysis of soil–pile–structure interaction system subjected to horizontal earthquake excitation. Soil Dyn. Earthq. Eng. **84**, 145–156 (2016)

Nguyen, Q.V., Fatahi, B., Hokmabadi, A.S.: The effects of foundation size on the seismic performance of buildings considering the soil-foundation-structure interaction. Struct. Eng. Mech. **58**, 1045–1075 (2016)

Nguyen, Q.V., Fatahi, B., Hokmabadi, A.S.: Influence of size and load-bearing mechanism of piles on seismic performance of buildings considering soil–pile–structure interaction. Int. J. Geomech. **17**, 04017007 (2017)

Shelke, A., Patra, N.: Effect of arching on uplift capacity of pile groups in sand. Int. J. Geomech. **8**, 347–354 (2008)

Sun, J.I., Golesorkhi, R., Seed, H.B.: Dynamic Moduli and Damping Ratios for Cohesive Soils, Earthquake Engineering Research Center, University of California Berkeley (1988)

Tabatabaiefar, H.R., Fatahi, B.: Idealisation of soil–structure system to determine inelastic seismic response of mid-rise building frames. Soil Dyn. Earthq. Eng. **66**, 339–351 (2014)

Vucetic, M., Dobry, R.: Effect of soil plasticity on cyclic response. J. Geotech. Eng. **117**, 89–107 (1991)

Remote Measurements and Vibration Analyses of Existing Wind Turbines

Chih-Hung Chiang[1(✉)], Keng-Tsang Hsu[1], Chih-Peng Yu[2], Chia-Chi Cheng[1], and Jie-Zhen Pan[1]

[1] Chaoyang University of Technology, Taichung, Taiwan
chiangc@cyut.edu.tw
[2] National Chung Hsin University, Taichung, Taiwan

Abstract. The service life of a wind turbine may be threatened by the collapse of the rotor blades or the supporting tower due to extreme weather. The change of dominant frequency may indicate that the stiffness of the turbine-tower system is deviated from its design. Dynamic measurements of lateral displacement of four 2-MW wind turbines of the same model have been achieved using a ground-based microwave interferometer. The frequency contents of a turbine-tower system can be measured when the rotor is stopped or operates at low to medium speed. The rotor blade exhibits multiple frequencies higher than 1 Hz that can only be measured when the wind turbine is shut down. Possible links between dynamic characteristics and structural integrity of wind turbine—tower systems are discussed. The results of current research are promising in terms of combining non-contact dynamic measurements and vibration analyses towards the condition assessment of wind turbine-tower systems.

1 Introduction

Wind energy has been drawn a lot of public attention as one of the source of renewable energy. Besides the concern over efficiency of power generation, issues such as noise and safety are raised by the news media and environmental groups from time to time. Threats to the safe operation of wind turbines may come from natural disasters such as lightening strikes, wind gust, and earthquakes. The supporting towers of wind turbines are usually well designed and properly built. However the structural integrity of the tower should not be overlooked since structural failures of wind turbines often involve collapsed towers.

Incidents of expected or unexpected wind turbine collapses have been reported in many parts of the world, including Europe, Japan, USA, and Taiwan. Chou and Tu investigated 715 incidents of wind turbines and found the top three causes were blade failures 23%, fire 13%, and structural failures 12% (Chou and Tu 2011). They also reported a total of 62 cases of tower collapse occurred from 1997 to 2009. More recent incidents include the severe damages to wind turbines caused by the extreme weather conditions at the Taichung's Gaomei wetlands in August 2015. Typhoon Soudelor`s powerful gusts of up to 220 km per hour blew six of 18 wind turbines at the Gaomei wind farm to the ground.

B. Fatahi et al. (eds.), *Sustainable Design and Construction for Geomaterials and Geostructures*, Sustainable Civil Infrastructures,
https://doi.org/10.1007/978-3-319-95753-1_13

The objective of this work is to look for distress signals of the supporting towers that may be used as damage indicators for wind turbines. Dynamic measurements of lateral displacement of four 2-MW wind turbines of the same model have been achieved using a ground-based microwave interferometer. One of them collapsed during the incident in August 2015. The results of vibration analyses will be closely examined, followed by some discussion on possible selection of damage indicator based on deflection curves.

2 Remote Measurements of Lateral Displacements of Wind Turbines

Dynamic measurements of lateral displacement of four 2-MW wind turbines of the same model have been achieved using a ground-based microwave interferometer. Figure 1 illustrates that the reflective signals from the wind turbine are received by the microwave interferometer IBIS. The displacement measurement is obtained by the phase shift of the same reflection point in two consecutive radar images recorded (Pieraccini et al. 2004). The expected accuracy is better than 1 mm at a signal-to-noise ratio of 40 dB or greater. The operating parameters of IBIS are listed in Table 1. The sampling frequency up to 200 Hz is automatically determined by the IBIS system.

Fig. 1. Non-contact measurement of a wind turbine (left) and the microwave interferometer (right)

Table 1. Operating parameters of the microwave interferometer IBIS-s

Center frequency (GHz)	Half-power beam width (°)	Bandwidth (MHz)	Number of transmitted frequencies	Single tone duration (μs)	Sweep duration (ms)
16.75	15	380	2502	12.3	30.76

An example of measurement results are shown as Fig. 2 in which the lateral displacements from five reflection points are recorded for approximately 30 min. Note that the middle part of data, starting from about 12 min, was recorded while the wind turbine is parked. The sharp increase in amplitude takes place when sudden movement is detected.

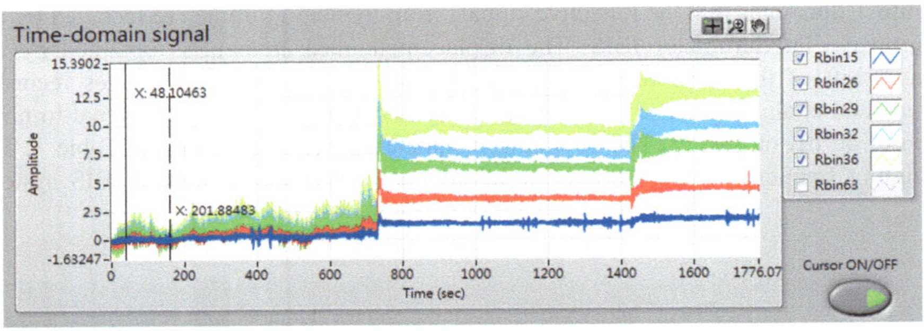

Fig. 2. Time-domain displacement signal for WT No. 9H, recorded on May 11 2015

3 Vibration Analyses

Vibration analyses in the current study refer to the methods of analyzing frequency contents of the wind turbines under scrutiny. Besides processing the transient data using fast Fourier transform, the signals obtained in the field testing of wind turbines are further analyzed using frequency domain decomposition and reassigned spectrogram. A brief account of the two methods is given in the following.

3.1 Frequency Domain Decomposition and Modal Analysis

Modal analysis that provides modal frequencies, mode shapes, and even damping ratios are frequently applied to system identifications and structural dynamics. Civil structures such as bridges and towers may sustain complex loading conditions that are difficult to reproduce in a full scale test. Operational modal analyses using ambient excitations thus have many practical applications in structural dynamics and health monitoring of civil structures. Frequency domain decomposition, or FDD, is one type of operational modal analyses. The basic idea is to take advantage of vibration measurements under a light loading condition (ambient excitation). The associate analysis assumes the input to be white noise and thus the power spectral density (PSD) matrix of the responses can be expressed as a function of the frequency response function matrix. The PSD matrix is then decomposed into a set of auto spectral density functions, each corresponding to a single-degree-of-freedom system (SDOF). Other assumptions include the mode shapes of close modes are geometrically orthogonal and the damping ratio of the structure is small. Even when these assumptions are not satisfied, the decomposition into SDOF systems is still a useful approximation. FDD and its extension EFDD have found great success in recent years for dynamic analysis of bridges and other large structures (Harte et al. 2012; Jacobsen et al. 2008; Magalhaes et al. 2010), partly due to the commercial software ARTeMIS[@] and similar products.

3.2 Reassigned Spectrogram for Time-Frequency Analysis

A spectrogram is a visual representation of the frequency spectrum in a transient (or spatial) signal. This can be estimated by computing the squared magnitude of the short-time Fourier transform of the signal. The type and size of the window and the frame advance of the adjacent window must be specified. The method of reassignment sharpens blurry time-frequency data and re-arranging the data according to local estimates of instantaneous frequency and group delay. Flandrin et al. (2003) described a spectrogram as a smoothed Wigner-Ville distribution, or a smoothed WVD. The value that a spectrogram takes at a given point (t, ω) of the plane cannot be considered as point-wise, but instead results from the summation of a continuum of WVDs within some time-frequency domain defined as the essential TF support of the short-time window. Assign the total mass to the center of gravity of the distribution within the domain, and this is precisely what reassignment does: at each TF point (t, ω) where a spectrogram value is computed, one also computes $t_x(t, \omega)$ and $\omega_x(t, \omega)$. The spectrogram value is then moved from the point (t, ω) to this centroid $(\hat{t}_x(t, \omega), \hat{\omega}_x(t, \omega))$, leading to the reassigned spectrogram as given by Flandrin et al. (2003):

$$\hat{s}_x^h(t,\omega) = \iint_{-\infty}^{+\infty} S_x^h(s,\zeta)\delta(t - \hat{t}_x(s,\zeta), \omega - \hat{\omega}_x(s,\zeta))\frac{dsd\zeta}{2\pi} \tag{1}$$

Detail information on a variety of algorithms for time-frequency reassignment can be found in the literature such as (Flandrin et al. 2003; Fulop and Fitz 2006). A computer code developed in the previous study (Chiang et al. 2015) will be employed to generate the reassigned spectrograms in the following sections.

3.3 Results of Vibration Analyses

The reassigned spectrogram offers the duration of a given frequency event that may be compared with the information recorded by the SCADA of the wind turbines. For example a rotor speed of 0.18 Hz can be found in the reassigned spectrogram in Fig. 3c.

The measurement result of wind turbine No. 4P had a dominant frequency of 0.44 Hz based on both reassigned spectrum and FDD. The same dominant frequency

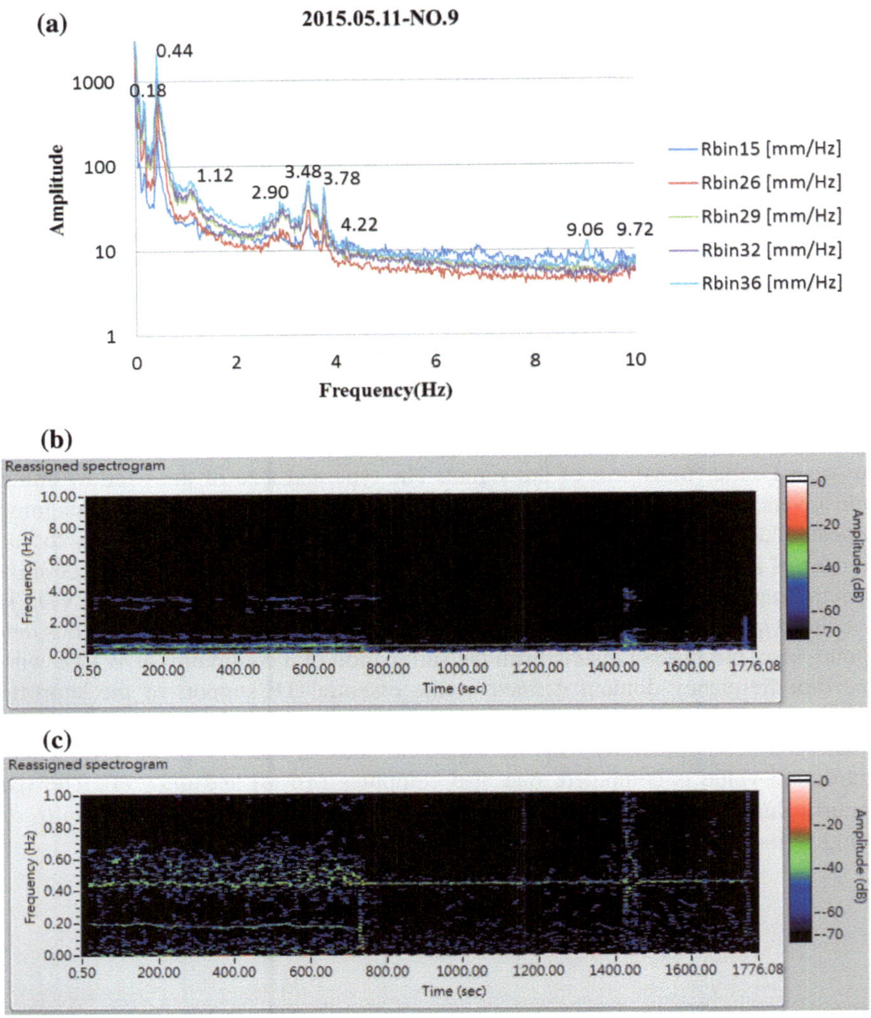

Fig. 3. Results of vibration analyses on the displacement record of WT#9H in Fig. 2. **a** FDD, **b** reassigned spectrum, and **c** reassigned spectrum for frequency below 1 Hz

was also detected for the rotor blade of wind turbine No. 5P. Additional five peak frequencies can be identified while the rotor of the wind turbine was parked, as shown in Fig. 4a.

The measurement results of wind turbine No. 18H had a dominant frequency at 0.35 Hz in both November and December 2013, lower than that at 0.44 Hz taken in 2012, as shown in Fig. 4b.

The measurement results of wind turbine No. 9H on the other hand, had a dominant frequency of 0.44 Hz, as shown in Fig. 4c, before and after the extreme wind gust of Typhoon Soudelor in 2015.

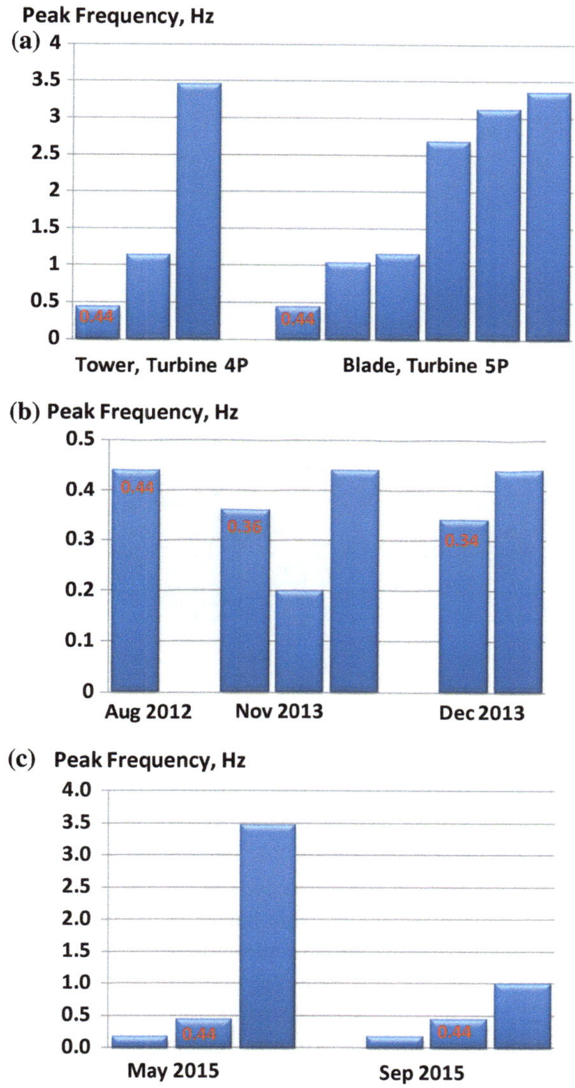

Fig. 4. Peak frequencies for **a** No. 4P tower and No. 5P rotor blade, **b** No. 18H tower in 2012 and 2013, and **c** No. 9H tower in 2015, respectively

4 Discussions

FDD may provide more peak frequencies than a reassigned spectrogram, as can be seen in Fig. 5. Frequencies higher than 3 Hz are related to higher modes of tower vibration, similar to those found by Smith and Mahmoud (2016). They identified the first three

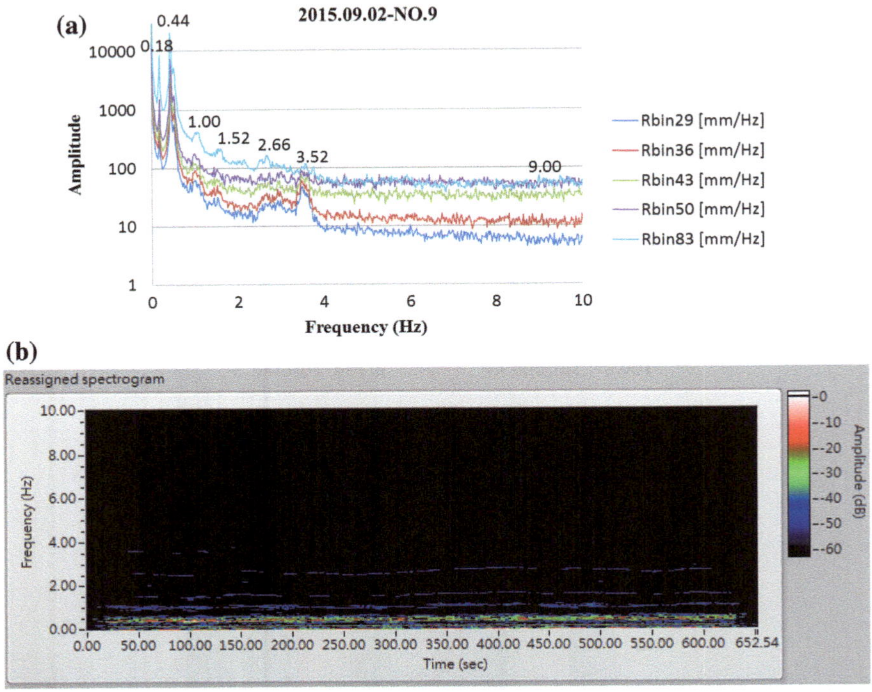

Fig. 5. **a** Results of FDD for five reflection points of the displacement signals recorded on September 2, 2015 **b** Reassigned spectrum for the reflection point Rbin83 of the same signals

modes of vibrations as 0.64, 3.33, and 9.09 Hz respectively for a 60 m-tall tower in the simulation model.

Decrease in the dominant frequency could be an indicator of distress for wind turbine No. 18H. A distress message was sent based on the findings in Fig. 4b in 2013. Further analysis on the turbine-tower structure found that change in dominant frequencies can be directly linked to the decrease in tower stiffness. Numerical modeling based on the Eq. (2) was then carried out using one-dimensional spectral elements derived from the linear continuous formulation (Chiang et al. 2014a)

$$\omega^2 = \frac{\int_0^L EI(y'')^2 dx + \int_{x_1}^{x_1+\Delta L} (\Delta EI)[y''(x_1)]^2 dx}{\int_0^L \rho A y^2 dx + \int (\Delta M) g(x_2)[y(x_2)]^2 dx} \tag{2}$$

ΔEI is the stiffness variation from position x_1 to $x_1 + \Delta L$. ΔM is the added mass at position x_2. The advantage for such a spectral model is that only a few discretized elements is required instead of a fine mesh usually employed by a time-domain finite element model.

The simplified case for the fixed foundation is given here. The lower sections are more sensitive than the top section of the tower that a 4% decrease in frequency is resulted from 20 to 30% decrease in *EI*, based on the numerical simulation (Chiang et al. 2014b). The effect of stiffness reduction is less detectable as the height of tower increases. Dominant frequency is independent of variation of *EI* at the top section of the tower, as shown in Fig. 6b. ***Besides the solid foundation considered in the current study, the influence of size and load-bearing mechanism of piles may also affect the natural frequency of the wind turbine system, as indicated by the work of Fitzgerald and Basu*** (2016) ***and Lombardi et al.*** (2013).

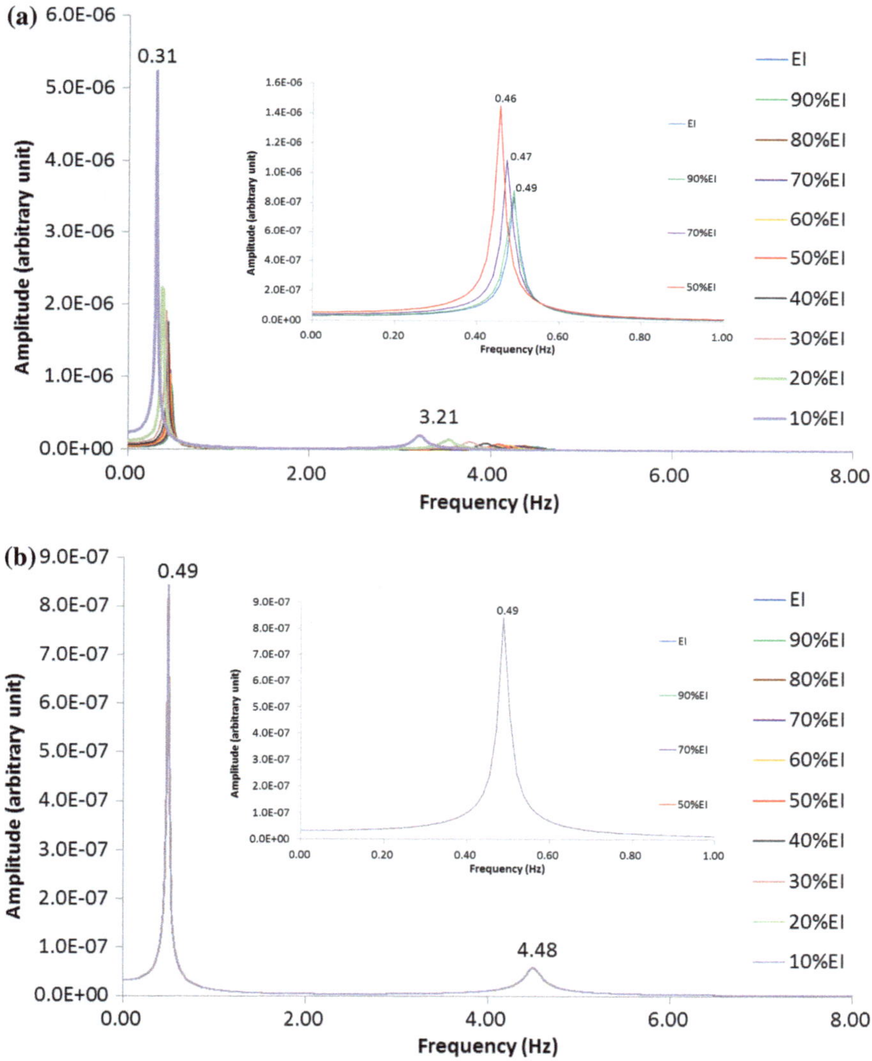

Fig. 6. Dominant frequency as a function of stiffness reduction at the lowest section of the tower and (**b**) at the top section of the tower

No severe weather associated with storms was recorded between January 2014 and July 2015. The decrease in the dominant frequency in Fig. 4b is then a ***possible*** indication that some damage ***may*** exist in the turbine-tower structure of No. 18H prior to the collapse in August 2015. The question then becomes whether or not the location ***and the cause*** of damage can be identified? ***Furthermore would the damage result in the reduction stiffness of the turbine-tower structure?***

The second derivative of measured deflection curve, or curvature is directly linked to the bending moment divided by the product of elastic modulus and moment-of-inertia (*EI*) of a beam member. Hence a finite element model has been applied to simulate the static deflection of the supporting tower as a vertical cantilever beam. The tower base is fixed while a concentrated load is exerted at the tower top in the lateral direction. Numerical simulation shows that a 30% decrease in stiffness *EI* of the section near the tower base will result in some change in the deflection curve. The deflection at the top of the tower increases as the lateral load increases, as shown in Fig. 7. The slope of the deflection curves also changes near the section of the tower base. Hence we propose that deflection curves can be used to investigate the trend of tower deflection between two different measurements, preferably on two separate dates for comparison.

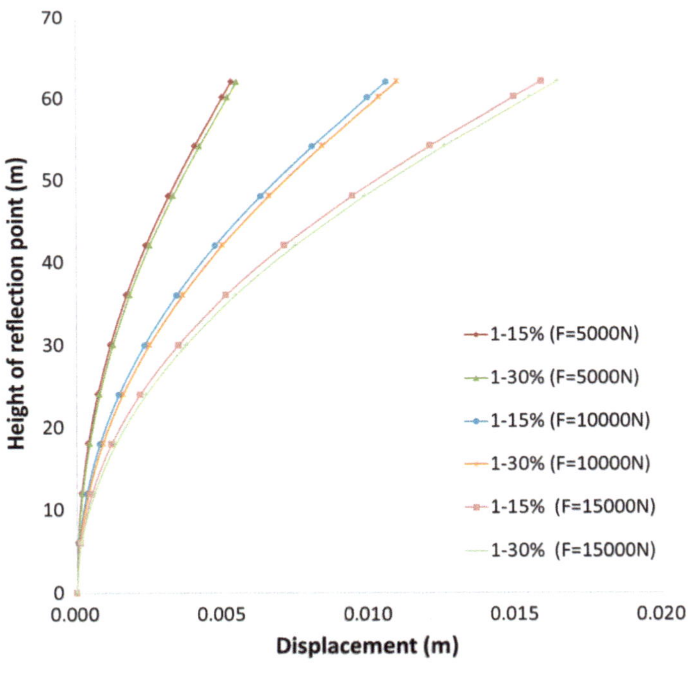

Fig. 7. Simulated deflection curves for a wind turbine tower

5 Concluding Remarks

The service life of a wind turbine may be threatened by the collapse of the rotor blades or the supporting tower due to extreme weather. Dynamic measurements of lateral displacement of four 2-MW wind turbines of the same model have been achieved using a ground-based microwave interferometer. The frequency contents of a turbine-tower system can be measured when the rotor is stopped or operates at low to medium speed. The rotor blade exhibits multiple frequencies higher than 1 Hz that can only be measured when the wind turbine is shut down. One wind turbine has shown a distress signal of decrease in the dominant frequency prior to its collapse in the 2015 incident. ***The change of dominant frequency indicates that some damage does exist in the turbine-tower structure. Further investigation should be conducted on whether the stiffness of the turbine-tower system is deviated from its design.***

This study shows the potential applications of combining non-contact dynamic measurements and vibration analyses towards the condition assessment of wind turbine-tower system. Numerical study found that the reduction in stiffness of the tower does lead to decrease in the dominant frequency of the turbine-tower system. If this were to be regarded as a plausible indicator of damage for existing wind turbines, then the deflection curves have to be measured effectively. Two alternatives can be explored in the future. First one is the direct measurements of deflection curves. Some form of normalization may be necessary to account for dissimilar loading conditions between two measurements. The other is modal analysis based on dynamic measurements of lateral displacements. Summation of the first three vibration modes could be a reasonable approximation to the deflection curve.

Acknowledgements. This research was made possible by the support of Ministry of Science and Technology Taiwan through project number MOST 103-2221-E-324-003 and NSC 102-2221-E-324-015. Additional support also came from the research grants of Chaoyang University of Technology.

References

Chiang, C-H., Yu, C-P., Hsu, K-T., Cheng, C-C., Ke, Y-T., and Shih, Y-R.: Remote monitoring and nondestructive evaluation of wind turbine towers. Proc. SPIE **8321**, 90631 (2014a)

Chiang, C.-H., Yu, C.-P., Hsu, K.-T., Cheng, C.-C., Shih, Y.-R., Pan, J.-Z.: Project report No. NSC 102-2221-E-324-015 (2014b)

Chiang, C.-H., Huang, C.-L., Hsu, K.-T., Cheng, C.-C., Lai, J.: "Time frequency analyses of vibrations of wind turbine towers. Proc. SPIE **9437**, 94371C (2015)

Chou, J.-S., Tu, W.-T.: Failure analysis and risk management of a collapsed large wind turbine tower. Eng. Fail. Anal. **18**, 295–313 (2011)

Fitzgerald, B., Basu, B.: Structural control of wind turbines with soil structure interaction included. Eng. Struct. **111**, 131–151 (2016)

Flandrin, P., Auger, F., Chassande-Mottin, E.: Time-frequency reassignment: from principles to algorithms. In: Papandreou-Suppappola, A. (ed.) Applications in Time-Frequency Signal Processing, ch. 5, pp. 179–203. CRC Press (2003)

Fulop, S.A., Fitz, K.: Algorithms for computing the time-corrected instantaneous frequency (reassigned) spectrogram, with applications. J. Acoust. Soc. Am. **119**(1), 360–371 (2006)
Harte, M., Basu, B., Nielsen, S.R.K.: Dynamic analysis of wind turbines including soil-structure interaction. Eng. Struct. **45**, 509–518 (2012)
Jacobsen, N.J., Andersen, P., Brincker, R.: Applications of frequency domain curve-fitting in the EFDD technique. In: Proceedings of the 18th International Modal Analysis Conference. Orlando, FL (2008)
Lombardi, D., Bhattacharya, S., Muir Wood, D.: Dynamic soil-structure interaction of monopole supported wind turbines in cohesive soil. Soil Dyn. Earthquake Eng. **49**, 165–180 (2013)
Magalhaes, F., Cunha, A., Caetano, E., Brincker, R.: Damping estimation using free decays and ambient vibration tests. Mech. Syst. Signal Process. **24**, 1274–1290 (2010)
Pieraccini, M., Luzi, G., Mecatti, D., Fratini, M., Noferini, L., Carissimi, L., Franchioni, G., Atzeni, C.: Remote sensing of building structural displacements using a microwave interferometer with imaging capability. NDT&E Int **37**(7), 545–550 (2004)
Smith, V., Mahmoud, H.: Multiharzard assessment of wind turbine towers under simultaneous application of wind, operation, and seismic loads. J. Perform. Constr. Facil. **30**(6), 04016043 (2016)

Parametric Selection for the Seismic Numerical Simulation of Geosynthetic Reinforced Slope

Sao-Jeng Chao(✉) and Tsan-Hsuan Yu

Department of Civil Engineering, National Ilan University, Ilan, Taiwan
chao@niu.edu.tw

Abstract. Reinforced soil structures encompass the advantage of easy and quick construction. In addition, they are highly resistant to stress and earthquakes. Conventional gravity retaining structures are thus being gradually replaced by geosynthetic reinforced soil structures. Bearing in mind that high frequency of earthquakes in Taiwan area, earthquake resistance should be taken into account in the design of reinforced soil structures to consider the effects of additional stress from earthquakes. This paper utilizes PLAXIS finite element program to simulate the seismic behavior of geosynthetic reinforced slope during the earthquake in FoGuang University. Since the repeated seismic reflections and magnified seismic response can occur within the model boundary when a seismic input is directly applied to the base location, the reflection of seismic input may affect the dynamic analysis simulation. Appropriate boundary condition and damping ratio are emphasized in this paper to facilitate further analyses to resolve the effects of seismic response. Although greater distance for model margins may also be employed without the boundary conditions to reduce the seismic reflections, more calculation for processing additional elements and storage space requirements may be demanded. Therefore, dynamic boundary condition provided in Plaxis can be applied to simplify the calculation. Furthermore, it is necessary to properly consider the soil damping ratio for a seismic simulation analysis to enhance accuracy of the predicted response.

1 Introduction

As it belongs to the seismically active region of the Circum-Pacific belt between the Eurasian and Philippine plates, developing effective means of preventing earthquake damage is a vital concern in Taiwan area. Consequently, this study investigates the effects of earthquakes on geosynthetic reinforced slope by installing a seismic recording device Palert on a geosynthetic reinforced slope to measure the magnitude of ground acceleration at the start.

By facilitating interactive behavior between the soil and the reinforcement, reinforced soil retaining structures stabilize the soil and resist earth pressure or additional sources of stress. Because of the advantages such as simplicity of application, low cost, short construction duration, adaptability, soil mass balance, and earthquake resistance, reinforced soil retaining structures have been widely applied in slope engineering in Taiwan in recent times, even gradually superseding the use of conventional gravity retaining structures (Chou et al. 1991; Chao 2004).

B. Fatahi et al. (eds.), *Sustainable Design and Construction for Geomaterials and Geostructures*, Sustainable Civil Infrastructures,
https://doi.org/10.1007/978-3-319-95753-1_14

However, reinforced soil retaining structures still displayed damage after the disastrous Chi-Chi earthquake (also known as the 921 earthquake). To examine the influence of seismic force on the geosynthetic reinforced slope, a seismic simulation was conducted through the tool of finite element method (FEM), with the help of an earthquake monitoring system deployed on the slope. The monitoring system was used to record the state of ground acceleration and displacement at various layers under actual seismic conditions. The earthquake data recorded at the demonstration site are used to validate the FEM analysis of Plaxis program (Kao 2014; Wu 2015). Subsequently, using the appropriate soil damping ratio and boundary condition in dynamic analysis, this paper explores the effects of earthquakes on geosynthetic reinforced slope with the aid of seismic numerical simulation.

2 Geosynthetic Reinforced Slope Demonstration Site and Monitored Data

FoGuang University is located in seismically active region of the Ilan area (Fig. 1), where slope maintenance is not easy at all. It is effective to use the Palert earthquake P wave sensor installing on the slope for the purpose of seismic data recording. The monitoring system was established to record the state of ground acceleration and displacement on various layers of the geosynthetic reinforced slope under actual seismic conditions. The earthquake data recorded at the demonstration site were used to compared and validated with a FEM analysis called Plaxis program. Subsequently, the

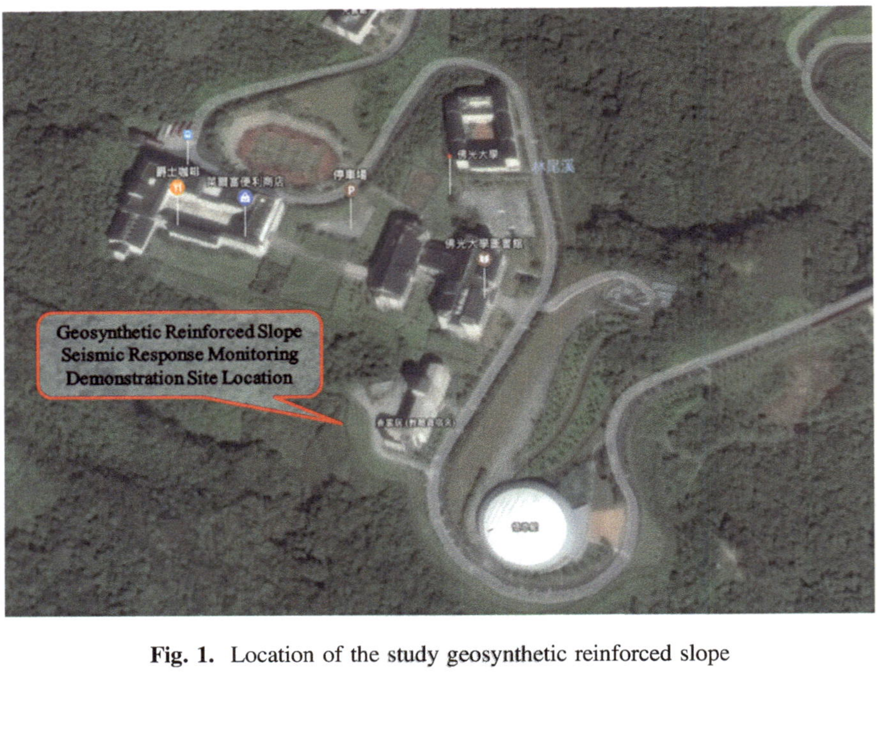

Fig. 1. Location of the study geosynthetic reinforced slope

boundary condition and soil damping ratio retrieved from the analysis were specifically investigated to improve the accuracy of seismic numerical simulation of geosynthetic reinforced slope.

2.1 Demonstration Site of the Geosynthetic Reinforced Slope Seismic Monitoring

The slope region of the adopted demonstration site was extremely unstable because of previous damage from debris flows caused by typhoons and severe rains. Considering the hazards of forthcoming and further damage, this study was conducted after the completion of restoration work. Several earthquake detectors (Palerts) were deployed at the top, middle, and ground layers of the geosynthetic reinforced slope, which is illustrated in Fig. 2, to explore the various conditions of the slope under a seismic occurrence.

Fig. 2. **a** Geosynthetic reinforced slope at FoGuang University; **b** Earthquake monitoring instrument Palert

The cross-sectional design of the adopted geosynthetic reinforced slope contains three layers, with the additional application of a 200 kN/m reinforced geogrid. The first layer (base structure) of the geosynthetic reinforced slope was designed to be 2.5 m high, with a six-layer reinforced geogrid was applied arranging as 0.5 m spacing; the second layer (middle structure) was 15.5 m high with a reinforced slope angle of 45°, as well as a 51-layer reinforced geogrid overlaid with 0.3 m spacing; the third layer (top structure) was 12.5 m high with a reinforced slope angle of 73°, and a 41-layer reinforced geogrid with 0.3 m spacing. Furthermore, an extra 2-m-high backfill formation was additionally designed on the top structure of the reinforced slope, which had been previously mixed with cement at an approximately 1.5–2% weight ratio and processed by compaction after filling up.

2.2 Seismic Monitoring Data of the Geosynthetic Reinforced Slope

The earthquake monitoring system in this study observed the seismic activity in the first, second, and third layers, and recognized it through built-in detection technology and recording the data on a host computer. The related seismograph software was used to instantly retrieve data content such as triaxial acceleration and displacement, as well as enabled a subsequent seismic simulation analysis.

According to the 11 data entries recorded in the earthquake monitoring system since October 2013, most maximum triaxial acceleration responses have different magnification ratios from the first layer (base structure) to third layer (top structure). Specifically, Data 0222 was adopted as the earthquake input for the later seismic simulation. The maximum acceleration of Data 0222 observed from the first to third layer was 0.092, 0.132, and 0.209 m/s^2, respectively; the magnification ratio detected from the first to second layer was 1.46, whereas the magnification ratio from the first to third layer was identified as 2.25. The earthquake acceleration time history observed from the field monitoring shown in Figs. 3, 4 and 5 represent the seismic activities in the three layers (where x = time, y = acceleration magnitude).

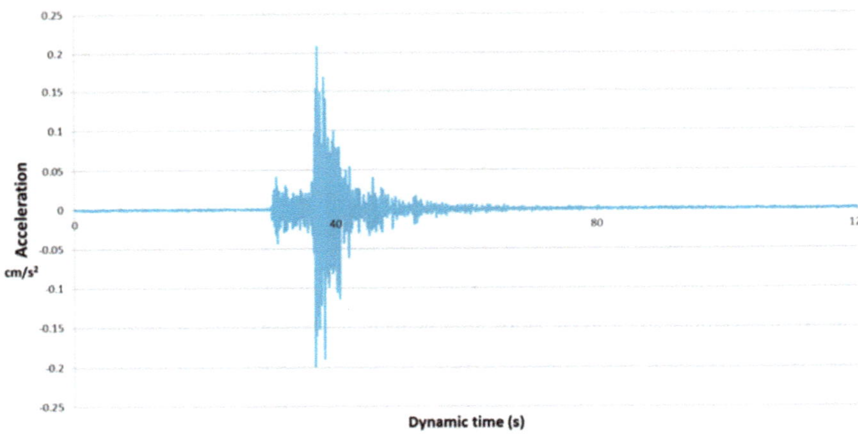

Fig. 3. Top structure earthquake acceleration time history observed in field monitoring

The use of recorded seismic data can also be used to validate for performing a seismic simulation of geosynthetic reinforced slope. The recorded data were converted and applied in Plaxis seismic simulation analysis to examine acceleration status under the effects of earthquake forces. To provide the basis for the design of the geosynthetic reinforced slope, it is expected to minimize the impact of the slope disaster and to build a stable and safe slope structure.

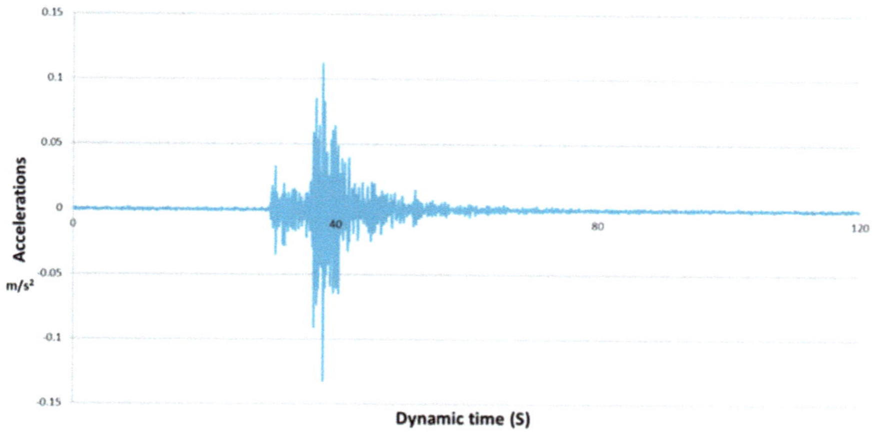

Fig. 4. Middle structure earthquake acceleration time history observed in field monitoring

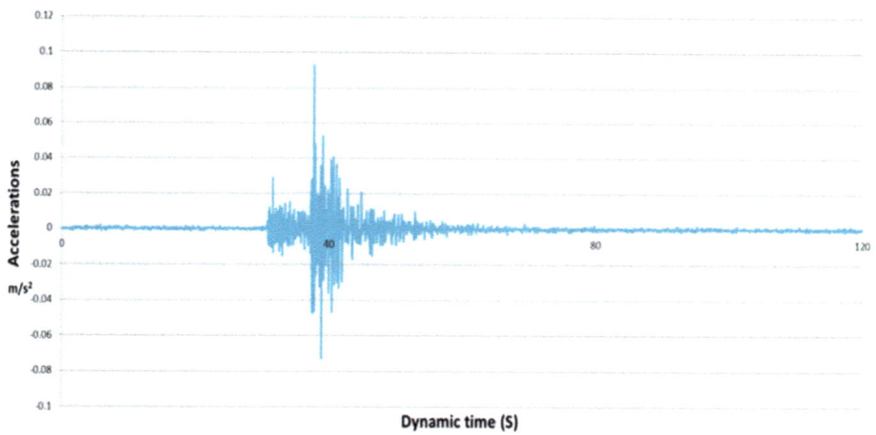

Fig. 5. Base structure earthquake acceleration time history observed in field monitoring

3 Seismic Numerical Simulation of Geosynthetic Reinforced Slope

Plaxis is a two-dimensional FEM analysis of deformation and stability for rock or soil that can construct a model using plane strain or axial symmetry. The computer program is able to instantly generate a finite element grid in the geometric model. Interactive behavior in various geotechnical and soil structures can be analyzed using a display window to input model conditions such as material parameters, boundary conditions, and seismic data; in addition to standard static analyses, earthquake data can be input for seismic analyses by the Plaxis (2015).

More specifically, because soil structures can be affected by dynamic loading in addition to static loading, vibrating behavior in the soil can also be detected through the dynamic simulation with Plaxis. Considering that earthquakes are the source of naturally generated vibrations, they can be calculated using dynamic analysis when the frequency of dynamic loading is equal or higher than the frequency of certain quality, whereas vibration lower than standard vibration frequency may be measured through a quasi-static calculation.

3.1 Effect of Boundary Condition on the Seismic Simulation of Geosynthetic Reinforced Slope

Plaxis seismic numerical analysis is used to perform the simulation using specific earthquake data given in the model. Inputting seismic waves directly to the base structural layer using fixed boundary conditions can lead to repetitive internal seismic reflection within the model boundary and magnified seismic response. By the way, in a static deformation analysis, boundary displacement can be simply generated at a completely free or fixed boundary of the FEM model in one or two directions.

In a dynamic calculation, the boundary distance should be greater than that in the static calculation, considering that the reflection of stress waves can result in imprecise calculations. Because of the additional elements computing time, and storage demanded in a remote boundary setting, the extraordinary dynamic boundary condition is required to simplify the calculation process. In this study, a dynamic simulation was employed using the free-field boundary through the Plaxis finite element program. The simulation results were subsequently compared with the corresponding field monitoring records for further verification.

In most simulations, seismic waves are often inputted at the base structural layer. In addition, they assume the independent reception of seismic wave but disregard other possible sources of vibration energy derived from the structure or external forces. Therefore, the free-field boundary condition can simulate seismic behavior in infinite far-field boundaries to decrease influence from the boundaries and represent actual conditions without generating additional calculations as a result of an enlarged model; thus, infinite far-field boundary conditions have been widely adopted in current numerical simulations. Moreover, in the Plaxis simulation, the free-field boundary conditions can only be applied in lateral boundary conditions (i.e. $x_{\min}$ and $x_{\max}$) to simulate the transmission of minimal seismic reflection at far-field boundaries (Fig. 6).

3.2 Effect of Soil Damping Ratio on the Seismic Simulation of Geosynthetic Reinforced Slope

Damping ratio is considered for viscidity soil and friction of the Plaxis numerical seismic analysis. Most soil models show pure elastic behavior during unloading and loading, thus the material properties considered in the Plaxis program alone do not accurately represent the damping condition of the real soil. Therefore, there is the need to add additional damping ratio to the increase dynamic simulation of the real soil characteristics, which are defined by Rayleigh damping in the Plaxis.

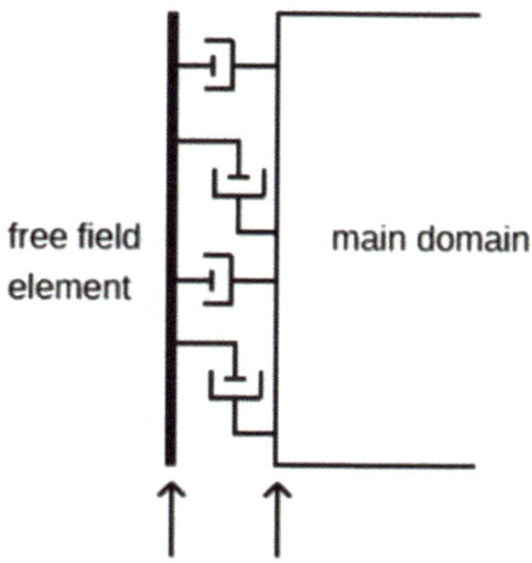

Fig. 6. Energy absorption of the free-field boundary

In this study, seismic simulation analysis achievement observed from the dynamic analysis results appear less accurate if the damping ratio is set incorrectly. For illustrating purpose, 0% damping ratio simulation results and 5% damping ratio comparison results are shown in Fig. 7. As we can tell that Fig. 7b is more consistent with the current earthquake monitoring results. As a result, the damping ratio is set to be 5% for dynamic simulation in this study.

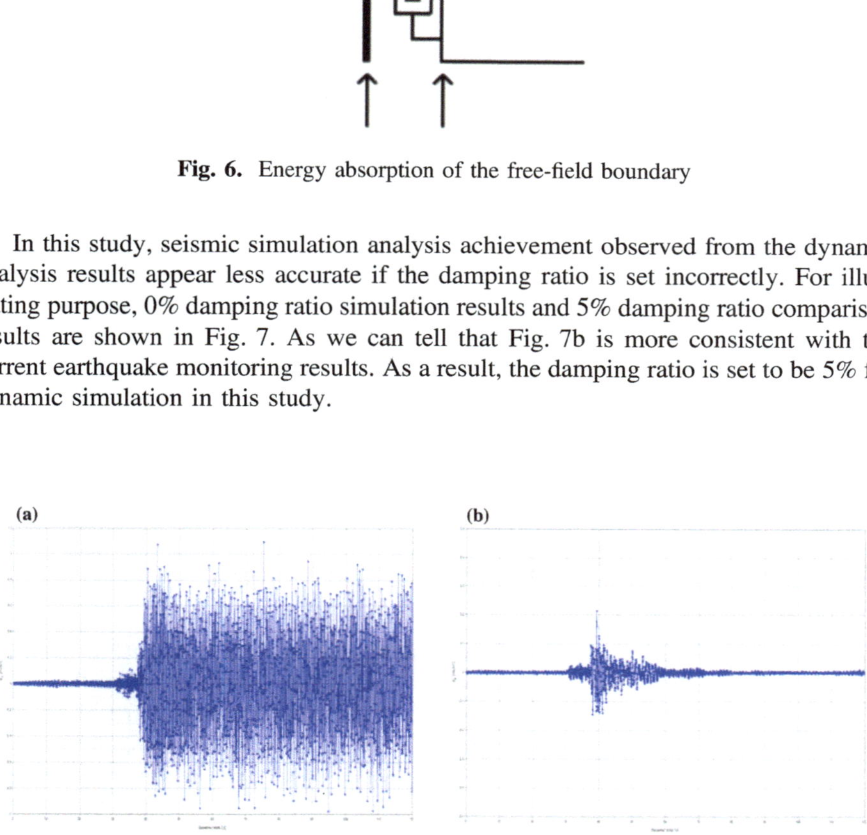

Fig. 7. The predicted acceleration time histories at top structure with **a** damp ratio 0%, **b** damp ratio 5%

4 Seismic Response of Geosynthetic Reinforced Slope

In this study, earthquake monitoring instruments (Palerts) were set up on the campus of FoGuang University to study the response of geosynthetic reinforced slope during the earthquake. However, the local equipment was only set up on the slope platforms. For more understanding, the boundary condition, the soil parameter, and the soil damping ratio can be numerically simulated to study the values of the external and internal maximum acceleration. The seismic data received show that the maximum acceleration value will be enlarged with height. The increase values by 3 m with the height of the geosynthetic reinforced slope, on the surface of the geosynthetic reinforced slope, in the structure inside of the geosynthetic reinforced slope, and in the soil contact interface of the geosynthetic reinforced slope, are drawn as the curve points (Fig. 8). Subsequently, it is expected that the completion of the calculation would determine whether the maximum acceleration value with the height of geosynthetic reinforced slope altering.

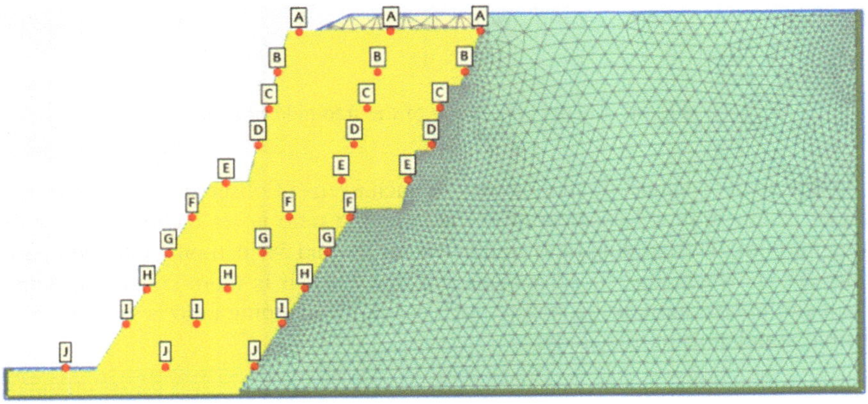

Fig. 8. Calculation locations for the marker positions

4.1 Seismic Numerical Simulation Results of Geosynthetic Reinforced Slope

According to the different height, the maximum acceleration is discussed here. The simulation results show that when the points of exploration are located on the surface of the geosynthetic reinforced slope, the maximum acceleration is 0.090 m/s^2 at the bottom of the model, maximum acceleration value as altitude increases, and when the height reaches 2.5 m maximum acceleration value is 0.098 m/s^2 on the first layer. When the height reaches 6 m maximum acceleration value is 0.098 m/s^2, 9 m maximum acceleration value is 0.098 m/s^2, 12 m maximum acceleration value is 0.11 m/s^2, 15 m maximum acceleration value is 0.12 m/s^2, 18 m (second layer) maximum acceleration value is 0.132 m/s^2, 21 m maximum acceleration value is 0.104 m/s^2, 24 m maximum acceleration value is 0.093 m/s^2, 27 m maximum acceleration value is 0.136 m/s^2, and 30 m (third layer) maximum acceleration value is 0.201 m/s^2.

The simulation results show that when the point of exploration is located in the structure inside of the geosynthetic reinforced slope, while the height reaches 2.5 m maximum acceleration value is 0.087 m/s^2, height reaches 6 m maximum acceleration value is 0.104 m/s^2, 9 m maximum acceleration value is 0.094 m/s^2, 12 m maximum acceleration value is 0.11 m/s^2, 15 m maximum acceleration value is 0.112 m/s^2, 18 m maximum acceleration value is 0.107 m/s^2, 21 m maximum acceleration value is 0.104 m/s^2, 24 m maximum acceleration value is 0.092 m/s^2, 27 m maximum acceleration value is 0.133 m/s^2, and 30 m maximum acceleration value is 0.201 m/s^2.

The simulation results show that when the point of exploration is located in the soil contact interface of the geosynthetic reinforced slope, while the height reaches 2.5 m maximum acceleration value is 0.124 m/s^2, height reaches 6 m maximum acceleration value is 0.125 m/s^2, 9 m maximum acceleration value is 0.097 m/s^2, 12 m maximum acceleration value is 0.087 m/s^2, 15 m maximum acceleration value is 0.097 m/s^2, 18 m maximum acceleration value is 0.109 m/s^2, 21 m maximum acceleration value is 0.112 m/s^2, 24 m maximum acceleration value is 0.105 m/s^2, 27 m maximum acceleration value is 0.111 m/s^2, and 30 m maximum acceleration value is 0.101 m/s^2.

In the geosynthetic reinforced slope maximum acceleration value is discussed in three places, which are chosen at the surface of the geosynthetic reinforced slope, in the structure inside of the geosynthetic reinforced slope, and in the soil contact interface of the geosynthetic reinforced slope, to understand the impact at different places and locations by the earthquake. The maximum acceleration values at different heights are summarized in Table 1 for the maximum acceleration value comparison for the three places according to different heights.

Table 1. Summarized maximum acceleration values at different heights

Elevation (m)	Location		
	Maximum acceleration (m/s^2)		
	Located on the surface of the geosynthetic reinforced slope	Located in the structure inside of the geosynthetic reinforced slope	Located in the soil contact interface of the geosynthetic reinforced slope
30	0.201	0.201	0.101
27	0.136	0.133	0.111
24	0.093	0.092	0.105
21	0.104	0.104	0.112
18	0.132	0.107	0.109
15	0.12	0.112	0.097
12	0.11	0.11	0.087
9	0.098	0.094	0.097
6	0.098	0.104	0.125
2.5	0.098	0.087	0.124

4.2 Discussion on Seismic Response of Geosynthetic Reinforced Slope

The seismic analysis calculates the maximum acceleration value of the simulation results at each point and the maximum acceleration value of each point is plotted according to height shown in Fig. 9. Figure 9 shows that the maximum acceleration value would be reduced when the point moves into the interior part of the geosynthetic reinforced slope. The simulation results show that the maximum acceleration value is

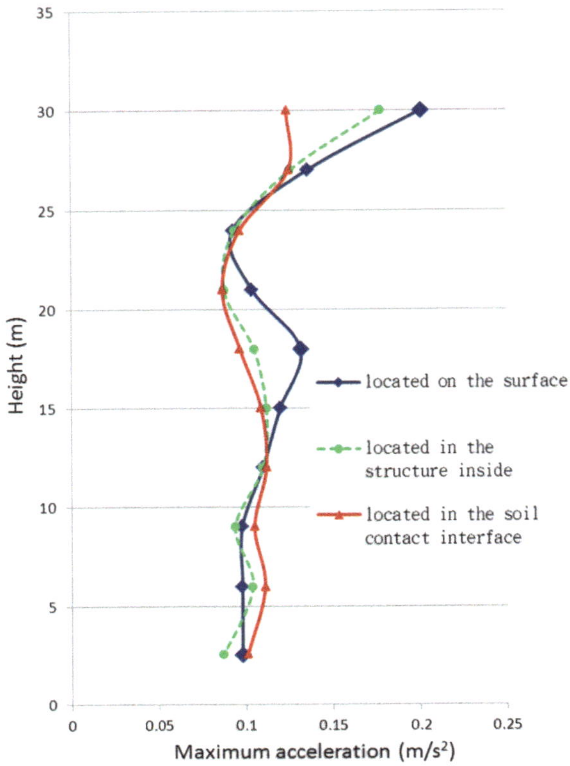

Fig. 9. Maximum acceleration value according to height

obvious on the slope surface during the earthquake. Because of the large degree of freedom on the geosynthetic reinforced slope platform, the maximum acceleration at the middle layer of the slope is not amplified immediately, but the acceleration value is gradually increased from the middle to the top of the slope. However, at the middle of the geosynthetic reinforced area and at the boundary of the soil, the seismic amplification reached the peak around overall height of 2/3. The seismic acceleration was significantly reduced compared with the maximum acceleration value on the surface.

The seismic magnification ratio at different place is plotted with height shown in Fig. 10. When the earthquake occurs, the acceleration in the soil contact interface of the geosynthetic would not show a significant amplification situation. Moreover, the simulation predicted the ratio of the maximum acceleration between the third and the first layer to be just 1.2. However, there are different maximum acceleration values in the structure inside and on the surface of the geosynthetic reinforced area, which have the almost same maximum acceleration between the third and the first layer to be 2.

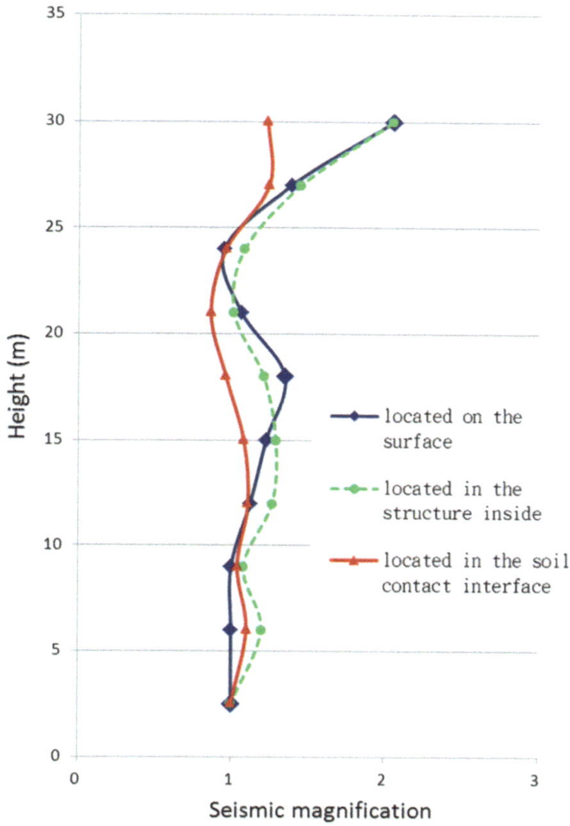

Fig. 10. Seismic magnification according to height

5 Conclusions

In this study, the monitoring system was set up to record the state of ground acceleration and displacement of various layers under actual seismic conditions. The earthquake data recorded at the demonstration site were used to validate a FEM analysis of Plaxis program.

The free-field boundary is used to simulate seismic behaviors in the Plaxis finite element program and compared well with the data obtained from the field monitoring. The simulation results demonstrated that the free-field boundary is quite appropriate for the seismic analysis of geosynthetic reinforced slope.

From the simulation results obtained from this study, it can also be seen that the damping parameter setting in the seismic analysis affects the simulation accuracy. The larger value of the damping ratio implement, the more reduction of acceleration value can be expected. When damping ratio set too large, anti-earthquake energy will be caused by improper absorption, which is not consistent with current monitoring data in this study.

The seismic simulation results show that the maximum acceleration value is obvious on the slope surface during the earthquake, and the amplification behavior reduces moving into the inner portion of the geosynthetic reinforced slope.

References

Chao, S.J: Finite element analysis of the I-Lan reinforced soil retaining walls. In: The Proceedings CD-ROM of the Sixth World Congress on Computational Mechanics in conjunction with the Second Asia-Pacific Congress on Computational Mechanics, Beijing, China, 2004

Chou, N.N.S., et al.: FEM Analysis of Staged Construction for a Reinforced Earth Wall. ASCE, Geotechnical Congress, Boulder, Colorado (1991)

Kao, C.H.: Seismic behavior monitoring and numerical simulation for the geosynthetic reinforced soil demonstration site. Master thesis, Department of Civil Engineering, National Ilan University (2014)

PLAXIS: Reference Manual, pp. 275–277 (2015)

Wu, R.H.: Study of appropriate seismic numerical simulation for the geosynthetic reinforced soil. Master thesis, Department of Civil Engineering, National Ilan University (2015)

Uncertainties of a Nuclear-Air-Blast Induced Ground Displacement Model

Shashank Pathak(✉) and G. V. Ramana

Civil Engineering, Indian Institute of Technology Delhi, New Delhi, India
shashankpathaks@gmail.com;
gvramanaiitdelhi@gmail.com

Abstract. Scenario based design of strategic underground structures is increasingly becoming important with increasing capacities of nuclear warheads. The ground displacement caused by nuclear-air-blast is an important design parameter for such structures. A simplified model to compute ground displacement, proposed earlier by authors, is recommended for preliminary designs in professional practice. This paper quantifies the uncertainties associated with this simplified model as well as its input parameters namely P-wave velocity, strain recovery ratio, velocity ratio, and constrained modulus, while estimating the nuclear blast induced ground displacement. The model uncertainty quantification is generally accomplished using sampling based Bayesian updating techniques. However, this is a computationally time-consuming process and not suitable with increasing number of model parameters. To circumvent this, an alternate simplified model uncertainty characterization approach is developed based on Taylor's series approximation. The validity of the proposed approach is established by comparing the results for a simple case of an earth pressure problem, routinely encountered in geotechnical engineering, against Bayesian approach. Subsequently, this simplified approach has been used to quantify the model uncertainties of air-blast induced ground displacement model and it is observed that this model is biased towards conservative side.

1 Introduction

Nuclear-air-blast induced ground displacement is an important parameter in the design of underground strategic structures (Brode 1964; Crawford et al. 1974). A pictorial representation of the loading caused by a nuclear blast at a height above ground level is shown in Fig. 1 and the point located on the ground surface vertically below the explosion centre is defined as Ground zero (GZ). Recently authors have proposed a simple model (Pathak and Ramana 2017) to estimate ground displacement at distances away from GZ caused by rapidly moving air-overpressure generated by nuclear explosion. This simplified model requires input parameters such as, peak overpressure, duration of loading, P-wave velocity, constrained modulus, velocity ratio, stress attenuation and strain recovery ratio. This model is based on simplifications and/or assumptions such as zero rise-time and linear decay of blast load time-history, bilinear behaviour of geomaterials, geometric stress attenuation and the effect of stress wave reflection in ground media is ignored. Like any other simplified model, this model would also have model uncertainties (Refsgaard

B. Fatahi et al. (eds.), *Sustainable Design and Construction for Geomaterials and Geostructures*, Sustainable Civil Infrastructures,
https://doi.org/10.1007/978-3-319-95753-1_15

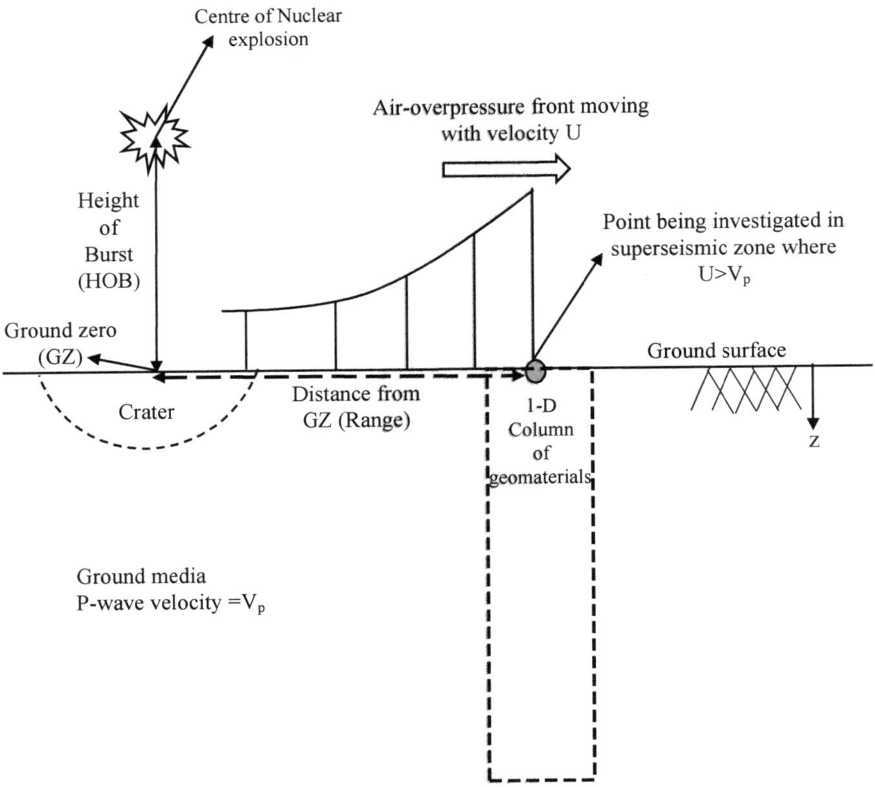

Fig. 1. Pictorial representation of the nuclear-blast-induced moving air-overpressure (shock wave) on ground surface

et al. 2007; Fenton and Neil 2012) and it is important to quantify these uncertainties to judge the credibility of model predictions.

There are two prominent sources that contribute to model uncertainties: (i) uncertainties of the input parameters (primarily of aleatoric nature) and (ii) uncertainties of the cause and effect relationship, i.e., of the model itself (primarily of epistemic nature) (Tang 1984; Phoon and Kulhawy 1999; Whitman 2000; Christian 2004; Honjo and Amatya 2005; Nadim 2007; Griffiths et al. 2011). Over the last several years, major efforts have been focused on the quantification of effect of input parameter uncertainties on the model predictions compared to the treatment of epistemic uncertainties. In geotechnical engineering, the model uncertainties have popularly been quantified as the ratio of model predictions to actual experimental observations (e.g., Sidi 1986). Different techniques such as statistical approaches (Tang 1989; Ronold and Bjerager 1992) and Bayesian updating approach (Kiureghian 1990) are adopted to analyze this ratio. However, there are two major complexities (Juang et al. 2004; Phoon and Kulhawy 2005; Zhang 2009; Zhang et al. 2009) associated with characterization of model uncertainty in geotechnical engineering: (i) insufficiency of experimental observations to compare with model predictions and (ii) the presence of uncertainties in input parameters and experimental observations. A Bayesian updating method can be used to characterize the model uncertainties that takes into account input and observation uncertainties using Markov Chain Monte Carlo (MCMC) simulation (Zhang 2009; Zhang et al. 2009, 2012). However, this approach is computationally very complex, time-consuming and not suitable for practicing engineers. In this paper, an alternate simplified model uncertainty characterization approach is developed based on Taylor's series approximation. The validity of the proposed approach is established by comparing the predicted model factor statistics for a simple case of an earth pressure problem against the results of Bayesian approach adopted from Zhang et al. (2012). Subsequently, the proposed model uncertainty characterization approach is then utilized to quantify the model uncertainty factor for Pathak-Ramana model.

2 Proposed Approach for Model Uncertainty Characterization

The model uncertainty is generally characterized (e.g., Ronold and Bjerager 1992) by evaluating the statistics of the factor β defined below:

$$\beta = \frac{Y_t}{f(\mu_{x_1}, \ldots, \mu_{x_n})} \tag{1}$$

where, Y_t is the true value of the experimental observation, $f()$ is the functional form of the model under investigation, $\{\mu_{x_1}, \ldots, \mu_{x_n}\}$ are the mean values of n model input parameters $\{x_1, \ldots, x_n\}$. If k number of observations are available for Y_t along with the corresponding input parameters, then the mean (μ_β), standard deviation (σ_β), and coefficient of variation (CV_β) of model uncertainty factor β are obtained as:

$$\mu_\beta = \frac{1}{k}\sum_{i=1}^{k} \beta_i \tag{2}$$

$$\sigma_\beta = \left[\frac{1}{k-1}\sum_{i=1}^{k} (\beta_i - \mu_\beta)^2\right]^{1/2} \tag{3}$$

$$CV_\beta = \frac{\sigma_\beta}{\mu_\beta} \tag{4}$$

where, β_i is the model uncertainty factor corresponding to the ith experimental observation. It should be noted that the above definition of β is based on mean values (or single representative value) of input parameters only and does not account for the uncertainty in input parameters.

If input parameters of a model are also uncertain, i.e., random variables with known statistical properties, the expected value of the model outcome $(E[f(x_1,\ldots,x_n)])$ can be approximated using truncated Taylor's series as below:

$$E[f(x_1,\ldots,x_n)] = f(\mu_{x_1},\ldots,\mu_{x_n}) + \frac{1}{2}\sum_{i=1}^{n}\sum_{j=1}^{n}\mathrm{cov}[x_i,x_j]\left.\frac{\partial^2 f(x_1,\ldots,x_j,\ldots,x_n)}{\partial x_i \partial x_j}\right|_{(\mu_{x_1},\ldots,\mu_{x_n})} \tag{5}$$

where, $\mathrm{cov}[x_i, x_j]$ is the co-variance of jth and ith input parameters and represents the uncertainties associated with these input parameters. If jth and ith input parameters are uncorrelated then Eq. (5) simplifies to

$$E[f(x_1,\ldots,x_n)] = f(\mu_{x_1},\ldots,\mu_{x_n}) + \frac{1}{2}\sum_{j=1}^{n}\mathrm{var}[x_j]\left.\frac{\partial^2 f(x_1,\ldots,x_j,\ldots,x_n)}{\partial x_j^2}\right|_{(\mu_{x_1},\ldots,\mu_{x_n})} \tag{5a}$$

where, $\mathrm{var}[x_j]$ is the variance of jth input parameter x_j. In geotechnical practice, coefficient of variation (CV_{x_j}) is used instead of $\mathrm{var}[x_j]$ to represent the uncertainty of the parameter. These two statistical parameters are related to each other as given below:

$$\mathrm{var}[x_j] = (\mu_{x_j} \times CV_{x_j})^2 \tag{6}$$

where, μ_{x_j} is mean value of the jth input parameter x_j. Therefore, a more general expression for model uncertainty factor β is proposed as below:

$$\beta = \frac{Y_t}{E[f(x_1,\ldots,x_n)]} \tag{7}$$

It should be noted that the first order approximation of Eq. (5) substituted in Eq. (7) leads to Eq. (1).

It is important to note that if there is a set of large number of random realizations of input parameters $\{x_1,\ldots,x_n\}$ then the corresponding outcomes $\{f(x_1,\ldots,x_n)\}$ to each realization and the expected value of the model outcome $E[f(x_1,\ldots,x_n)]$ can be obtained. This expected model outcome when multiplied with μ_β can be shown to be equal to the expected true outcome as below:

$$\begin{aligned}\mu_\beta \times E[f(x_1,\ldots,x_n)] &= E\left[\frac{Y_t}{E[f(x_1,\ldots,x_n)]}\right] \times E[f(x_1,\ldots,x_n)] \\ &= \frac{E[Y_t]}{E[f(x_1,\ldots,x_n)]} \times E[f(x_1,\ldots,x_n)] = E[Y_t]\end{aligned}$$

In this way the proposed model factor can also be used to treat the model uncertainties.

3 Comparison with Bayesian Approach

To compare the model factor statistics obtained from the proposed approach with Bayesian approach, the examples considered by Zhang et al. (2012) are utilized in this study. The models of Jacky (1944) (Eq. 8) and Wenklow (1979) (Eq. 9) for predicting the at-rest earth pressure coefficient k_0 of normally consolidated soils based on the friction-angle ϕ are investigated for their model uncertainty.

$$k_0 = 1 - \sin\phi \tag{8}$$

$$k_0 = \tan(45^\circ - \phi/2) \tag{9}$$

The uncertainty of the input parameter is represented as coefficient of variation of ϕ, i.e., CV_ϕ. The true values of k_0 along with respective ϕ values are shown in Fig. 2 for 25 laboratory tests on at-rest pressure coefficient originally compiled by Burland and Federico (1999) and adopted here from Zhang et al. (2012). The model uncertainty characterization is carried out for the six cases of input parameter uncertainty, i.e., $CV_\phi = 0, 0.05, 0.10, 0.15, 0.20, 0.30$. The variation in statistics (μ_β, CV_β) of model uncertainty factors corresponding to Jacky's model and Wenklow's model are shown in Fig. 3a–d, respectively, using the proposed approach and the results of Bayesian approach (adopted from Zhang et al. 2012). It is observed that the predicted μ_β using the proposed approach deviates marginally from the μ_β predicted using the Bayesian

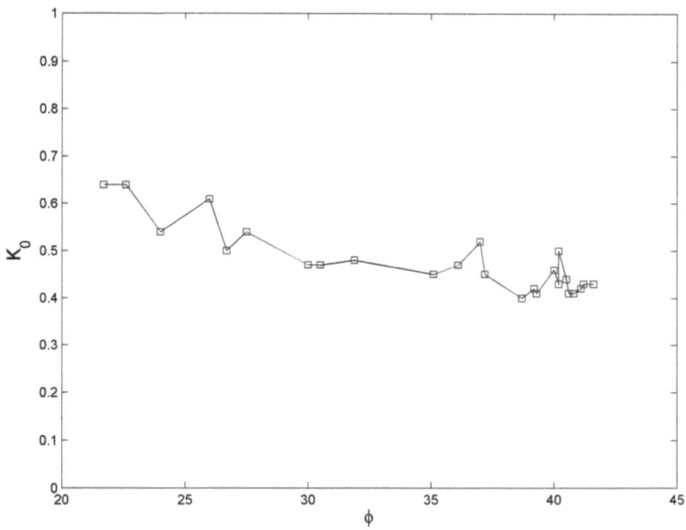

Fig. 2. Measured earth pressure coefficient at rest (adopted from Burland and Federico 1999)

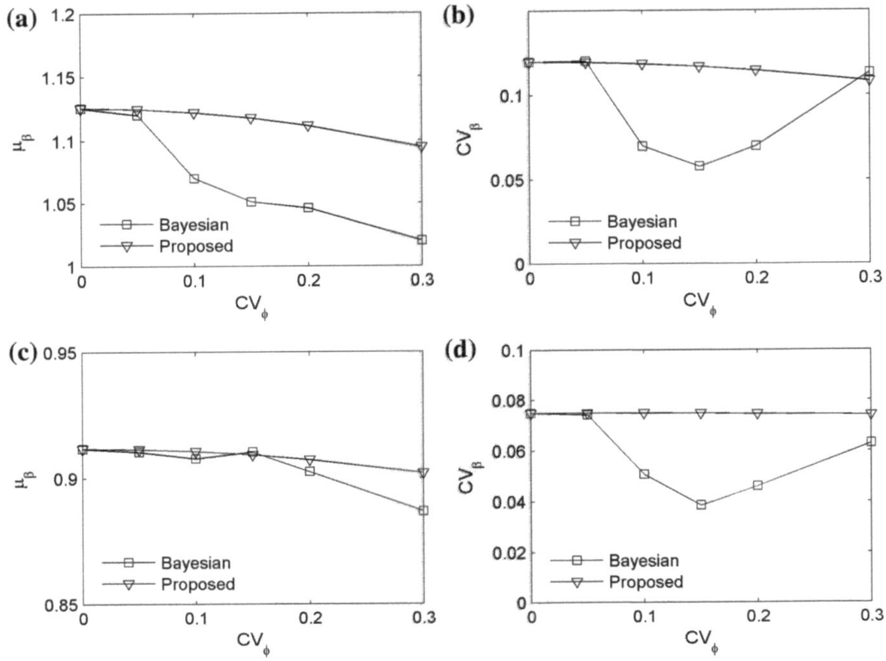

Fig. 3. Variation of μ_β (**a**) and CV_β (**b**) for Jacky's model and μ_β (**c**) and CV_β (**d**) for Wenklow's model with input uncertainty (CV_ϕ)

approach (by not more than 7%). However, the proposed method predicts higher CV_β compared to the Bayesian approach.

Furthermore, to demonstrate the effect of model uncertainty in design applications, an example of a retaining wall (H = 6 m high) that retains dry sand ($\phi = 30°$ and γ = 1620 kg/m^3) is considered. Assuming the condition of earth pressure at-rest, the thrust per meter run of the wall (F) can be written as,

$$F = \frac{1}{2} k_0 \gamma H^2 \tag{10}$$

where, k_0 is earth pressure coefficient at rest and can be evaluated as a function of ϕ using any of the previously mentioned models. Thus, the design parameter F depends upon two geotechnical parameters ϕ and γ. To explicitly highlight the effect of the model uncertainty associated with k_0 only, ϕ is assumed to be a log-normally distributed random variable with a mean value of 30° and coefficient of variation of 0.1 (Fig. 4a), whereas, uncertainty of γ and correlation between ϕ and γ is ignored. This uncertainty of ϕ would be propagated to k_0 and then to F (Eq. 10). The model uncertainties (including uncertainties of parameter ϕ) in k_0 and F can be accounted for by multiplying factor β. From Fig. 3a and b, statistics of the log-normally distributed model factor β for $CV_\Phi = 0.10$ is obtained as $\mu_\beta = 1.12$ and $CV_\beta = 0.12$ (using proposed method) and $\mu_\beta = 1.07$ and $CV_\beta = 0.07$ (using Bayesian approach) in case of Jacky's model of k_0. The probability distribution of factor β is shown in Fig. 4b, for Bayesian and proposed approaches. Using the Monte-Carlo simulation (MCS) technique, the distribution of k_0 (from Jacky's model) and F are also obtained as shown in Figs. 4c and d, respectively, with and without applying the model uncertainty factor. From Fig. 4d, the uncorrected expected mean and standard deviation of F are obtained as 146 and 13 kN/m, respectively. As these values do not account for model uncertainty associated with Jacky's model of k_0, henceforth, referred as 'uncorrected'. The corrected k_0, i.e., βk_0 and corrected F, i.e., βF for model uncertainty in k_0 can be represented by their respective probability distributions as shown in Fig. 4c and d for Bayesian and the proposed approach. The expected mean and standard deviation of F are obtained as 156 and 11 kN/m, respectively, after using Bayesian model uncertainty factor and 163 and 20 kN/m, respectively, after using proposed model uncertainty factor. It indicates that ignoring the model uncertainty factor may underestimate

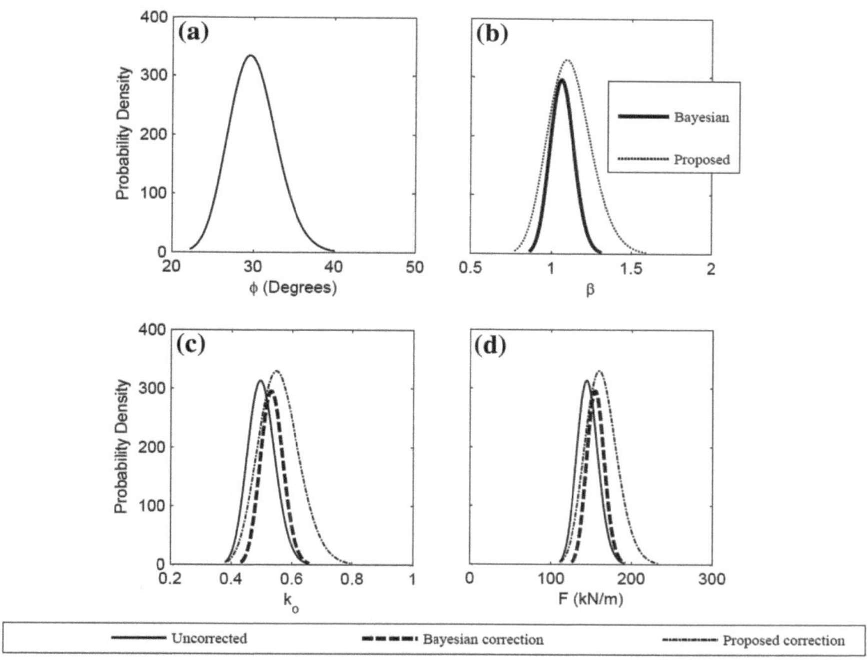

Fig. 4. Probability densities of: **a** ϕ, **b** β (Bayesian and proposed), **c** k_0 and **d** F (with and without treatment of model uncertainties)

the design forces and the proposed method for accounting model uncertainty is slightly conservative (by 4.5%) compared to the Bayesian approach.

4 Problem of Air-Blast Induced Ground Displacement

From the above demonstration, it is clear that the newly proposed simplified method of representing model uncertainty can be used to judge the credibility of any model with uncertain input parameters. In view of this, the proposed method is applied to the Pathak-Ramana model (Eq. 11) used for estimation of induced ground displacement due to nuclear air-blast.

$$S(t) = \frac{P_0 t_p V_p}{2fM_L}\left[\frac{2fL_W}{t_p V_p}\left(1 - \frac{rt}{t_p} - \frac{rfL_W}{t_p V_p}\right)\ln\left(1 + \frac{V_p t}{fL_W}\right) + \frac{2frtL_W}{t_p^2 V_p} + \frac{t(f-1)}{t_p\left(1 + V_p t/fL_W\right)}\right] \quad (11)$$

where, $S(t)$, P_0, t_p, V_p, M_L, f, L_W, r are the ground displacement at time-instant t, peak overpressure, duration of the air-overpressure time-history (or blast loading), P-wave velocity of ground, constrained modulus of geomaterials, ratio of the P-wave velocity to the wave propagation velocity of peak stress wave, i.e., velocity ratio, stress attenuation parameter, and ratio of the recovered strain (upon unloading) to the peak strain (upon loading), i.e., strain recovery ratio, respectively. For the purpose of calibration of model uncertainty factor, experimental data (Perret 1960) of an atmospheric nuclear test (Shot Priscilla, 37 kt yield at height of burst of 213 m) conducted at Frenchman Flat (Nevada, USA) is used.

5 Model Parameters and Associated Uncertainties

Pathak-Ramana model (2017) uses an equivalent linearly decaying blast load of the energy content or impulse (defined as the area under overpressure time-history) equal to the experimentally recorded impulse. Ignoring the finite rise-time and nonlinear decay of the blast load are the first source of model uncertainty. The actual air-overpressure recorded at a distance of 198 m from GZ and the idealized linearly decaying equivalent loading are shown in Fig. 5. The recorded peak positive overpressure and impulse are

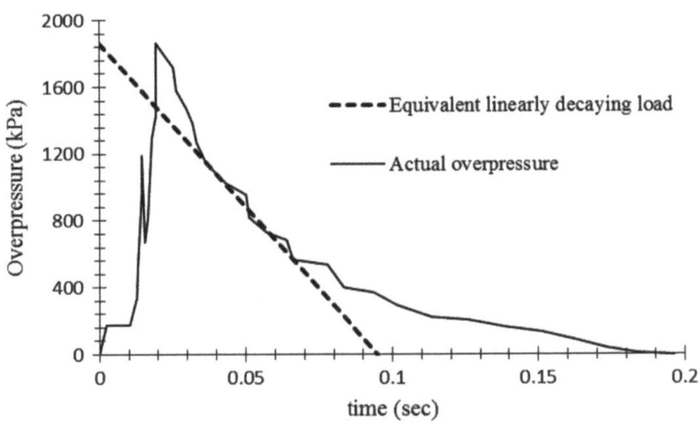

Fig. 5. Equivalent linearly decaying blast load (Pathak and Ramana 2017) and actual overpressure time-history (Perret 1960)

1863 kPa and 88.32 kPa-s, respectively. The duration of the equivalent linearly decaying overpressure is $2 \times 88.32/1863 = 0.095$ s. Further, the uncertainty associated with blast load parameters (P_0 and t_p) itself are not taken into account.

The nuclear test site consists of Frenchman Flat silts and the measured V_p varied between 306.75 and 952.57 m/s over a depth of about 60 m (Fig. 6). The representative P-wave velocity is estimated as 721.53 m/s (Pathak and Ramana 2017). The standard deviation of V_p is estimated using six-σ rule (see Baecher and Christian 2005) as (952.57−306.75)/6 = 107.64 m/s and hence, the coefficient of variation of V_p is 0.15.

The value of attenuation parameter L_W is adopted as 14.22 m (Pathak and Ramana 2017). The attenuation parameter is evaluated as a function of peak overpressure and yield of explosion (Newmark 1964). As the uncertainties associated with blast parameters are not considered, the parameter L_W is deterministic. It should be noted that even though L_W is assigned a representative value with zero coefficient of variation, the uncertainties associated with the model (Newmark 1964) used to estimate L_W will be propagated to the Pathak-Raman model.

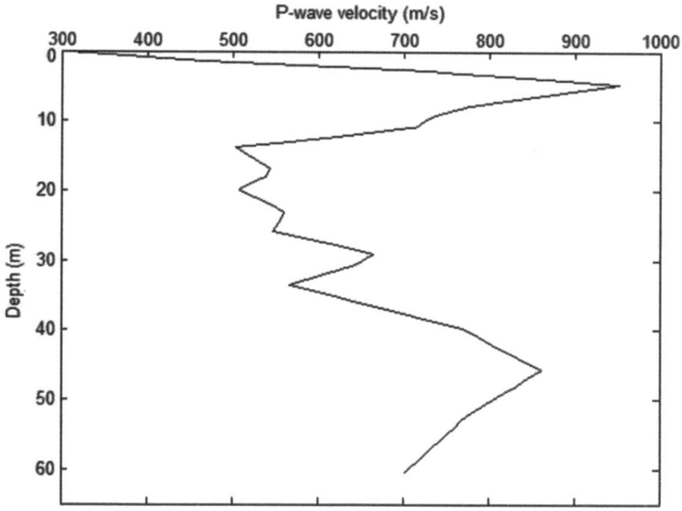

Fig. 6. Variation of P-wave velocity at Frenchman Flat nuclear test site (Perret 1960)

Based on the typical ranges of *r* and *f* as 0.3–1.0 and 1.5–2.0, respectively (Wilson and Sibley 1962), the values of *r* and *f* are adopted as 0.6 and 2, respectively by Pathak and Ramana (2017) for Frenchman Flat silts. Using six-σ rule, the coefficients of variation of *r* and *f* can also obtained as 0.19 and 0.08, respectively.

Reported values of constrained modulus for Frenchman Flat silts based on various methods such as, triaxial tests, dynamic 1-D compression test, deduced from observed ground motion, calculated from seismic velocity, and resonant column test can be found in Whitman (1970). Pathak and Ramana (2017) judiciously selected the representative constrained modulus and modelled the ground as a double layered media with a representative constrained modulus of 69 MPa (M_{L1}) for depths lesser than 18 m and 759 MPa (M_{L2}) for depths more than 18 m. This engineering judgment involved in the selection of moduli value is another source of the epistemic uncertainty. The parametric

Table 1. Mean values and coefficient of variation of parameters for Pathak and Ramana (2017) model at Frenchman Flat nuclear test site

X	μ_X	CV_X
P_0	1863 kPa	0
t_p	0.095 s	0
V_p	721.53 m/s	0.15
L_W	14.22 m	0
r	0.6	0.19
f	2	0.08
M_{L1}	69 MPa	0.10
M_{L2}	759 MPa	0.10

uncertainty in constrained moduli are caused by the measurement errors of these test methods. A coefficient of variation of M_{L1} and M_{L2} is adopted as 0.10 to examine the influence on model uncertainty. The list of model parameters along with their respective mean values and coefficient of variation are summarized in Table 1.

6 Model Uncertainty Factor

To obtain the statistics of the model uncertainty factor, the experimental displacement time-history (at a distance of 198 m from GZ) measured during an actual nuclear-air-blast at Frenchman Flat are compared with those obtained by Pathak-Ramana model. To obtain the model estimate of the ground displacement time-history, the original form of the model (Eq. 11) is further expanded (Pathak and Ramana 2017) to accommodate the two layers of different constrained moduli such that Eq. (12a) is applicable up to t = 0.025 s, Eq. (12b) is applicable from t = 0.025 s to t = 0.050 s and Eq. (12c) is applicable beyond t = 0.050 s, in this particular case.

$$S(t) = \frac{P_0 t_p V_p}{2fM_{L1}}\left[\frac{2fL_W}{t_p V_p}\left(1 - \frac{rt}{t_p} - \frac{rfL_W}{t_p V_p}\right)\ln\left(1 + \frac{V_p t}{fL_W}\right) + \frac{2frtL_W}{t_p^2 V_p} + \frac{t(f-1)}{t_p\left(1 + V_p t/fL_W\right)}\right] \tag{12a}$$

$$S(t) = \frac{P_{0}t_pV_p}{2fM_{L1}}\left[\begin{array}{l}\frac{2fL_W}{t_pV_p}\left(1-\frac{rt}{t_p}-\frac{rfL_W}{t_pV_p}\right)\ln\left(1+\frac{V_pt}{fL_W}\right)+\frac{2frtL_W}{t_p^2V_p}+\\ \frac{t}{t_p(1+V_pt/fL_W)(f-1)}\left(\left(\frac{18f}{V_pt}-1\right)\left(2f-\frac{18f}{V_pt}-1\right)+\frac{M_{L1}}{M_{L2}}\left(f-\frac{18f}{V_pt}\right)^2\right)\end{array}\right] \tag{12b}$$

$$S(t) = \frac{P_{0}t_pV_p}{2fM_{L1}}\left[\begin{array}{l}\frac{2fL_W}{t_pV_p}\left(1-\frac{rt}{t_p}-\frac{rfL_W}{t_pV_p}\right)\ln\left(1+\frac{18}{L_W}\right)+\frac{18\times 2f^2rL_W}{t_p^2V_p^2}\\ +\frac{M_{L1}}{M_{L2}}\left(\frac{2fL_W}{t_pV_p}\left(1-\frac{rt}{t_p}-\frac{rfL_W}{t_pV_p}\right)\ln\left(\frac{L_W+V_pt/f}{L_W+18}\right)\right.\\ \left.+\frac{2(V_pt/f-18)f^2rL_W}{t_p^2V_p^2}+\frac{t(f-1)}{t_p(1+V_pt/fL_W)}\right)\end{array}\right] \tag{12c}$$

Using Eqs. (12a)–(12c), estimated displacement time-histories (using the mean values of model parameters) are plotted along with the experimental data as shown in Fig. 7. In Fig. 7, time is normalized with t_p as positive phase durations of actual and equivalent overpressure time-histories are different.

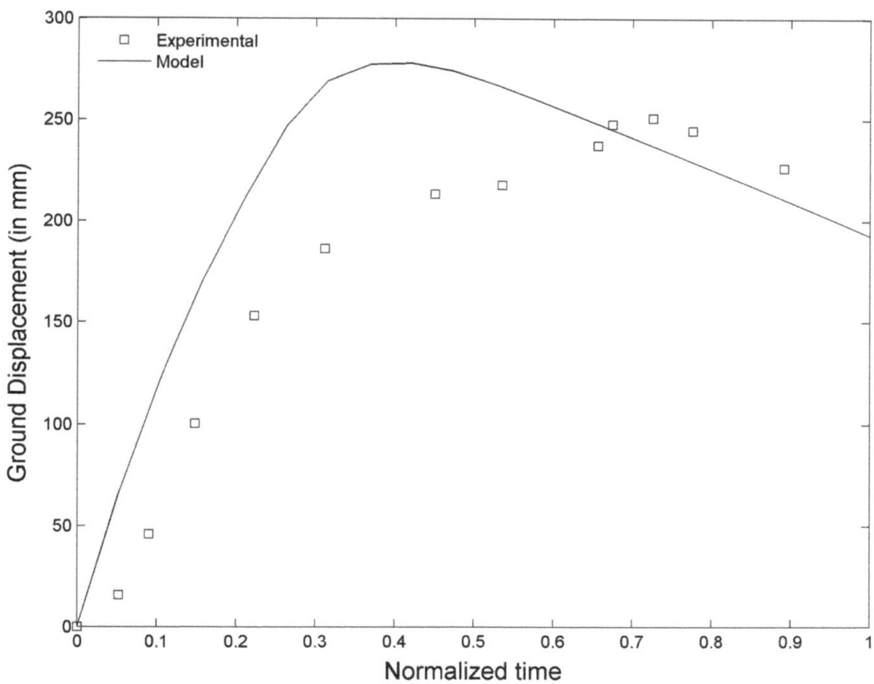

Fig. 7. Nuclear-air-blast induced ground displacement time-history as predicted by model and actual experimental records [adopted from Perret (1960)]

First the expected value of ground displacement at a time-instant 't' is evaluated using Eq. (5a). As the model is in terms of three piece-wise closed-form solutions (Eqs. 12a–12c), the analytical evaluation of partial double derivative (in Eq. 5a) is difficult. Therefore, a simple numerical technique such as finite difference method can be used to obtain the partial derivatives (Eq. 13),

$$\left.\frac{\partial^2 f(x_1,..,x_j,..,x_n)}{\partial x_j^2}\right|_{(\mu_{x_1},..,\mu_{x_n})} = \frac{f(\mu_{x_1},..,\mu_{x_j}-h,..,\mu_{x_n}) - 2f(\mu_{x_1},..,\mu_{x_j},..,\mu_{x_n}) + f(\mu_{x_1},..,\mu_{x_j}+h,..,\mu_{x_n})}{h^2} \tag{13}$$

where, $\left.\partial^2 f \middle/ \partial x_j^2\right|_{(\mu_{x_1},..,\mu_{x_n})}$ is the partial double derivative corresponding to the jth model parameter evaluated at the mean values of model parameters and h is the step size of the finite difference method. Substituting the mean values and coefficient of variations of model parameters (Table 1) sequentially in Eqs. (13), (5a) (assuming uncorrelated parameters), and (7), provides the model uncertainty factors for the available experimental records. The model uncertainty factors corresponding to the experimental records are listed in Table 2 and thus, μ_β and CV_β are found to be 0.78 and 0.34,

Table 2. Model uncertainty factors corresponding to measured records

Normalized time	Observed ground displacement (mm)	Model uncertainty factor β
0.05	15.75	0.23
0.09	45.97	0.42
0.15	100.33	0.61
0.22	153.16	0.69
0.31	186.44	0.70
0.45	213.36	0.77
0.54	217.93	0.81
0.66	237.49	0.94
0.67	247.90	1.00
0.73	250.95	1.04
0.78	244.86	1.05
0.89	226.57	1.06
Mean model uncertainty factor, μ_β		0.78
Coefficient of variation of model uncertainty factor, CV_β		0.34

respectively. As the mean model factor is less than unity, the Pathak-Ramana model overestimates the air-blast induced ground displacement on an average. In other words, the Pathak-Ramana model is biased on the conservative side. It is worth noting that the estimates of the model uncertainty factor in this case correspond to a particular test site and therefore, the estimates of μ_β and CV_β can be regarded as site-specific model uncertainty factor statistics. To obtain a more general statistics of model uncertainty factor, test data from different types of geologies and blast yield are required.

7 Conclusions

Model uncertainties are generally associated with any engineering model and need to be quantified to judge the credibility of the model. In this paper, a simple and intuitive approach is proposed to characterize the model uncertainty in cases where input parameters of model are also uncertain. The proposed approach is compared with the computationally advanced Bayesian approach for the two models related with the estimation of static earth pressure coefficients in terms of friction angle of soil. It is found that the proposed method is in close agreement with the Bayesian approach (variation less than 7%) for prediction of mean model factors. The proposed model uncertainty characterization approach is then applied to the Pathak-Ramana model that estimates the nuclear-air-blast induced ground displacement. A site-specific statistics of model factor for Pathak-Ramana model is obtained with mean model factor of 0.78 and coefficient of variation of 0.34 which indicates that the model is slightly biased on conservative side.

References

Baecher, G.B., Christian, J.T.: Reliability and Statistics in Geotechnical Engineering, Chapter 6. John Wiley & Sons (2005)

Brode, H.L.: A Review of Nuclear Explosion Phenomena Pertinent to Protective Construction, Vol. 425, No. PR. Rand Corporation, Santa Monica, CA (1964)

Burland, J., Federico, A.: At-rest earth pressure coefficient K0(NC) from Cam Clay model. In: Hong, S.W., Kim, M.M., Yang, G.S., Lee, S.R., Chung, S.S., Ihm, C.C., Kim, H.T., Park, J. B. (eds.) Proceedings of the 11th Asian Regional Conference on Soil Mechanics and Geotechnical Engineering, pp. 15–18. Balkema, Rotterdam (1999)

Christian, J.T.: Geotechnical engineering reliability: how well do we know what we are doing? J. Geotech. Geoenviron. Eng. **130**(10), 985–1003 (2004)

Crawford, R.E., Higgins, C.J., Bultmann, E.H.: The Air Force Manual for Design and Analysis of Hardened Structures. Final report Jun 71-May 74 (No. AD-B-004152/5). Civil Nuclear Systems Corp., Albuquerque, NM (USA) (1974)

Fenton, N., Neil, M.: Risk Assessment and Decision Analysis with Bayesian Networks. CRC Press (2012)

Griffiths, D.V., Huang, J., Fenton, G.A.: Probabilistic infinite slope analysis. Comput. Geotech. **38**(4), 577–584 (2011)

Honjo, Y., Amatya, S.: Partial factors calibration based on reliability analyses for square footings on granular soils. Géotechnique **55**(6), 479–491 (2005)

Jacky, J.: The coefficient of earth pressure at rest. J. Soc. Hung. Archit. Eng. **7**, 355–358 (1944)

Juang, C.H., Yang, S.H., Yuan, H., Khor, E.H.: Characterisation of the uncertainty of Robertson and Wride model for liquefaction potential evaluation. Soil Dyn. Earthq. Eng. **24**(5), 771–780 (2004)

Kiureghian, A.D.: Bayesian analysis of model uncertainty in structural reliability. In: Proceedings of Third IFIP Working Conference on Reliability and Optimization of Structural Systems. Springer-Verlag, New York, NY (1990)

Nadim, F.: Tools and strategies for dealing with uncertainty in geotechnics. In: Griffiths, D.V., Fenton, G.A. (eds.) Probabilistic Methods in Geotechnical Engineering, pp. 71–96. Springer Wien, New York (2007)

Newmark, N.M.: The basis of current criteria for the design of underground protective construction, opening address-session I. In: Proceedings of the Symposium on Soil-Structure Interaction, University of Arizona, Tucson, pp. 1–24 (1964)

Pathak, S., Ramana, G.V.: Air-blast induced ground displacement. Procedia Eng. (11th International Symposium on Plasticity and Impact Mechanics, 2016) **173**, 555–562 (2017)

Perret, W.R.: Ground motion studies at high incident overpressure. Operation PLUMBBOB-Project 1.5, WT-1405 for DASA (1960)

Phoon, K.K., Kulhawy, F.H.: Characterisation of geotechnical variability. Can. Geotech. J. **36**, 612–624 (1999)

Phoon, K.K., Kulhawy, F.H.: Characterisation of model uncertainties for laterally loaded rigid drilled shafts. Géotechnique **55**(1), 45–54 (2005)

Refsgaard, J.C., Van der Sluijs, J.P., Højberg, A.L., Vanrolleghem, P.A.: Uncertainty in the environmental modelling process—a framework and guidance. Environ. Model. Softw. **22**, 1543–1556 (2007)

Ronold, K.O., Bjerager, P.: Model uncertainty representation in geotechnical reliability analyses. J. Geotech. Eng. **118**(3), 363–376 (1992)

Sidi, I.D.: Probabilistic prediction of friction pile capacities. PhD thesis, University of Illinois, Urbana-Champaign (1986)

Tang, W.H.: Principles of probabilistic characterization of soil properties. In: Bowles, D.S., Ko, H.Y. (eds.) Probabilistic Characterization of Soil Properties, Bridge Between Theory and Practice, pp. 74–89. ASCE, New York (1984)

Tang, W.H.: Uncertainties in offshore axial pile capacity. In: Kulhawy, F. (ed.) Foundation Engineering: Current Principles and Practices, vol. II, pp. 833–847. ASCE, New York, NY (1989)

Wenklow, W.: An apparatus for the evaluation of the coefficient of earth pressure at rest. In: Design Parameters in Geotechnical Engineering, Proceedings of the 7th European Conference on Soil Mechanics and Foundation Engineering. British Geotechnical Society, London, pp. 171–174 (1979)

Whitman, R.V.: Organizing and evaluating uncertainty in geotechnical engineering. J. Geotech. Geoenviron. Eng. **125**(6), 583–593 (2000)

Whitman, R.V. The response of soils to dynamic loadings. Report 26, Contract No. DA-22-079-eng-224, MIT, USA (1970)

Wilson, S.D., Sibley, E.A.: Ground displacements from air-blast loading. Soil Mech. Found. Div., ASCE **88** (SM-6), 1–31 (1962)

Zhang, J. Characterizing geotechnical model uncertainty. PhD thesis, Hong Kong University of Science and Technology, Hong Kong (2009)

Zhang, J., Zhang, L.M., Tang, W.H.: Bayesian framework for characterizing geotechnical model uncertainty. J Geotech. Geoenviron. Eng. **135**(7), 932–940 (2009)

Zhang, J., Tang, W.H., Zhang, L.M., Huang, H.W.: Characterising geotechnical model uncertainty by hybrid Markov Chain Monte Carlo simulation. Comput. Geotech. **43**, 26–36 (2012)

Development of Mobile Application to Assess and Enhance Earthquake Preparedness Level of Individuals and Community in India

S. Kolathayar(✉), V. Karan Kumar, V. R. Rohith, K. Priyatham, S. Nikil, and S. A. Anupa

Department of Civil Engineering, Amrita School of Engineering Coimbatore, Amrita Vishwa Vidyapeetham, Coimbatore, India
sreevalsakolathayar@gmail.com

Abstract. Earthquake is one of the natural hazards that affected the mankind the most with large scale destruction and loss of lives. According to IS 1893:2002, 60% Indian landmass (Zone V = 12%, Zone IV = 18%, Zone III = 26% and Zone II 44%) is vulnerable to earthquakes and several studies have shown that highly populated states of North and North Eastern India, belong to seismically active regions in the world. There is a need to bring the spirit of togetherness and alertness well before a disaster so that a community can be prepared well to face the disaster and thus reduce the loss of life and properties. This involves building a community that is aware of the earthquake hazards and prepared for it. There is a dire need for effective earthquake preparedness and an assessment tool that can assist individuals and communities to prepare, respond and recover from earthquakes. With the latest technological developments, the effective way of communicating the information required for improving the preparedness levels of individuals towards earthquakes is through mobile based applications. Hence development of a mobile APP is need of the hour. This study focuses on developing java based application which serves as an assessment tool to measure the preparedness level of an individual and community. The information required to be prepared for an earthquake such as latest information about earthquakes, actions to be taken during an earthquake, seismic zones in India are also provided through this tool. This paper also presents a preliminary case study of applying the developed app among the community from south Indian city of Coimbatore to understand the status of earthquake preparedness and awareness among the community.

1 Introduction

Earthquake has proved to be one of the most devastating natural disasters in a lot of cases. Millions of quakes are happening around the world each year, some of which may cause enormous personnel casualties and economic loss and even lead to a devastating blow in some areas. Even though it remains impossible to predict when and where an earthquake will occur, its impact can be reduced by taking several personal, community and building safety measures. This is where preparedness plays a major role. Earthquake preparedness consists of survival measures, preparation that will

B. Fatahi et al. (eds.), *Sustainable Design and Construction for Geomaterials and Geostructures*, Sustainable Civil Infrastructures,
https://doi.org/10.1007/978-3-319-95753-1_16

improve the chances of survival during the earthquake. Therefore it is vital to measure the preparedness level of individuals towards earthquakes. These measurements can be done through a suitable scale or an index. The creation and use of an "index" has been a popular methodology for evaluating relative levels of some state of being, whether economic health, quality of life, or something similar.

Indices are attractive because of their ability to summarize a considerable amount of technical information in a way that is easy for lay persons to understand (Davidson and Lambert 2001). Some indexes, such as the United Nations Human Development Index (HDI), do not provide the entire picture, but they are more beneficial than just examining one indicator such as income because they will incorporate into the index other variables, such as education and health (United Nations 2005). By examining multiple indicators, the level of bias can be reduced and the picture that is provided by investigating just one indicator is much clearer, providing a wider perspective (Cobb and Rixford 1998). Thus there is a need to develop a fast and efficient tool which involves in measuring the preparedness index of individuals during an earthquake. The indices which has been used in this study are termed as 'Earthquake Readiness Indices' (Kolathayar et al. 2017).

Figure 1 shows a graph which depicts the statistical details of the increase in the number of smart phone users in India. This shows that smart phone applications are one of the most efficient ways to communicate any kind of information to the people. As we are in need of an efficient mode of communication of Earthquake relevant information, a java based mobile application has been developed. A combination of information technology, community based disaster preparedness and collaborative relationships between community-based organisations, government and Non-Government

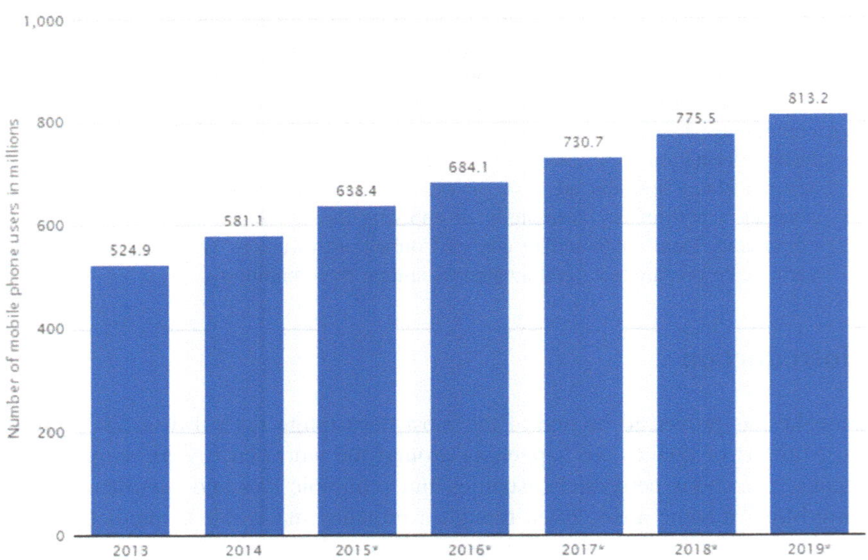

Fig. 1. Number of mobile phone users in India from 2013 to 2019 (in millions) (Ref: statista.com)

Organizations and can strengthen disaster awareness level to great extent (Kangabam et al. 2012). In this context, this study aims to develop an android based application with below mentioned specific objectives

- To disseminate latest knowledge on seismicity of India with digitized seismic zoning maps and links to important seismic design codes
- To give information on safety and preparedness measures to face earthquake using infographics
- To present an user friendly tool to evaluate earthquake preparedness level of individual and community
- To use the data collected from the surveys for research purposes to assess the preparedness level of community as a whole. This data can also be used by the government to streamline disaster mitigation plans.

The Sendai Framework for Disaster Risk Reduction (2015–2030), adopted at the third United Nations World Conference in Sendai, Japan, on March 18, 2015, highlights the need for disaster preparedness as one of its four priorities for global risk reduction. Past studies shows that disaster preparedness is vital for effective response and to "Build Back Better" in recovery, rehabilitation and reconstruction in regard to the tremendous growth in population and assets exposure. Empowering women and promoting the weaker sections of the society form the key approach to attain preparedness at national and community levels. It elucidates the various measured to be followed at national and local levels, and global and regional levels to achieve the expected goal. The present work is a step taken towards earthquake risk reduction in line with UN Sendai framework.

2 The Purpose of Mobile Apps

The purpose of mobile apps developed for disaster risk reduction can be different for different apps based on the context for which they were developed. Tan et al. (2017) carried out a detailed study reviewing the existing mobile apps developed for disaster prevention. The purpose of the majority of apps is to facilitate crowdsourcing during disaster events with an aim to organise crowdsourcing processes. The crowdsourcing process starts with a call for specific actions where participants are requested to perform tasks to tag the relevance of tweets and images of the disaster. These apps can be used for post-disaster damage assessment and pre-disaster risk assessment. Purpose for some disaster apps is to act as a collaborating platform that provide an open avenue for communities to work together during disaster events. Apps are used as platforms to nurture self-help communities helping each other like where donation seekers can link-up with suppliers. Providing alerts and information is another common purpose for a built-for-disaster purpose app. The objective of such apps is to disseminate authentic information. Here, the primary interaction is one-way; usually originating from the authorities to the public. Few apps serve the purpose of gathering information. These apps collate publicly available social media data, without necessarily engaging the public, and then repackage the information for easier consumption. Another purpose is more localised to the user networks where one can notify others of their situation during a disaster.

3 Methodology

3.1 Selection of Contents of the Application

The contents of the application were decided after deciding the appropriate deliverables of the application to the public. After various group discussions, the contents of the applications were shortlisted to: Basic Information, Do's and Don'ts, Seismic map of India and Survey-Questionnaire. This format of the application also ensures the delivery of necessary information about Earthquakes to the public.

3.2 Layout of the Application

In order to create a user-friendly application, the layout was decided as shown in Fig. 2, according to the function of the application.

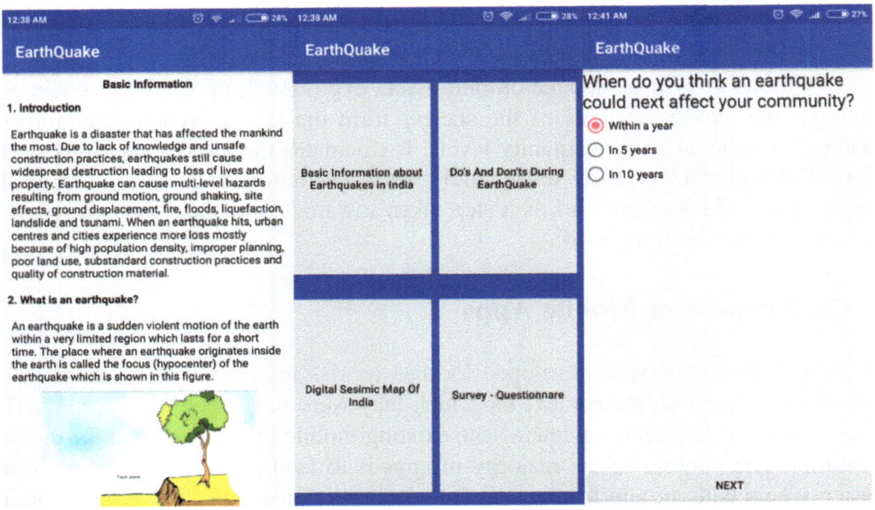

Fig. 2. Snapshot of 'BhookampRaksha' App

3.3 Coding and Running of the Application

The application was decided to be compatible to Android versions of Jelly Bean and above. This application is XML and Java-based. Thus it was programmed using Android Studio. All the pages of the application were designed as separate layouts using XML and they are connected using java programs. The response from the surveys are stored in a third-party server so that the response from any mobile phone around the world in this application can be viewed by the administrator. Test run of different layouts and the whole application together is run by Android Viewer which comes along with Android Studio.

The Android development tools incorporated are listed below.

- Product name: Android studio 2.2.1
- eXtended Markup Language (XML) for developing User Interface
- JAVA language for developing Android structure and functions
- Virtual Android emulator for debugging and running the app.

3.4 Testing of the Application

The application was shared to a few people and their survey responses were collected. After a few upgrades, the final version of the application is to be shared through cloud storages and some websites to the public.

4 Features of the Application

The mobile based application has been coined with the name '**BhookampRakhsha**' which literally means 'Safety from earthquakes'. The snapshot of the developed mobile application is shown in Fig. 2. The app has been provided with a single login page through which both the common users and the administrator can login. The admin login page has been provided in order to view and collect the survey results of the users who attend the survey. Figure 3 shows the contents of the app in the form of a framework.

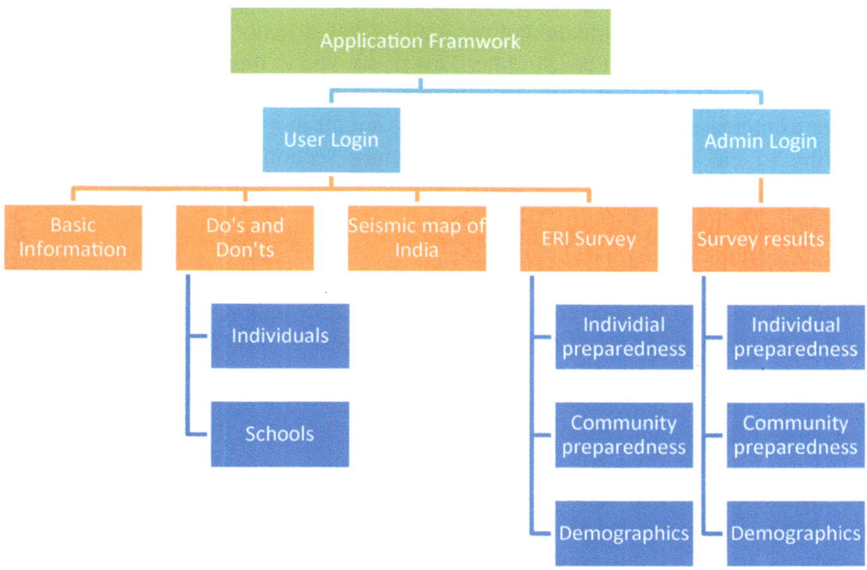

Fig. 3. Framework of the application

The application contains the Basic information of earthquake which has some content regarding what an earthquake is, causes of earthquakes, ways to measure the impact of earthquakes and so on. The application also shows basic Do's and Don'ts which involves measures to be taken to minimise the effects of an earthquake. These measures are classified as actions to be taken as an individual as well as in a community place like schools. The individual preparation measures include actions to be taken before, during and after an earthquake. A seismic map showing the different earthquake zones of India is included as a part of the application. The ERI questionnaire consists of three sets of questionnaire which involves Individual preparedness, Community preparedness and Demographics.

This application mainly focuses on how to increase the awareness of earthquake disaster among the public. Thus the main framework has to be neat, simple and elegant, so that it would be easy even for a layman to understand the facts of earthquakes. The information given in the application has been divided into 4 modules.

The first three modules comprises of the general information required for the public to be known in order to increase their awareness. The first module, "**Basic Information about Earthquake**" provide necessary basic information about earthquake, its occurrence & measurement of its magnitude. It is provided with infographics representing how an earthquake occurs i.e. the underlying geological phenomenon.

The second module, "**Do's and Don'ts**", conveys the mandatory action to be taken before, during and after an earthquake as an individual and also as a community. Various literature were examined in order to obtain precise and accurate information on these actions, which includes the guidelines from American Red Cross, FEMA, NDMA, WEMO, several disaster management agencies and Sitharam and Kolathayar (2018). These actions are broadly classified into two categories, '**Individuals**' and '**Community**'. Both Individuals and community has a major role to play in order to maintain safety during such catastrophe. This reason led to the split of information into such categories. And also, in terms of community, where a group of people act together in order to increase their safety efficiency. But, not all in a group would be able to act according to the others. Thus in the community section, the actions included are exclusively for school students, where someone should lead the students and guide them to safety, in order to protect them. The information provided in this module has been devised in such a way that all the necessary survival guidelines are included. In each individual section, further three divisions i.e. 'before', 'during' and 'after' have been created in order to have an organized form of information. These three categories give the actions to be taken at the stipulated time period. And also, these information are provided in the form of infographics, so that even people whose literacy rate are less would be able to understand the message to be conveyed just by looking into those infographics.

The third module "**Digital seismic map of India**", has a digitalized seismic map developed by Geological Survey of India (G.S.I). This can be used to identify the zone under which any particular place has been categorized. The last module "**Readiness assessment tool**", is developed to incorporate a questionnaire based earthquake readiness assessment tool. This module has 2 sets of questionnaires, one for assessing the individual earthquake readiness and the second is to assess community readiness towards earthquake. And the user has a choice of undertaking both or any one of the surveys. These tools are developed after an elaborate study is carried out in creating the questionnaires, validating for consistency and reliability through pilot surveys and the questionnaires have been subjected to many revisions before the final form. Questionnaire for evaluating Earthquake readiness of individuals is a 27 item scale and the questionnaire for community readiness is a 20 item scale. The questionnaires also collect demographic information which can be used for interpretative studies. These questionnaires try to evaluate the individual's level of intention to prepare against a major earthquake. Three point Likert scale was incorporated. In the community readiness questionnaire the 20 items try to assess community participation, collective efficacy along with trust and empowerment. From the scores generated based on the responses of the user, one can assess their present levels of preparedness and take respective actions to improve, by actually just moving to the other modules of the app, where the basic information is elucidated in a simple and elegant way.

Another important feature of the app is **daily notification and inspirational quotes** highlighting important aspect of disaster risk reduction at personal and household level. The reminder notification is stranded in the trigger construct persuasion theory by Fogg (2009) which suggests that trigger is necessary to engage in target behavior even when target behavior is easy to do and incentivized. The daily inspirational quote is based on intrinsic motivation theories similar to inspirational memes (Zichermann and Christopher 2011). The daily inspirational quotes are given irrespective of participants' involvement with a hope that they will like the quote and will respond by taking the survey and required tasks. This kind of behavioral influence is referred as reciprocity in the social influence literature (Cialdini 1987).

5 Case Study of Coimbatore

5.1 Seismic Background of Coimbatore City

Until 1967, most of the engineers and scientists in the country were under the impression that peninsular India was free from seismic activity. But the earthquakes at Koyna (1967) and Latur (1993) changed their perception. Tamil Nadu is thought to be mostly free from seismic activity. But an earthquake rocked Coimbatore as far back as 1900. Even though Coimbatore has experienced an earthquake of moment magnitude 6.3 in 1900 (the epicentre of the earthquake was located at 10° 45′ North and 76° 45′ East), it was placed in seismic zone Zero, in the first version of the IS 1893 (BIS 1962). Presently, Coimbatore city is placed in Zone III as per the latest release of IS 1983 (BIS 2002). Coimbatore is located on thin lithosphere, part of Gondwanaland (Kumar et al. 2007) [30]. Geologically, this area is oldest sedimentary and is called as area of Dharwar (Gupta 2006) (Table 1).

Statistical analysis indicates that such seismological events might recur in 100 years, plus or minus about 30 years. Although such statistical forecasts are probabilistic, they give a reasonable idea about the seismogenic potential of the region. It could be said that seismic activity in this region has been on the rise for the past 15 years, (Panjamani et al. 2012). There were two earthquakes, each with a magnitude of around 5.0 on the Richter Scale, in Idukki and Coimbatore districts on the Kerala-Tamil

Table 1. Statistics and data for Coimbatore City

Parameter	Value
Expanse (km^2)	105.6
Population Density (per km^2)	11,234
Local co-ordinates	10.96°N, 76.9°E–11.06°N, 77.06°E
Seismic zones	III (as per BIS 1893 2002)
Reason to be at risk	Though it falls in the Peninsular Shield region, studies show that seismogenic activities have higher chances to recur at 100 ± 30 years
Literacy rate	89.23%

Nadu border in December 2000 and January 2001 respectively. It could be said that seismic activity began sometime in 1988 when the Idukki area was shaken by an earthquake of a magnitude of 4.8 on the Richter scale.

5.2 Sample Survey in Coimbatore

The sample questionnaire was pretested by sending it for pilot survey using Google Forms. The sample size obtained for pilot survey was 132, which fulfilled the bare minimum criteria of three respondents against one item in both the scales. Construct validity was determined by exploratory factor analysis. It was done by clustering items into common factors. Principal component analysis (PCA) and varimax methods were used for the undertaking FEA. To ensure an appropriate sample size for running factor analysis, the Kaiser-Meyer-Olikin (KMO) sampling adequacy on the scale was computed. The accepting values of ≥ 0.5 were recommended. The KMO sampling adequacy of the tool was 0.75 indicating the sample size of 132 had been adequate for performing factor analysis. The observations of the survey are tabulated in Appendix. Figure 4 presents the response analysis of the pilot survey. Internal consistency check produced alpha co-efficient (Cronbach's alpha) greater than 0.7, indicating acceptable correlation between the items of the questionnaire. Figure 5 gives a clear picture regarding the ways to disseminate earthquake information to people. Television, Newspapers and Internet have a fair share of passing information to the people. Hence, it highlights the importance of mass media during the event of an earthquake. Care must be taken to by the concerned authorities and broadcasting organisations to ensure only reliable information reach the people.

6 Conclusion

The android based application that is developed would assist individuals to assess their preparatory levels towards earthquake through the unique ERI questionnaire based survey that is incorporated. Respondents with less preparatory levels can go through the information presented in the first three modules of the app and gain the required disaster knowledge. The response information that is collected from various respondents is stored in a separate server and the use of this information can be immense.

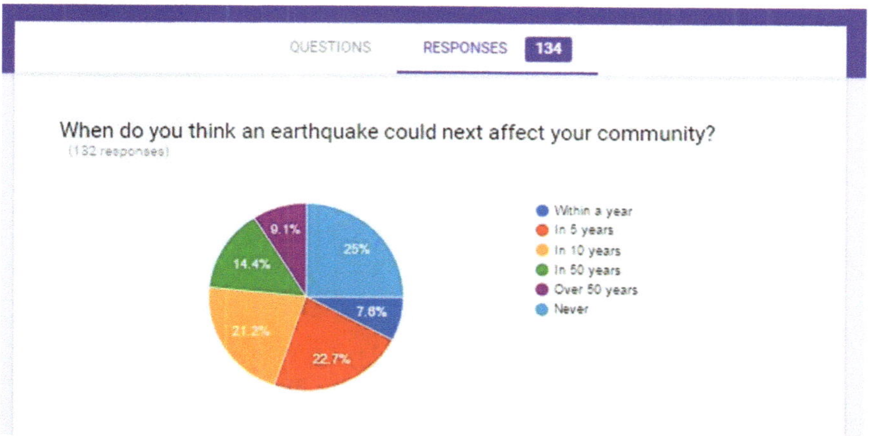

Fig. 4. Response analysis of online pilot survey

Once sufficient number of respondents use the application and take up the surveys, we can use all the collected response information and do interpretative studies, from which we can understand the influence of demographic factors on preparedness levels etc. The application can be used to the full potential, if maximum number of people take up the assessment surveys and this can be achieved by popularising the application through disaster awareness campaigns, social media tools, government disaster initiatives etc.

The case study of Coimbatore city reveals that Television, Newspapers and Internet have a fair share of passing information to the people. Hence, it highlights the importance of mass media to disseminate earthquake information to people. Results of the online pilot survey indicates that the earthquake preparedness level among the respondents is very less. The case is more pathetic when it comes to community preparedness level. Lack of awareness, skills and knowledge play a crucial role in preparedness level independent of the demographic and economic backgrounds.

Summary of Online Pilot Survey Results

Total no: of respondents = 132
C—Count, N%—Count in percentage

Demographics

1. Are you male or female? (131 responses)

	C	N%
Male	72	45
Female	59	55
Other	0	0

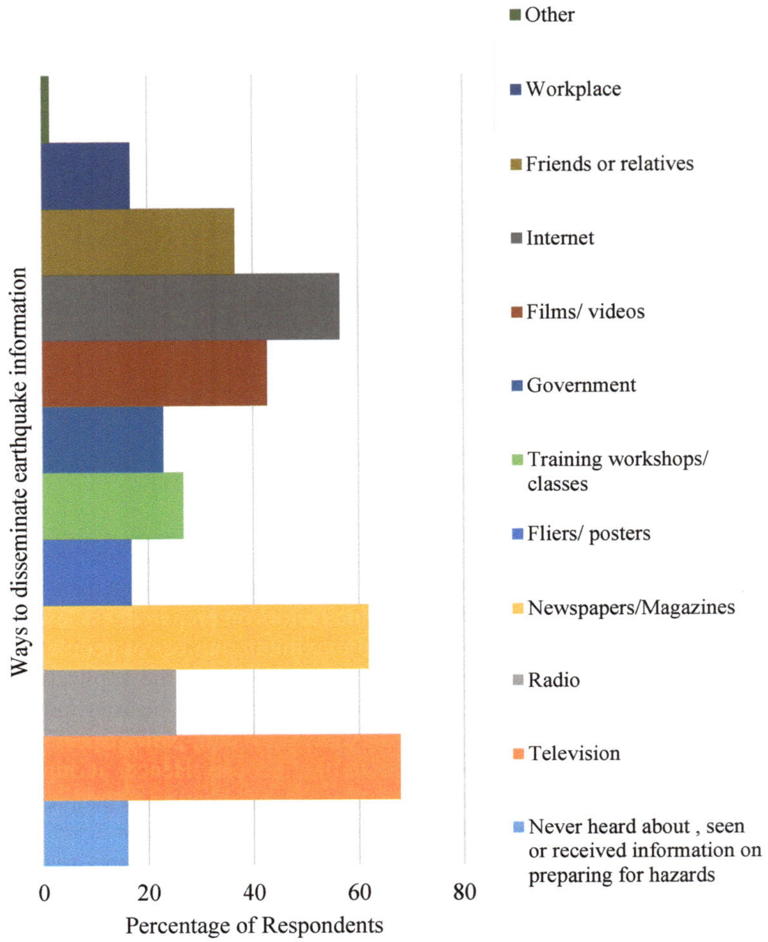

Fig. 5. Percentage of Respondents against the various way to disseminate earthquake information

2. Which category below includes your age? (132 responses)

	C	N%
17 or younger	2	1.5
18–20 years	19	14.4
21–29 years	105	79.5
30–39 years	1	0.8
40–49 years	1	0.8
50–59 years	4	3
60 years or older	0	0

3. Which of the following categories best describes your employment status? (132 responses)

	C	N%
Employed (full-time)	33	25
Employed (part-time)	0	0
Self-employed	5	3.8
Student	88	66.7
Not employed	4	3
House person	2	1.5
Retired	0	0
Disabled, not able to work	0	0

4. Are you an Indian resident? If yes, mention your native state in the other options tab (132 responses)

	C	N%
Yes	130	98.5
No	2	1.5

5. What is the highest level of your educational qualification? (132 responses)

	C	N%
Less than high school	0	0
High school	4	3
Some college but no degree	10	7.5
Associate degree	0	0
Bachelor degree	60	45.5
Graduate degree	58	44

6. How much total combined money did all members of your HOUSEHOLD earn in 2015? (Annual income of your Household) (131 responses)

	C	N%
Less than ₹72,000	6	4.6
₹72,000–₹143,999	16	12.2
₹144,000–₹239,999	25	19.1
₹240,000–₹719,999	44	33.6
₹720,000–₹2,879,999	37	28.2
₹2,880000–₹14,399,999	2	1.5
₹14,400,000 or More	1	0.8

7. How long have you lived in your current house? (in years) (132 responses)

Value	C	N %
0.5	3	2.3
1	11	8.3
2	5	3.8
3	10	7.6
4	4	3.0
5	5	3.8
6	2	1.5
7	3	2.3
8	5	3.8
9	2	1.5
10	15	11.4
11	3	2.3
12	4	3.0
13	3	2.3
14	1	0.8
15	10	7.6
16	2	1.5
17	3	2.3
18	5	3.8
20	8	6.1
21	1	0.8
22	3	2.3
23	11	8.3
24	4	3.0
25	1	0.8
27	3	2.3
28	3	2.3
50	1	0.8
57	1	0.8

8. Including yourself, how many people are there in your household? (128 responses)

Value	C	N%
0	1	0.8
1	3	2.3
2	2	1.6
3	18	14.1
4	68	53.1
5	26	20.3
6	4	3.1
7	5	3.9
20	1	0.8

9. Which of the following describes your family status? (127 responses)

	C	N%
Married or partnered with children at home	32	25.2
Married or partnered with NO children at home	4	3.1
Single with children at home	7	5.5
Single with NO children at home	84	66.1

10. Which of the following describes the ownership of the house/apartment you live in? (131 responses)

	C	N%
Owned or buying, to live in it	95	72.5
Owned or buying as a holiday home	3	2.3
Rent, to live in it	31	23.7
Rent, as a holiday home	2	1.5

Earthquake Awareness and Preparedness Related Questions

11. Have you ever experienced an earthquake? (130 responses)

	C	N%
Yes, direct experience (e.g.: damage, injury, tremors)	21	16.2
Yes, indirect experience (e.g.: was inconvenient, couldn't travel)	17	13.1
No	92	70.8

12. When do you think an earthquake could next affect your community? (130 responses)

	C	N%
Within a year	9	6.9
In 5 years	30	23.1
In 10 years	27	20.8
In 50 years	19	14.6
Over 50 years	12	9.2
Never	33	25.4

13. Please rate the extent to which you perceive the preparedness level of the following (132 responses)

	Very much prepared		Prepared		Somewhat prepared		May not be prepared		Not at all prepared	
How prepared do you think you are for a major earthquake?	9	6.8	16	12.1	45	34.1	34	25.8	28	21.2
How well prepared do you think other members of your community are for a major earthquake	6	4.5	16	12.1	43	32.6	34	25.8	33	25.0
How well prepared do you think your local government is for a major earthquake?	7	5.3	28	21.2	46	34.8	28	21.2	23	17.4

14. Have you ever heard, seen or received information on earthquake preparedness from the following? (Tick all that apply) (131 responses)

Value	C	N%
Never heard about, seen or received information on preparing for hazards	21	16.0
Television	89	67.9
Radio	33	25.2
Newspapers/Magazines	81	61.8
Fliers/Posters	22	16.8
Training workshops/classes	35	26.7
Government	30	22.9
Films/videos	56	42.7
Internet	74	56.5

(continued)

(*continued*)

Value	C	N%
Friends or relatives	48	36.6
Workplace	22	16.8
Other	2	1.5

15. Rate the importance of each of the following mentioned below in earthquake preparedness (132 responses)

	Very important		Somewhat important		Has no role		May be important		Not at all important	
	C	N%	C	N%	C	N%	C	N%	C	N%
Individuals	91	68.9	32	24.2	2	1.5	5	3.8	2	1.5
Community group	102	77.3	25	18.9	0	0.0	4	3.0	1	0.8
Governmental organizations	112	84.8	16	12.1	1	0.8	2	1.5	1	0.8
Agencies	70	53.0	47	35.6	6	4.5	9	6.8	0	0.0
Schools	104	78.8	23	17.4	2	1.5	3	2.3	0	0.0
Workplace	98	74.2	27	20.5	2	1.5	5	3.8	0	0.0
Emergency services (e.g. Police, fire)	120	90.9	9	6.8	2	1.5	1	0.8	0	0.0
Insurance companies	83	62.9	41	31.1	2	1.5	6	4.5	0	0.0

16. To what extent do each of the following prevent you from preparing for earthquakes? (132 responses)

	Very much		Somewhat		Has no role		May be		Not at all	
	C	N%	C	N%	C	N%	C	N%	C	N %
The cost	24	18.2	58	43.9	29	22.0	16	12.1	5	3.8
The skill or knowledge required	54	40.9	58	43.9	12	9.1	5	3.8	3	2.3
The time needed to prepare	48	36.4	57	43.2	9	6.8	11	8.3	7	5.3

17. Please describe the extent to which you agree or disagree with each of the following statements (132 responses)

	Strongly agree		Agree		Neither agree nor disagree		Disagree		Strongly disagree	
	C	N%	C	N%	C	N%	C	N%	C	N %
In community meetings, I am often a leader	12	9.1	36	27.3	42	31.8	33	25.0	9	6.8
In community meetings I prefer to be a leader rather than a follower.	16	12.1	41	31.1	40	30.3	29	22.0	6	4.5
People around here will express opinion for relevant issues	21	15.9	82	62.1	20	15.2	5	3.8	4	3.0
People in this community can express their needs to others easily	17	12.9	71	53.8	31	23.5	10	7.6	3	2.3
The talks of community depends on residents' interests	27	20.5	71	53.8	30	22.7	3	2.3	1	0.8
Difficult situations determine what issues this community should focus on	42	31.8	66	50.0	21	15.9	2	1.5	1	0.8
Peoples perceptions of community problems control the measures adopted	29	22.0	72	54.5	25	18.9	5	3.8	1	0.8
I trust my Local government to respond to meet the needs of its residents	16	12.1	52	39.4	44	33.3	14	10.6	6	4.5
I trust the media (newspapers, TV, radio) to report fairly	23	17.4	55	41.7	42	31.8	11	8.3	1	0.8
I have confidence in the law to protect and maintain order in my community	19	14.4	60	45.5	41	31.1	8	6.1	4	3.0

18. With regard to your participation in this community, record your responses (132 responses)

	Yes		May do This		Will not do this	
	C	N%	C	N%	C	N %
I have worked with others to improve community life	49	37.1	79	59.8	4	3.0
I participate in local activities or events (e.g., festivals, fairs)	66	50.0	60	45.5	6	4.5
I have contributed money, local causes, charities, or to others in my community	61	46.2	64	48.5	7	5.3
I have attended public meetings on community issues	42	31.8	78	59.1	12	9.1
I have been involved in volunteer activities beneficial to my community (e.g., fund-raising, clean-up days, local groups, Scouts)	59	44.7	64	48.5	9	6.8

19. Thinking about earthquake preparedness issues would you be willing to (132 responses)

	Yes		Possibly		No	
	C	N%	C	N%	C	N%
Participate regularly, on an on-going basis (e.g. belong to a group; attend monthly meetings)?	39	29.5	83	62.9	10	7.6
Participate for specific reasons or events (e.g. attend a one-off community meeting; be involved in a preparedness fair)?	50	37.9	74	56.1	8	6.1
Pass on information about earthquake and preparing to other community members?	75	56.8	52	39.4	5	3.8
Encourage other people in your community to get prepared for earthquakes?	78	59.1	53	40.2	1	0.8
Take a training course to be qualified to help your community recover from earthquakes?	54	40.9	63	47.7	15	11.4

20. The following are things that can be done to minimize damage and disruption if an earthquake occurs. Record your responses (132 responses)

	Yes		May do this		Will not do this	
	C	N%	C	N%	C	N%
I have considered the chance of an earthquake when I decided to live in my present house	48	36.4	70	53.0	14	10.6
I have fixed tall furniture to the wall	32	24.2	79	59.8	21	15.9

(*continued*)

(*continued*)

	Yes		May do this		Will not do this	
	C	N%	C	N%	C	N%
I have ensured that heavy utility hardware will not fall down in a major earthquake (e.g.: water heaters, chimney)	46	34.8	76	57.6	10	7.6
I have ensured that my roof will probably not collapse in a major earthquake	52	39.4	70	53.0	10	7.6
I have arranged the cupboards so that heavy objects are stored at ground level	53	40.2	67	50.8	12	9.1
I have securely fastened cupboards with latches	45	34.1	74	56.1	13	9.8
I have ensured that objects that contain water have not been stored on top of electrical equipment (e.g., a pot plant or fishbowl on top of the television)	56	42.4	69	52.3	7	5.3
I have ensured that heavy objects are stored on the floor	62	47.0	65	49.2	5	3.8
I know the locations of utility shut-off valves and switches (e.g., water shut-off valve, gas shut-off valve, electric power shut-off switch)	65	49.2	62	47.0	5	3.8
I know how to operate utility shut-off valves and switches	59	44.7	66	50.0	7	5.3
I have stored water for survival	45	34.1	80	60.6	7	5.3
I have put aside spare plastic bags and toilet paper for use as an emergency toilet	38	28.8	85	64.4	9	6.8
I have accumulated enough tools to make minor repairs to the house following a major earthquake	44	33.3	81	61.4	7	5.3
I have obtained a supply of tinned or dehydrated food that could be used in an emergency	35	26.5	77	58.3	20	15.2
I have purchased or put together a first aid kit	45	34.1	75	56.8	12	9.1
I have a supply of essential medicines for illness or allergies	49	37.1	73	55.3	10	7.6
I have obtained a working battery radio (or solar/dynamo equivalent)	31	23.5	86	65.2	15	11.4
I have secured movable objects in my home (e.g., TV, computer)	42	31.8	76	57.6	14	10.6
I have access to an alternative cooking source.	33	25.0	81	61.4	18	13.6
I have a household earthquake emergency plan	33	25.0	78	59.1	21	15.9
My plan covers where the family should meet if an earthquake occurs during the day	32	24.2	81	61.4	19	14.4
I have contacts of emergency services (e.g.: medical, police, support groups)	62	47.0	59	44.7	11	8.3
I have obtained a working fire extinguisher	32	24.2	81	61.4	19	14.4

(*continued*)

(*continued*)

	Yes		May do this		Will not do this	
	C	N%	C	N%	C	N%
I have an earthquake emergency plan at work place	41	31.1	75	56.8	16	12.1
I have taken some other preparedness steps at work (knowing emergency exits, where to be in case of an earthquake)	45	34.1	74	56.1	13	9.8
I have obtained spare batteries for the appliances I might need to use	36	27.3	79	59.8	17	12.9
I have specifically put together an emergency kit	32	24.2	85	64.4	15	11.4
I check the contents/operation of my emergency supplies at least every six months	36	27.3	79	59.8	17	12.9
I know the location of a medical emergency center near to my house	57	43.2	66	50.0	9	6.8

21. In the next month or so, do you intend to do the following?

	Yes		Possibly		No	
	C	N%	C	N%	C	N%
Check your level of preparedness for earthquakes	47	35.6	69	52.3	16	12.1
Increase your level of preparedness for earthquakes	54	40.9	63	47.7	15	11.4
Become involved with a local group to discuss how to reduce earthquake damage or losses	40	30.3	68	51.5	24	18.2
Seek information on earthquake risk	61	46.2	58	43.9	13	9.8

References

BIS-1893: Indian Standard criteria for earthquake resistant design of structures. Part 1—General provisions and buildings. Bureau of Indian Standards, New Delhi (2002)

Cialdini, R.B.: The psychology of influence. Currents **13**, 48–51 (1987)

Cobb, C.W., Rixford, C.: Competing paradigms in the development of social and economic indicators. In: CSLS Conference on the State of Living Standards and the Quality of Life in Canada, pp. 30–31. Ottawa (1998, October)

Davidson, R.A., Lambert, K.B.: Comparing the hurricane disaster risk of US coastal counties. Nat. Hazards Rev. **2**(3), 132–142 (2001)

Fogg, B.J.: A behavior model for persuasive design. In: Proceedings of the 4th international Conference on Persuasive Technology, p. 40. ACM (2009)

Kolathayar, S., Anupa, S.A., Lalith Prakash, E.: Development of earthquake readiness index tool to assess individual earthquake preparedness level. In: ASCE India Conference (2017)

Tan, M.L., Prasanna, R., Stock, K., Hudson-Doyle, E., Leonard, G., Johnston, D.: Mobile applications in crisis informatics literature: a systematic review. Int. J. Disaster Risk Reduct. (2017). https://doi.org/10.1016/j.ijdrr.2017.06.009
Sitharam, T.G., Kolathayar, S.: Do's and don'ts before, during and after an earthquake. In: Preparing for Earthquakes: Lessons for India, pp. 17–28. Springer, Cham (2018)
Zichermann, G., Christopher, C.: Gamification by design: implementing game mechanics in web and mobile apps. O'Reilly Media, Inc. (2011)

On Optimal Reinforcement Layout for Underground Openings

Kazem Ghabraie(✉)

School of engineering, Deakin University,
Waurn Ponds, Australia
k.ghabraie@deakin.edu.au

Abstract. The problem of finding the optimal reinforcement layout for underground openings is considered. The rock material is assumed to sustain only compressive stresses while the reinforcements are in pure tension. It is shown that in this case, the optimal reinforcement layout can be achieved by solving a truss optimisation problem. The theory of truss optimisation is briefly presented. It is shown that the optimal layout of reinforcing bars is independent of the material properties of the reinforcing bars and the ground material.

1 Introduction

The most common type of reinforcement system used in underground excavations is rock bolting. Designing a rock bolt system requires selection of a layout for the rock bolts and then specifying the diameter and length of each bolt. By layout here, we practically mean the location and orientation of bolts around the opening. In many cases a pattern is selected for bolts where spacing between the individual bolts is also considered as a design variable specifying which the layout of the bolts is identified.

In available design schemes for rock bolts, the pattern and/or layout is mostly predefined intuitively or based on previous experience. Design parameters such as length, spacing and diameter have been well studied but finding the optimal pattern/layout of rock bolts seemed to have only received attention recently.

The problem of finding the optimal pattern of reinforcements can be classified as a topology optimisation problem. Using different numerical topology optimisation techniques, Yin et al. (2000), Yin and Yang (2000a; 2000b) and later Ghabraie (2009), Ghabraie et al. (2010) tried to find the optimal distribution of reinforced area around openings in isotropic linear elastic rock media. In these works, a homogenised, isotropic, linear elastic material model with a larger Young's modulus is used to model the reinforced areas of host rock. Imposing a constraint on the volume of the reinforced material around the opening, the optimal distribution of the reinforced material is found using topology optimisation. The simplified material models used in these studies significantly limits their applicability. Nguyen et al. (2014) extended these applications to non-linear

B. Fatahi et al. (eds.): *Sustainable Design and Construction for Geomaterials and Geostructures*, Sustainable Civil Infrastructures,
https://doi.org/10.1007/978-3-319-95753-1_17

materials by presenting a sensitivity analysis independent of material model. However, this study also optimised the distribution of a homogenised reinforced material rather than the rockbolt layout. This limitation was removed by Nguyen et al. (2015, 2016) where the rock bolts were explicitly included in the model and their length, orientation, and cross section were optimised.

Despite these works, however, As far as the author is aware, there is no theoretical study on the optimality of reinforcement patterns around underground openings. In this paper, the basics of pin-joined truss optimisation theory is represented, expanded and used to reach insightful conclusions in tunnel reinforcement design.

2 Problem Statement

Prior to the excavation, assume that the stress distribution in the ground is known at least in the vicinity of where the opening is to be excavated. This in-situ stress field is denoted by the tensor σ_{ij}. After the excavation is completed, surface tractions vanish along the opening boundary (Fig. 1a). Prior to the excavation, the in-situ stresses along the boundary of the opening were balanced by the surface traction $t_i = \sigma_{ij} n_j$ where n_j is the unit normal vector of the opening boundary pointing outwards (Fig. 1b). Assuming that the deformations are small, the post-excavation stress field around the opening can be obtained by superimposing the negated surface traction $-t_i$ over the in-situ stresses (Fig. 1). Whether the deformations are small or not, the surface tractions form a *self-equilibrated load system.*

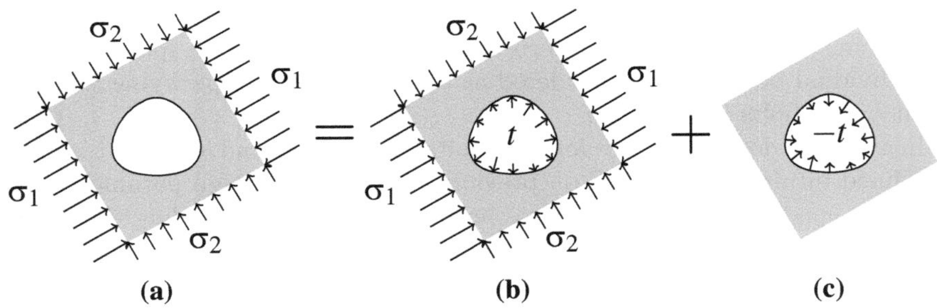

Fig. 1. The post-excavation stresses **a** can be considered as the superposition of pre-excavation (in-situ) stresses (**b**) and the surface traction due to stress relief (**c**)

The problem of concern in this paper is finding the lightest reinforcement system to withstand the self-equilibrated surface tractions $-t_i$ around an underground opening preventing opening collapse. By lightest reinforcement system, we mean the system with the least amount of reinforcement material. Assuming that the reinforcements are made of homogeneous materials, in the sequel we may use the terms lightest, least volume, and optimal interchangeably. It should be noted that in practice, other parameters such as the amount of drilling need to be considered in finding the most cost-effective design. Hence the most cost-effective design is not necessarily equivalent to the least volume design. Nevertheless, in most cases knowing the the least volume design is insightful in finding the most cost-effective design.

It is assumed that the rock and reinforcement materials are elastic-perfectly plastic and that reinforcements are in form of one-dimensional inclusions with negligible flexural strength. It is also assumed that the rock material can only withstand compressive stresses up to the known limit of σ_c and that the reinforcement material can withstand tensile stresses up to the known limit of σ_t. Compressive stresses are considered positive.

In a region of space, let's denote by $\mathscr{C}(\mathrm{F}, \sigma_t, \sigma_c)$ the set of *safe truss-continua configurations*, which are truss-continua combinations capable of carrying a self-equilibrated load system F[1] such that the truss elements are all in tension with tensile stresses $|\sigma| \leq \sigma_t$, and the continuum regions are completely in compression with (compressive) principal stresses $0 \leq \sigma_3 \leq \sigma_2 \leq \sigma_1 \leq \sigma_c$. Given our assumption on elastic-perfectly plastic material behaviour, based on the limit design theorems (Drucker et al. 1952), if such a safe truss-continua configuration can be found for an excavation problem (i.e. if $\mathscr{C}(-t_i, \sigma_t, \sigma_c) \neq \emptyset$), the truss elements in that configuration will provide a safe reinforcement design for the opening.

For any configuration like $\mathsf{C} \in \mathscr{C}(\mathrm{F}, \sigma_t, \sigma_c)$ we denote the volume of its tensile (truss) members by $V_t(\mathsf{C})$. We may write $\mathsf{C}^* = \min_t\{\mathscr{C}(\mathrm{F}, \sigma_t, \sigma_c)\}$ to express that the configuration C^* has the least volume of tensile (truss) members among all other safe configurations, that is $\forall \mathsf{C} \in \mathscr{C}(\mathrm{F}, \sigma_t, \sigma_c): \; V_t(\mathsf{C}) \geq V_t(\mathsf{C}^*)$. Now, the problem of concern can be stated as finding the tensile (truss) elements in the configuration $\mathsf{C}^* = \min_t\{\mathscr{C}(-t_i, \sigma_t, \sigma_c)\}$, given $-t_i$, σ_t, and σ_c.

3 Optimal Reinforcement Systems and Optimal Trusses

Similar to the definition of safe truss-continua configuration, let's use $\mathscr{T}(\mathrm{F}, \sigma_t, \sigma_c)$ to denote the set of trusses capable of safely carrying the self-equilibrated load system F with tensile and compressive stresses limited to σ_t and σ_c respectively. It is clear that $\mathscr{T} \subset \mathscr{C}$.

For any truss like $\mathsf{T} \in \mathscr{T}(\mathrm{F}, \sigma_t, \sigma_c)$ we denote the volume of its tensile and compressive members by $V_t(\mathsf{T})$ and $V_c(\mathsf{T})$, respectively. Total volume of this truss is denoted by $V(\mathsf{T}) = V_t(\mathsf{T}) + V_c(\mathsf{T})$. We may write $\mathsf{T}^* = \min\{\mathscr{T}(\mathrm{F}, \sigma_t, \sigma_c)\}$ to express that the truss T^* has the least volume among all trusses in $\mathscr{T}(\mathrm{F}, \sigma_t, \sigma_c)$, that is $\forall \mathsf{T} \in \mathscr{T}(\mathrm{F}, \sigma_t, \sigma_c): \; V(\mathsf{T}) \geq V(\mathsf{T}^*)$.

In this section, we want to show that in order to find the tensile (truss) elements in $\mathsf{C}^* = \min_t\{\mathscr{C}(\mathrm{F}, \sigma_t, \sigma_c)\}$, it is enough to find the tensile elements in $\mathsf{T}^* = \min\{\mathscr{T}(\mathrm{F}, \sigma_t, \sigma_c)\}$. In other words, we want to show that by replacing

[1] It should be noted that by F, we identify not only the load vectors but also their points of action.

compressive members of a least weight truss by compressive only continua, one cannot reduce its tensile volume any further.

We first recall the following important corollary stated by Cox (1958) and Hemp (1958) following a theorem initially proved by Maxwell (1864).

Corollary 1. *The least volume truss has the least volume of tensile members and also the least volume of compressive members among all other safe trusses, i.e.,*

$$\mathsf{T}^* = \min\{\mathscr{T}(\mathrm{F}, \sigma_t, \sigma_c)\} \Leftrightarrow V_t(\mathsf{T}^*) \leq V_t(\mathsf{T}) \text{ and } V_c(\mathsf{T}^*) \leq V_c(\mathsf{T}), \forall \mathsf{T} \in \mathscr{T}(\mathrm{F}, \sigma_t, \sigma_c).$$

Now consider $\mathsf{C}^* = \min_t\{\mathscr{C}(\mathrm{F}, \sigma_t, \sigma_c)\}$ and let's refer to the collection of its (not necessarily connected) tensile truss members by T^t. The force system acting on C^* is also considered to be decomposed into two sets: one maintaining stress boundary conditions in members of T^t and one maintaining stress boundary conditions in the in-compression continua. We denote the former by F^t and the latter by F^c.

Let T^c be a truss in $\mathscr{T}(\mathrm{F}_c, \sigma_t, \sigma_c)$. It is noted that one can always find such a truss as there is no limitation on the amount of material used. Clearly, T^c has only compressive members. Superposing T^c and T^t, we define $\mathsf{T}^* = \mathsf{T}^\mathsf{t} \cup \mathsf{T}^\mathsf{c}$. Note that $\mathsf{T}^* \in \mathscr{T}(\mathrm{F}, \sigma_t, \sigma_c)$.

Given that $\mathscr{T}(\mathrm{F}, \sigma_t, \sigma_c) \subset \mathscr{C}(\mathrm{F}, \sigma_t, \sigma_c)$, one cannot find any truss in $\mathscr{T}$ with smaller tensile volume than $V_t(\mathsf{T}^*)$, so we have $V_t(\mathsf{T}^*) \leq V_t(\mathsf{T}),\ \forall \mathsf{T} \in \mathscr{T}(\mathrm{F}_c, \sigma_t, \sigma_c)$. Following Corollary 1 above, this means that $\mathsf{T}^* = \min\{\mathscr{T}(\mathrm{F}, \sigma_t, \sigma_c)\}$. In other words, the layout of the tensile (truss) members of the optimal truss-continua configuration is identical to that of the optimal truss.

The practical significance of this result can be expressed as the following statement: *In order to find the least volume of tensile bars, instead of searching for the optimal configuration in $\mathscr{C}$ one can find the optimal truss in $\mathscr{T}$.*

4 Finding the Least Weight Truss

We now focus on the truss optimisation problem. In a seminal contribution, Michell (1904) proved the following theorem which expresses the sufficient condition for optimality of a truss.

Theorem 1. (Michell's theorem) *In a region of space $\mathscr{S}$, the truss $\mathsf{T} \in \mathscr{T}(\mathrm{F}, \sigma_t, \sigma_c)$ is optimal if a kinematically admissible virtual deformation on $\mathscr{S}$ could be found to generate the strain $\pm\varepsilon$ along all members of T where ε is a small positive number and the sign of virtual strain in each member agrees with that of the stress caused by* F, *and that the normal strain value in no directions in $\mathscr{S}$ is numerically greater than ε.*

A proof similar to the original Michell's proof but using contemporary notation can be found in Hemp (1958, 1973).

The Michell's virtual deformation can assumed to be caused by a virtual (dual) load system subject to kinematic admissibility. For a self-equilibrated load system with no supports which is the case for our reinforcement design, kinematical admissibility is trivial for any virtual deformation. Hence choosing the dual load system equal to the actual (primal) load system and setting the truss elements along the trajectories of the principal stresses would automatically satisfy the requirements of the Michell's theorem. This leads to the following corollary.

Corollary 2. *For a self-equilibrated system of forces with no supports, the Michell optimal truss layout matches the trajectories of principal stresses (or strains) and therefore is independent of the values of* σ_t *and* σ_c.

It should be reminded that this statement only concerns the optimal *layout* of reinforcements. The required cross section and length of rock bolts *will* depend on the strength of the materials involved.

Earlier results obtained by the author's team confirm this conclusion even for optimised solutions obtained under additional constraints. To illustrate this, a numerical example included in Nguyen et al. (2016) is reused here in Fig. 2. This figure shows the results of a cross section optimisation of rock bolts around a circular tunnel in plane strain regime under different in-situ stresses. Here both

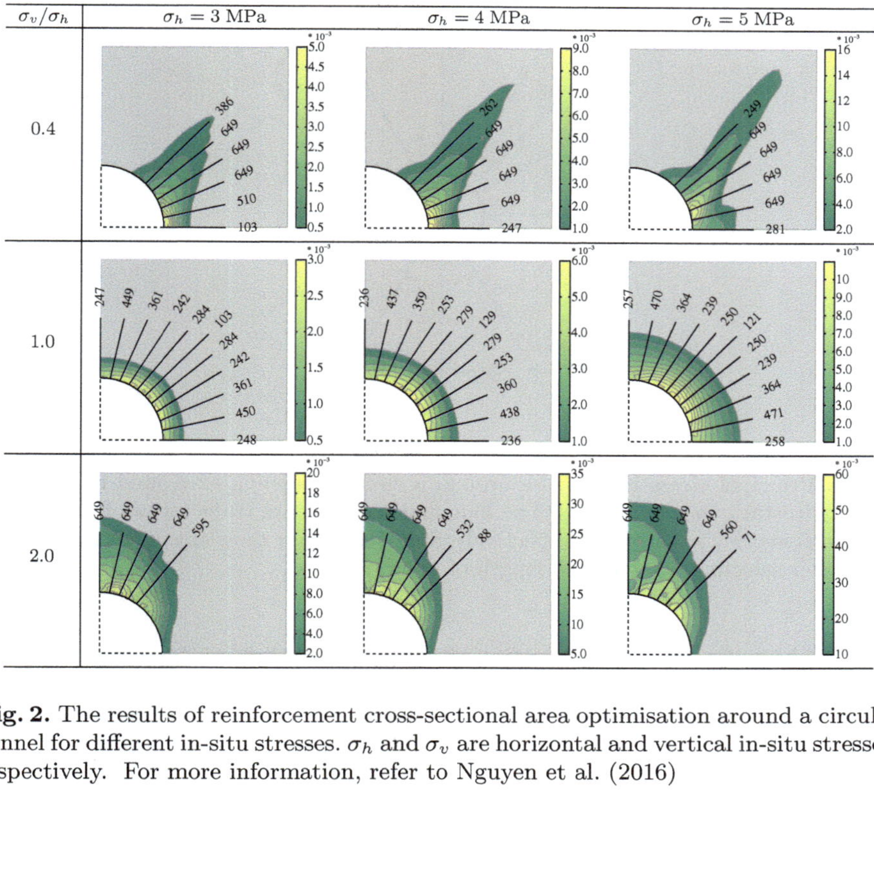

Fig. 2. The results of reinforcement cross-sectional area optimisation around a circular tunnel for different in-situ stresses. σ_h and σ_v are horizontal and vertical in-situ stresses, respectively. For more information, refer to Nguyen et al. (2016)

length and orientation of rock bolts are fixed. The numbers at the end of each bolt show the cross section of that bolt in mm^2 per unit meter of the tunnel. Bolts with cross sectional areas smaller than 65 mm^2/m are removed to make the patterns clearer. The contours in this figure show a measure of the plastic strain magnitude defined as $\sqrt{\frac{2}{3}\varepsilon^p : \varepsilon^p}$ where ε^p is plastic strain tensor.

Looking at Fig. 2, it is clear that the optimised results are not very sensitive to stress magnitudes, but change with the ratio of vertical to horizontal in-situ stress.

5 A Simple Example

Consider the simplest problem of its class: a circular opening in a plane strain setting excavated in a ground with hydrostatic in-situ stresses. In this case, $-t_i$ is a uniform pressure around the circular opening. The principal stress trajectories around the opening are shown in Fig. 3. The blue straight lines are in tension and the red circles are in compression. Based on these trajectories, the least weight reinforcement design can be achieved by putting rock bolts in radial directions.

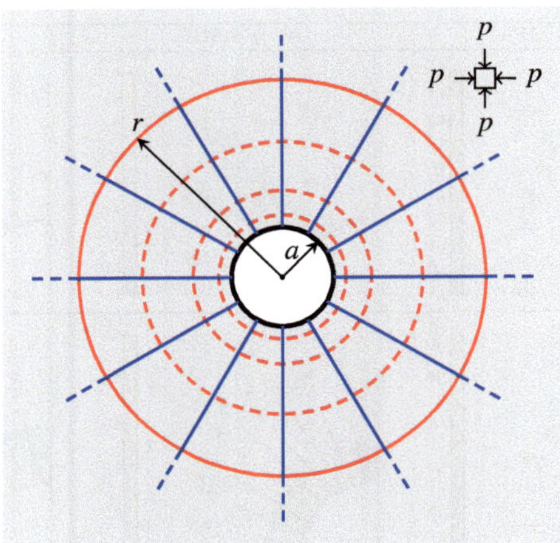

Fig. 3. Principal stress trajectories around a circular opening excavated in a ground with hydrostatic in-situ stresses. The blue and red lines are trajectories of tensile and compressive principal stresses respectively. Solid lines show the elements of the optimal truss after selecting the required length of bolts

Based on the corollaries mentioned above, the radial reinforcement pattern shown in Fig. 3 is optimal irrespective of the in-situ stress magnitude p and material properties of the rock and reinforcement material. However, the cross section area, spacing, and length of the bolts should be determined based on material constants (σ_c and σ_t) and in-situ stress magnitude (p). For example, the length of the bolts are generally found such that they are anchored in the undamaged part of the rock. As the plastic (damaged) zone depends on the material model and stress magnitude (see e.g. the second row in Fig. 2), so does the length of the bolts.

It should be noted that for other in-situ stresses and opening shapes, the optimal layout of reinforcements obtained based on Mechell trusses can be curved. Discussion on how to use such solutions to obtain practical results will be left to another paper.

6 Conclusion

The reinforcement layout optimisation problem around an underground opening is considered assuming that the reinforcements only work in tension. Extending the theories of structural optimisation, the following two conclusions are achieved:

1. The reinforcement layout optimisation problem can be solved as a truss optimisation problem.
2. The optimal layout of reinforcements is independent of the strength of rock or reinforcement material.

References

Cox, H.L: The theory of design. Technical Report 19791, Aeronautical research council (1958)

Drucker, D., Prager, W., Greenberg, H.: Extended limit design theorems for continuous media. Q. Appl. Math. **9**(4), 381–389 (1952)

Ghabraie, K.: Exploring topology and shape optimisation techniques in underground excavations. Ph.D. thesis, School of Civil, Environmental and Chemical Engineering Science, RMIT University, Melbourne, Australia (2009)

Ghabraie, K., Xie, Y.M., Huang, X., Ren, G.: Shape and reinforcement optimization of underground tunnels. J. Comput. Sci. Technol. **4**(1), 51–63 (2010). https://doi.org/10.1299/jcst.4.51

Hemp, W.S.: Theory of structural design. Technical Report 115, The College of Aeronautics, Cranfield (1958)

Hemp, W.S.: Optimum Structures. Oxford University Press, Oxford (1973)

Maxwell, J.C.: On reciprocal figures and diagrams of forces. Philos. Mag. Ser. **4**, **27**(182), 250–261 (1864). https://doi.org/10.1080/14786446408643663

Michell, A.G.M.: The limits of economy of material in frame-structures. Philos. Mag. Ser. **6**, **8**(47), 589–597 (1904). https://doi.org/10.1080/14786440409463229

Nguyen, T., Ghabraie, K., Tran-Cong, T.: Applying bi-directional evolutionary structural optimisation method for tunnel reinforcement design considering nonlinear material behaviour. Comput. Geotech. **55**, 57–66 (2014). https://doi.org/10.1016/j.compgeo.2013.07.015

Nguyen, T., Ghabraie, K., Tran-Cong, T.: Simultaneous pattern and size optimisation of rock bolts for underground excavations. Comput. Geotech. **66**, 264–277 (2015). https://doi.org/10.1016/j.compgeo.2015.02.007

Nguyen, T., Ghabraie, K., Tran-Cong, T., Fatahi, B.: Improving rock bolt design in tunnels using topology optimisation. Int. J. Geomech. **16**(1), (2016). https://doi.org/10.1061/(ASCE)GM.1943-5622.0000488

Yin, L., Yang, W.: Topology optimization for tunnel support in layered geological structures. International Journal for Numerical Methods in Engineering **47**(12), 1983–1996 (2000a). https://doi.org/10.1002/(SICI)1097-0207(20000430)47:12⟨1983::AID-NME863⟩3.0.CO;2-N

Yin, L., Yang, W.: Topology optimization to prevent tunnel heaves under different stress biaxialities. International Journal for Numerical and Analytical Methods in Geomechanics **24**(9), 783–792 (2000b). https://doi.org/10.1002/1096-9853(20000810)24:9⟨783::AID-NAG96⟩3.0.CO;2-X

Yin, L., Yang, W., Guo, T.: Tunnel reinforcement via topology optimization. International Journal for Numerical and Analytical Methods in Geomechanics **24**(2), 201–213 (2000). https://doi.org/10.1002/(SICI)1096-9853(200002)24:2⟨201::AID-NAG64⟩3.0.CO;2-P

Comparison Between Various Multiple Linear Regression Model for Prediction of TBM Performance

Hang-Lo Lee, Won-Suh Sung, and Ki-Il Song(✉)

Inha University, Incheon, South Korea
ksong@inha.ac.kr

Abstract. Reliable estimation of shield tunnel boring machine performance is very important for accurate construction period and cost. In this study, comprehensive statistical procedure for prediction of tunnel boring machine performance was established to develop multiple linear regression model. Also, comparison between various multiple linear regression model was performed based on the database obtained from cable tunnels in South Korea. This paper will be useful in selecting multiple linear regression techniques in terms of explanatory and predictive performance.

1 Introduction

A tunnel boring machine (TBM) is a machine used to excavate tunnels with a circular cross section through a variety of soil and rock mass (Fig. 1). TBM which is used as an alternative to drilling and blasting methods have the advantages of limiting the disturbance to the surrounding ground and constructing a smooth tunnel wall. This makes them suitable to use in heavily urbanized areas. For this reason, the use of shield TBM is gradually increasing due to the urbanization of social infrastructures. Reliable estimation of shield TBM performance is very important for accurate construction period and cost. Generally, there are various parameters for prediction of TBM performance related to machine, ground and utilization (Bruland 1998). Simplified empirical models have been proposed by many researchers since it is difficult to consider all of parameters relevant to TBM performance. Farmer and Glossop (1980) and Hughes (1986) predicted the net penetration rate using cutting force and rock mass strength. Bruland (1998) developed the empirical model which can estimate the TBM performance using various parameters obtained by using special experiments. Yagiz (2008) and Hassanpour et al. (2009) proposed multiple linear regression (MLR) model using stepwise based on database obtained from field. The main advantage of MLR is easily interpretable mathematical expression for its simple form.

In this study, comprehensive statistical procedure for prediction of TBM performance was established to develop MLR model. Also, comparison between various MLR model was performed based on the database obtained from cable tunnels in South Korea. Various MLR model such as lasso, ordinary least squares (OLS), ridge and best subset selection (BSS) were developed by applying training dataset. And these were finally evaluated by using testing dataset.

B. Fatahi et al. (eds.), *Sustainable Design and Construction for Geomaterials and Geostructures*, Sustainable Civil Infrastructures,
https://doi.org/10.1007/978-3-319-95753-1_18

Fig. 1. Tunnel boring machine (EPB type)

2 Various Multiple Linear Regression

2.1 Ridge Regression

The ridge regression proposed by Hoerl and Kennard (1970) is a shrinkage method limiting the coefficient estimates for linear model including all of input parameters. Shrinkage of coefficient estimates toward to zero have been found to reduce the variance of coefficient estimates (James et al. 2014).

Unlike the least squares that minimizes the residual sum of squares (RSS) as shown in Eq. (1), coefficient estimates of ridge regression can be estimated by minimizing the following Eq. (2).

$$RSS = \sum_{i=1}^{n}\left(y_i - \beta_0 - \sum_{j=1}^{p}\beta_j x_{ij}\right)^2 \tag{1}$$

$$RSS + \lambda\sum_{j=1}^{p}\beta_j^2 \tag{2}$$

where, n and p is the number of datasets and number of input parameters respectively. In particular, $\lambda \geq 0$ of Eq. (2) has the effect on making the coefficient estimate to zero.

2.2 Lasso Regression

Although ridge regression may not be a problem in terms of predictive performance, it can lead to difficulties in interpreting a model due to consideration of all input parameter. Lasso regression is a technique to overcome this limit of ridge regression. The coefficient estimate of lasso regression can be calculated by minimizing following Eq. (3).

$$RSS + \lambda \sum_{j=1}^{p} |\beta_j| \tag{3}$$

Lasso regression also shrink the coefficient estimate to zero, as that of ridge regression. However, as $\lambda \geq 0$ increases, some of the coefficient estimate is exactly converged to zero. That means it has a same effect on selecting input parameters and may be easier to interpret the model than that of ridge regression.

2.3 Best Subset Selection

Best subset selection (BSS) is a simple technique to select the best model by using least squares to all possible combinations of input parameters. Only one model is selected as best model with the lowest prediction error among $\sum_{i=1}^{p} \binom{p}{i} = 2^p - 1$ models. However, as the number of input parameter to consider increases, a computing time increases sharply. Therefore, this technique may have more computational constraints than lasso and ridge regression techniques (James et al. 2014).

3 Statistical Procedure for Prediction of TBM Performance

To build a prediction model for TBM performance, the construction of statistical procedures must be preceded. As shown in Fig. 2, the overall procedure for prediction of field penetration index (FPI) was established by utilizing the python programming language (Sanner 1999). It can be divided into two stages such as a preprocessing for processing dataset and a model establishment. The preprocessing stage includes diagnosis of abnormal dataset and normalization which should be preceded for construction of a multiple linear model. Then, various multiple linear models are established and verified.

3.1 Preprocessing Part

Abnormal dataset can have a large effect on variations of the coefficient estimates. It should be removed for its subsequent analysis. Elimination of dataset corresponding to outliers with influential point is one of the statistical method for diagnosis of abnormal dataset (Cousineau and Chartier 2010). Outliers are defined as dataset which is far from the regression model and can be diagnosed by using externally studentized residual (ESR) with an absolute value higher than 2. Influential point is also defined as dataset which great affect the variance of coefficient estimates and can be calculated using DFFITS suggested by Welsch (1980). This formula can be derived in combination of ESS and hat matrix value (HMV). Where, HMV means that the value of the input parameter x is far from the mean of x.

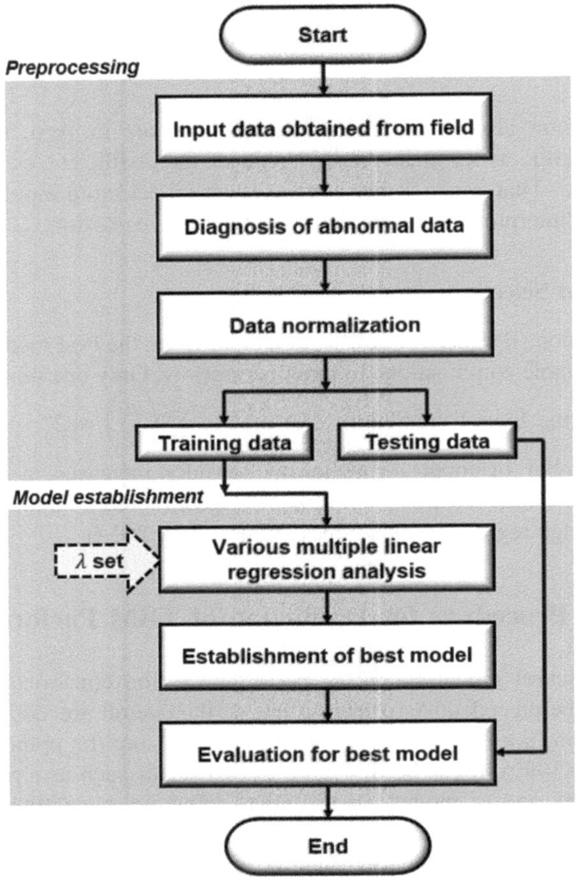

Fig. 2. Statistical procedure for prediction of TBM performance

After diagnosing abnormal dataset, it is followed by normalization of dataset using Eq. (4).

$$\bar{x} = \frac{x - x_{min}}{(x_{max} - x_{min})} \quad (4)$$

x is a parameter before normalization. As x approaches x_{min}, $\bar{x}$ converges to zero. Conversely x converges to one as it approaches x_{max}.

The normalized dataset are then randomly divided into training (80%) and testing (20%) data sets. Training dataset is used to build the model, and the testing dataset can be used to validate the constructed model.

3.2 Model Establishment Part

In this study, cross validation was selected as a criterion for model selection. In cross validation procedure, the dataset is clustered into several folds of same size. Sequentially, each fold is considered as the validation dataset for the model constructed by the all of rest fold respectively. Mean squared error (MSE) obtained from this procedure is called a cross validation mean square error (cv. MSE). It can directly evaluate the prediction performance of model. And then, the constructed model is validated by using testing dataset.

4 Implementation of Statistical Procedure

4.1 Description of Data and TBM

The database obtained from the shield TBM field in South Korea was applied to the statistical procedure to verify the predictive performance of the model. This field is TBM cable tunnel for electricity supply between substations, and the section where shield TBM is applied is 2258 m long. As a result of laboratory tests on samples collected from field, uniaxial compressive strength (UCS), total core recovery (TCR), rock quality designation (RQD), rock mass rating (RMR), Lugeon value (Lv), absorption ratio (Ar) and deformation modulus (Dm) were obtained in the design stages. As output parameter, FPI was analyzed to have the highest r squared between

Table 1. Descriptive statistics for dataset obtained from cable tunnel in South Korea

Parameter	N	Min	Max	Mean	Std. dev	Cov
UCS (MPa)	25	20.15	95.26	51.81	21.06	40.65
TCR (%)	25	94	100	99.56	1.53	1.54
RQD (%)	25	47	100	77.16	18.28	23.69
RMR (%)	25	54	84	68.6	8.78	12.80
Lv	25	0.01	15.89	1.95	3.37	172.82
Ar (%)	25	0.12	2.91	0.54	0.66	122.22
Ev (m/s)	25	3748	5252	4629	456	9.85
Dm (MPa)	25	14,821	77,036	41,445	17,852	43.07
Thrust (kN)	25	1532	3454	2718	528	19.42
RPM (rev/min)	25	7.07	7.26	7.17	0.04	0.56
FPI (kN/cutter/mm/rev)	25	12.47	68.74	33.38	15.49	46.41

Std. dev Standard deviation, *Cov* Coefficient of variation, *Lv* Lugeon value, *Ar* Absorption ratio, *Ev* Elastic wave velocity, *Dm* Deformation modulus

various rock mass parameters based on Hassanpour et al. (2009). For this reason, FPI was selected as output parameter in this study instead of penetration rate and penetration per revolution. As more details for field, the descriptive statistics except for two

Table 2. Specification of shield TBM

Specification of TBM	
Type	EPB type shield TBM
Machine diameter (mm)	3330
Cutter diameter (mm)	350
Number of disc cutter	26
Total thrust (kN)	9600
RPM (rev/min)	9

abnormal datasets was summarized in Table 1, and the specifications of TBM was represented in Table 2.

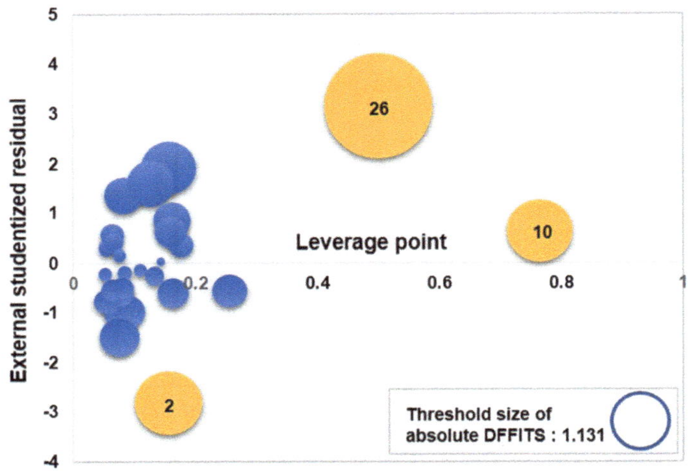

Fig. 3. Diagnosis for abnormal dataset

Diagnosis of abnormal dataset was performed with 27 datasets, and the results was shown in Fig. 3. The vertical axis of the graph represents ESR for outliers, and the size of the circle represents the magnitude of the absolute value of DFFITS. Dataset satisfying outliers and influential points simultaneously were number 2 and 26, which were eliminated. It would seem that this abnormal dataset was induced by intentional decrease of advance rate due to construction process.

4.2 Evaluation of Various Linear Model

As outlined in the introduction, various techniques, such as OLS, lasso, ridge and BSS, were applied for the purpose of comparison with several models. Subsequently, it was evaluated by using testing dataset.

Table 3. Comparison between various multiple linear model using testing dataset

Case	Lasso		OLS		Ridge		BSS	
	R squared	MAPE (%)	R squared	MAPE (%)	R squared	MAPE (%)	R squared	MAPE (%)
1	0.636	19	0.356	24.7	0.561	18.1	0.481	23
2	0.662	30.9	0.312	35.1	0.595	29.7	0.569	34.2
3	0.867	15.4	0.913	14.3	0.82	15.8	0.808	20.1
4	0.934	8.8	0.937	9.3	0.806	13.3	0.972	5.8
5	0.721	36.1	0.53	46.2	0.581	42.6	0.843	27.1
6	0.718	22.9	0.091	34.1	0.557	26.5	0.789	19.5
7	0.453	29.8	0.15	28.6	0.432	31.2	0.574	20.4
8	0.316	30.2	0.148	28	0.323	26.4	0.126	32.7
Mean	0.663	24.1	0.430	27.5	0.584	25.5	0.645	22

Meanwhile, when splitting dataset repeatedly into training and testing set, its characteristics can be changed, which can lead to make the biased results of the analysis. In order to prevent this, eight times were repeated and analyzed. Table 3 shows the results of the prediction performance for several models using the testing dataset. Where, r squared explain the relationship between the observed FPI and the predicted FPI, and the mean absolute percentage error (MAPE, %) can be expressed as the following Eq. (5).

$$MAPE = \frac{|predictedvalue - observedvalue|}{observedvalue} \times 100 \quad (5)$$

The lasso model had the highest R squared of 0.663 which is similar with that of BSS model. Also, BSS model had the lowest MAPE of 22% among the rest of models. That means that these two model can be applicable in terms of predictive power.

Table 4. Number of input parameter applied to each model

	Technique	Case 1	Case 2	Case 3	Case 4	Case 5	Case 6	Case 7	Case 8	Mean
Number of applied input parameter	Lasso	4	3	4	6	4	5	6	4	4.5
	OLS	8	8	8	8	8	8	8	8	8
	Ridge	8	8	8	8	8	8	8	8	8
	BSS	2	3	2	3	3	3	3	3	2.8

Table 4 shows the number of input parameters applied to model respectively. The BSS model used approximately 3 input parameters, which means the best model in terms of explanatory power. On the other hand, OLS and ridge regression model were determined as the poorest explanatory power due to the consideration of all input variables.

5 Conclusions

Comprehensive statistical procedure for prediction of TBM performance was established to develop MLR model. Also, comparison between various MLR model was conducted based on the database obtained from cable tunnels in South Korea.

The lasso model had the highest R squared of 0.663 which is similar with that of BSS model. Also, the BSS model had the lowest MAPE of 22% among the rest of models. That means these two model could be applicable in terms of predictive power. If more input parameters are expected to be used in the future than the number of input parameters used in this study, BSS techniques should be considered carefully because its computing time can take longer than that of lasso regression techniques.

BSS model used approximately 3 input parameters, which means the model with good explanatory power. On the other hand, OLS and ridge model were represented as the poorest explanatory power due to the consideration of all input variables.

The limitation of this study is that only twenty-five small number of dataset are used. The results derived through this process does not provide reliable value due to lack of dataset. The reliability of statistical analysis result can be enhanced with more dataset. Therefore, this paper will be useful in selecting MLR techniques roughly in terms of explanatory and predictive performance.

Acknowledgements. This research was supported by a grant (15SCIP-B105148-01) from the Construction Technology Research Program funded by the Ministry of Land, Infrastructure, and Transport of the Korean government.

References

Bruland, A.: Hard rock tunnel boring. Ph.D. thesis, Norwegian University of Science and Technology (1998)

Cousineau, D., Chartier, S.: Outliers detection and treatment: a review. Int. J. Psychol. Res. **3**(1), 58–67 (2010)

Farmer, I.W., Glossop, N.H.: Mechanics of disc cutter penetration. Tunn. Tunn. **12**(6), 22–25 (1980)

Hamidi, J.K., et al.: Performance prediction of hard rock TBM using Rock Mass Rating system. Tunn. Undergr. Space Technol. **25**(4), 333–345 (2010)

Hassanpour, J., et al.: Developing new equations for TBM performance prediction in carbonate-argillaceous rocks: a case history of Nowsood water conveyance tunnel. Geomech. Geoeng. **4**(4), 287–297 (2009)

Hoerl, A.E., Kennard, R.W.: Ridge regression: biased estimation for nonorthogonal problems. Technometrics **12**(1), 55–67 (1970)

Hughes, H.M.: The relative cuttability of coal-measures stone. Min. Sci. Technol. **3**(2), 95–109 (1986)

James, G., et al.: An Introduction to Statistical Learning: With Applications in R. Springer, New York (2014)

Sanner, M.F.: Python: a programming language for software integration and development. J. Mol. Graph. Model. **17**(1), 57–61 (1999)

Tibshirani, R.: Regression shrinkage and selection via the lasso. J. R. Stat. Soc. Ser. B Methodol. **58**, 267–288 (1996)
Welsch, R.E.: Regression sensitivity analysis and bounded-influence estimation. In: Evaluation of Econometric Models, pp. 153–167. Academic Press, New York (1980)
Yagiz, S.: Utilizing rock mass properties for predicting TBM performance in hard rock condition. Tunn. Undergr. Space Technol. **23**(3), 326–339 (2008)

The Effect of Degree of Fouling of Coal Dust on Railway Foundation Stability

F. N. Okonta(✉)

Department of Civil Engineering Science, University of Johannesburg, Johannesburg, South Africa
fnokonta@uj.ac.za

Abstract. Railway ballasts and large aggregates can be fouled internally by inter particle weathering and externally by soil pumping and deposition of rail fright particles. The gradual accumulation of soil fines and coal particles on road pavements, rail ballast and sleepers affect the aggregate interlocking properties. Track maintenance and ballast replacement schedule are dependent on the different classification methods used to evaluate the degree of fouling. Uniformly graded large aggregates were fouled with different percentages of residual fines of coal dusts that were prepared at different moisture content (w). The degree of fouling of the coal dust fouled aggregates (CDFA) were classified according to the percentage of fouling (PF) and Relative ballast fouling ratio (RBFR). The interlocking properties of the fouled aggregates were investigated in a large direct shear device and the Intergranular Friction Angles (IFA) were determined. Degree of fouling up to 8% had no effect on the dry density of the aggregate, between 8 and 16% the effect of fouling was marginal, however linear relationship was indicated beyond 16% fouling. The rate of decrease of IFA was found to be dependent on the degree of fouling and moisture content of the coal material. The IFA of the CDFA tend towards the IFA of the wet coal materials as the degree of fouling increase to 20% of the dry mass of the aggregate. 20% fouling represent moderately fouled and fouled conditions in accordance with PF and RBFR fouling classifications respectively. RBFR is based on the inter aggregate void volume occupied by the fouling material and specific gravity of coal. It was found to be more sensitive to the changes in IFA than PF. Regression models of the effect of coal dust fouling on aggregate integranular and interface friction based on IFA, MC, PF and RBFR were formulated and proposed for the development of track maintenance scheme, ballast replacement schedule and standard track inspection routines.

Keywords: Angle of friction · Gravels · Direct shear strength
Interface shear strength · Fouling · Railway ballasts · Percentage of fouling (PF) and relative ballast fouling ratio (RBFR) · Shear box apparatus
Optimum moisture content · Maximum dry unit weight · Coal dust

B. Fatahi et al. (eds.), *Sustainable Design and Construction for Geomaterials and Geostructures*, Sustainable Civil Infrastructures,
https://doi.org/10.1007/978-3-319-95753-1_19

1 Introduction

Mining and Agriculture are the major employment sector in Mpumalanga, providing jobs and contributing to over one fifth of Mpumalanga's Gross Geographic Product (GGP) (Thys 2003). Extensive coal beds are situated in many parts of South Africa with significant concentrations in Mpumalanga Province and Gauteng Province (Thys 2003). The Ermelo railway siding shown in Fig. 1, links major industrial railways used by Eskom power stations to transfer coal. Derailment in Emerlo Industrial sidings has been reported and linked to poor signals, settlement of rail tracks and slippages. Most of the accidents were recorded in wet weathers. Emerlo Industrial sidings is used for storing, loading and unloading and store stationary rolling stock, especially for loading and unloading coal materials from mines and quarries to factories, wharves, warehouses. The Ermelo rail sidings was also used for maintenance of railway equipment to allow trains to pass and also used to store alternate or spare engines.

Fig. 1. The Ermelo railway siding. The coal investigated by the author was taken from the siding at Ermelo in Mpumalanga Province

Railway foundation engineers often describe good quality ballasts as broken pieces of mainly igneous and sedimentary rocks of average particle size range of 25–60 mm, over which the railway tracks are laid. They are subangular and angular shaped rocks of high toughness and hardness, high resistance to weathering, minimum fouling from aggregate breakdown, of rough surfaces and minimum hairline cracks (Rottcha 2011; Tutumluer et al. 2009). The sub angular particle structure with predominant sharp corners enhance ballast-ballast and ballast-sleeper interlocking, which in turn ensures efficient transfer of train load, rail foundation stability and proper vertical and horizontal alignment of rail track. The filling of the open graded ballast with fine materials (fouling) from soils, coal etc. is known to affect morphological properties, static strength of ballast and the life span of rail and sleepers as these properties depend on the interlocking ability of the ballast. In South Africa, more than 50% of the total power generated for industrial and domestic use is derived from coal. Rail is a common means of coal transportation and the associated problems are the gradually deposition of fine coal particles on rail ballast materials. Conventional methods employed to determine the most economical time to clean or replace polluted and degraded ballast entail visual evaluation for evidence of fouling, pumping and ponding at ditches and shoulders, as well as ballast sampling and testing for fouling by laboratory sieve analyses of the bulk aggregate particle.

The study of the effect of wetting on the interlocking properties of aggregate - coal mixture in unbound matrix will facilitate the prediction of the service performance of railways foundation aggregates, because of the tendency of coal to exhibit significant decrease in strength in the wet state. As ballast ages, it can be progressively fouled by numerous fine materials, whose accumulation in the voids of ballast can result in a decrease in shear strength with reduced resiliency and drainage capability of the ballast (Janardhanam and Desai 1983). An increase in moisture in fouled ballast and infiltration of water to the subgrade can promote soft soil pumping or excessive deformation of the formation.

Sources of ballast fouling consist of ballast particle degradation, infiltration of fine foreign particles deposited on the track surface, sleeper wear, as well as sub-ballast and subgrade infiltration (Selig and Waters 1994). The common parameters used for evaluating ballast fouling can be divided into two groups, fouling index (FI) and percentage fouling (PF) which are based on textural properties and mass of the fouling material in relation to the mass of the dry ballast and percentage void contamination (PVC) and the relative ballast fouling ratio (RBFR) which are based on based on the void ratio and specific gravities of the fouling materials and ballasts. Because of the different material physical properties on which they are based, degree of correlation of the different fouling classification parameters and interlocking properties induced by fouling have not been investigated. Classification of fouling by mechanical interlocking properties is vital as it is indicator of rail road degradation. The effect of fouling on the intergranular friction differ depending on the loading mode. It was noted by Marto et al. (2009) and Tutumluer et al. (2009). Change in intergranular friction due to monotonic loading represent the upper bound of friction resistance associated with optimal interlocking. Other properties of rail ballast especially the permeability of the ballast bed also decreases as a result of increase in fouling. The percentage decrease in permeability is dependent on the grading of the ballasts and optimum for well graded ballasts. A number of investigations on the static and dynamic properties of ballasts and the effect of geofabric on rate of ballast degradation rates are well documented (Gentzis 2007), studies on the geotechnical properties of coal in relation to engineering applications was detailed (Indraratna et al. 2006; Jasinge 2009; Marto et al. 2009; Tutumluer et al. 2009). Also as noted by Indraratna and Salim (2005), the most importance factors of the ballast are the parent rock and the particle shapes. Recently, Pen et al. (2013) tested railway ballasts with different particle sizes varying from 9.5 to 62.5 mm to study the angularity of particles. As reported in Pen et al. (2013), due to a negligible variation in angularity, scaled-down materials are appropriate substitutes for testing purposes. Cyclic triaxial testing conducted using three parallel gradations of ballast (with the maximum size of 63.5, 38.0 and 19.0 mm) by Sevi and Ge (2012) reported that the particle shape and relative density are the most important parameters but the evaluation of geotechnical parameters as optimal indicators of coal dust fouling in relation to ballast interlocking properties have not been investigated in detail.

The effect of fine coal particles on the interlocking strength of uniformly graded gravel in direct shear was investigated. The objectives are to evaluate the effect of coal fines on the performance of railways especially during the rainy seasons and to identify and recommend physical and mechanical indicators that may better inform ballast screening and cleaning methods for fouled railways.

1.1 Fouling Parameters

The fouling index parameter (FI) was proposed by Selig and Waters (1994) for the description of ballast fouling based on gradations. Ballast fouling based on gradations obtained from representative ballast samples in North America was represented by Eq. (1)

$$FI = P_4 + P_{200} \tag{1}$$

where P_4 and P_{200} are percentages of ballast particles passing the number 4 sieve (4.75 mm) and number 200 sieve (0.075 mm), respectively. The categories of fouling based on FI as defined by Selig and Waters (1994) are listed in Table 1.

Table 1. Categories of fouling based on the fouling index, percentage of fouling, and relative ballast fouling ratio

Category	Fouling index (%) Selig and Waters (1994)	Percentage of fouling (%)	Relative ballast fouling ratio (%) Indraratna et al. (2011)
Clean	<1	<2	<2
Moderately clean	1 to <10	2 to <9.5	2 to <10
Moderately fouled	10 to <20	9.5 to <17.5	10 to <20
Fouled	20 to <40	17.5 to <34	20 to <50
Highly fouled	≥ 40	≥ 34	≥ 50

An index related to FI is the percentage of fouling (PF) (% fouling), which is the ratio of the dry weight of material passing a 9.5 mm sieve to the dry weight of the total sample.

By comparing the PF and the average void content of large aggregates fouled by different materials, Indraratna et al. (2011) formulated another fouling parameter, the relative ballast fouling ratio, (*RBFR*). It is a ratio between the solid volumes of fouling particles (passing a 9.5 mm sieve) and ballast particles (particles being retained on a 9.5 mm sieve).

The relative ballast fouling ratio was expressed by Eq. (2)

$$RBFR = \frac{M_f\left(\frac{G_{b-f}}{G_{s-f}}\right)}{M_b} \times 100\% \quad (2)$$

where M and Gs are the dry mass and specific gravities, respectively, and subscripts f and b represent the fouling materials and ballast, respectively. G_{b-f} and G_{s-f} are the specific gravities of the ballasts and foiling material. In Eq. (2), only the mass of the ballast and the mass and specific gravity of the fouling material need to be measured.

2 Materials and Test Methods

The coal materials were collected from an open cast coal mine close to Nelspruit, in Mpumalanga Province, South Africa. The coal material was oven dried at temperature of 60 °C to constant mass over three days, grounded to fine particles and sieved through the 0.075 mm mesh and sealed in airproof bags. The coal material could not be oven dried at 100 °C because the organic structure of the coal is distructured and contain small quantities of ash. The particle size curve of the residual granitic filler soil that is commonly used for road construction is shown in Fig. 2.

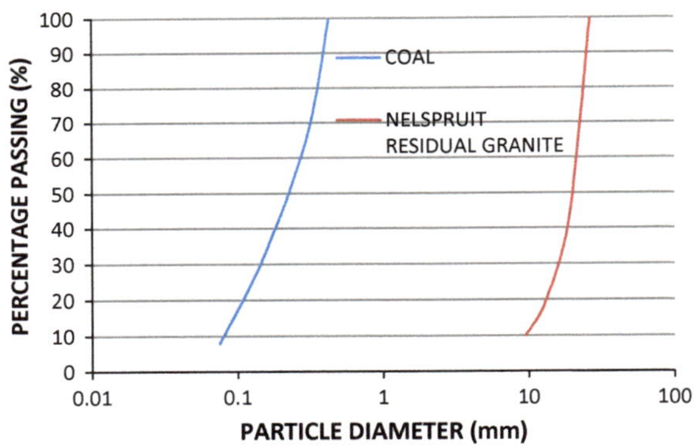

Fig. 2. The particle size curves of the granite aggregate and the coal were presented

The determinations of physical properties were in accordance to TMH (1996) and TRH 4 (1994). The soil material used for the strength properties tests shown in Fig. 2 were gravel fractions with particle distribution between 26.5 and 13.2 mm, with more than 65% larger than 19 mm.

The maximum density of the aggregate was determined by the use of dead weights on the samples placed on a vibrating table. The maximum densities and optimum moisture content of the dust fouled aggregates were determined by the modified proctor compaction method. Two major sets of specimens were investigated; specimens that were compacted to 90% maximum dry density and OMC, and dry specimens that were compacted to 90% maximum dry density. An initial concern was the non uniform distribution of dry coal dust within the gravel for specimens with low degree of fouling (Kim et al. 2005). However visual observation after compaction revealed uniform distribution of coal particles at the mid section. It was also decided to average the results of three tests.

2.1 Direct Shear Test

The direct shear properties of mixtures of coal and the granitic stones were studied in a fabricated 200 mm square shear box apparatus. The device was driven by electronic motor and equipped with boundary load cell on the expanded track rail of the conventional square box. The specimens were compacted to height of 150 mm and sheared at a strain rate of 0.5 mm/min. Although the selected strain rate does not simulate the field condition of interest, it does not allow for the build up of excess pore pressure especially within the wet specimens. Different percentages of coal fines per dry mass of the aggregates of 0, 5, 10, 15 and 20% were tested, at OMC and in dry state at applied normal stresses of 100, 200 and 300 kPa.

The series of tests conducted are:

- Series 1: Direct shear test on dry gravel and coal fines at different moisture contents.
- Series 2: Direct Shear test on gravel mixed with increasing percentage of coal fines at 95% dry densities and optimum moisture contents.
- Series 3: Direct shear test on gravel mixed with increasing percentage of dry coal fines at 95% maximum dry densities.

3 Results and Discussion

3.1 Physical Properties

The major compounds identified by XRD in the coal samples are silicon oxide, copper aluminium sulphate, benzene, Iron arsenide sulphate, Aluminium silicate (Rottcha 2011). The presence of arsenides sulphate and benzene makes the coal an environmentally toxic material (Karim et al. 1997; Kim et al. 2006). The particle size constitution of the as dug coal were gravel size = 75%, sand size = 15%; Fines (>0.425) = 10%. Only the fines shown in Fig. 2 with particle size range of 100% < 0.425 mm and 8% < 0.075 mm were used for this study. The liquid limit of the fresh coal was 16% and the plastic limit was 13%. The specific gravity was **1.25**. The maximum dry density of the coal is 1238 kg/m^3 and the optimum moisture content was 13%. The particle size curves of the granite aggregate and the coal were presented Fig. 2. The particle size distribution are gravel size = 68%, sand size = 32%; Fines = 15%. The fractions held between the 13.2 and 26.5 mm diameter sieves were used for this study. Granite is an important source of ballasts for railway foundations and coarse aggregate for road construction. The maximum dry density of the granites was 1585 kg/m^3 and the average specific gravity was **2.61**. The effect of coal fines on the compaction properties of the uniform aggregate is shown in Fig. 3. Very low and insignificant increase in optimum moisture content and a decrease in maximum unit weight are indicated with increase in the percentage of coal fines fouling, however the changes become significant as the fouling increases beyond 8%. Degree of fouling up to 8% have no effect on the dry density of the aggregate, between 8 and 16% the effect of fouling was marginal. However linear relationship between dry density and percentage fouling was indicated

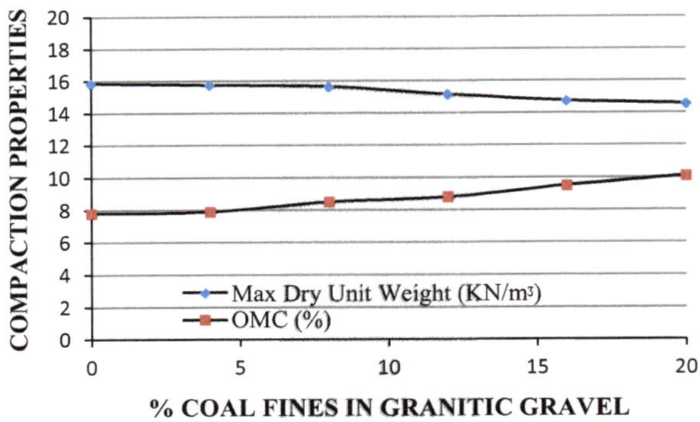

Fig. 3. The effect of coal fines on the compaction properties of the uniform aggregate

beyond 16% fouling. The change in the unit weight was due to the marginal displacement of the aggregate by the increasing volume of the coal fines.

3.2 Direct Shear Strength of Coal Fines and Gravel Aggregates

The effect of moisture content on the direct shear strength parameters of coal fines and gravel aggregates were investigated. The coal fines were compacted at the dry density and moisture content ranges from dry state of the OMC. The stress and deformation curves of compacted coal fines are shown in Fig. 4. The stress strain curves are elasto plastic and minimal strain hardening behaviour, typical of soft soil material. The coal fines exhibit shear induced compressive deformation in proportion to the magnitude of applied normal stress. The coal fines that were tested dry at zero moisture content indicated stress strain curves with weak trend of strain softening behaviour that decreased with increasing normal stress and exhibited marginal contraction that increased with increase in the applied normal stress. The direct shear envelopes of the coal fines and gravel aggregates are shown in Fig. 5. The direct shear strength envelope showed that the gravel aggregate has a frictional angle of 35° [$\tan^{-1}$ 0.688]. An increase in moisture content of the coal fines from the dry state to the OMC for 20% fines fouling which is a moisture content of 10%, resulted in increase in friction angle from 19° [$\tan^{-1}$ 0.354] to 6° [$\tan^{-1}$ 0.11]. The friction angles of gravel and coal fouled gravel were presented in Table 1.

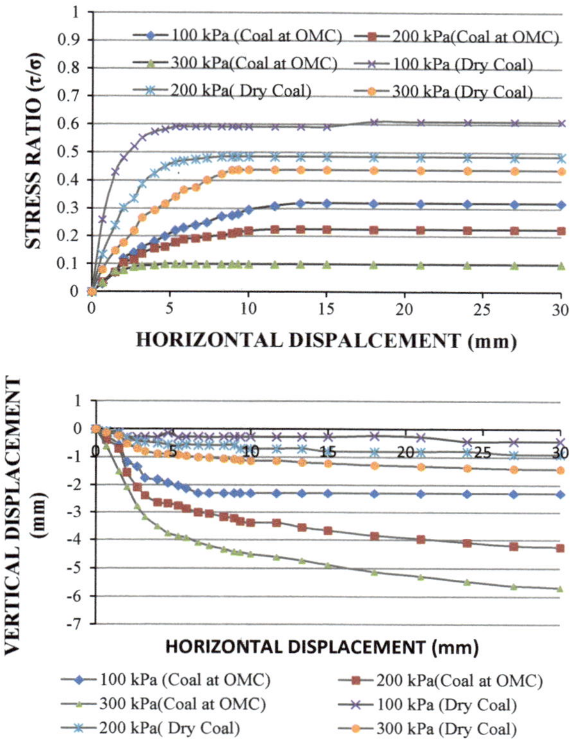

Fig. 4. The stress and deformation curves of coal fines at OMC and dry state

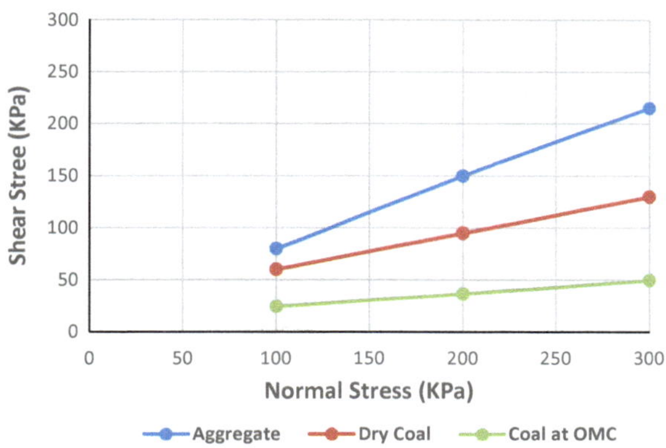

Fig. 5. Strength envelopes of coal and aggregate

3.3 Effect of Dry Coal Fines on Gravel

Series of tests were conducted on compacted mixtures of dry coal fines and gravel. The stress mobilized by the granitic material due to increasing normal stress from 100 to 300 kPa in direct shear is shown in Fig. 6. The maximum shear stress increased with increase in normal stress. For the degree of fouling investigated, no defined maximum peak stress ratio was indicated. At 4% coal fouling, the stress strain behaviour is general strain hardening elasto plastic response at high normal stress, however there is evidence of yielding and straining hardening at lower normal stress of 100 kPa. At 20% fouling the stress strain response is straining hardening response. The stress ratio at 4% fouling tends to a constant value at large imposed horizontal displacement. Increased degree of fouling resulted in decreased mobilized stress ratio for the range of normal stress applied. The effect of increase in the amount of coal fines on the intergranular friction angle of the granitic stones was presented in Tables 2 and 3. As the percentage of dry coal fines increases to 20%, the angle of friction of the granitic stones tends towards the angle of friction of the coal fines at the optimum moisture content, implying that the granite–granite intergranular friction is significantly reduced and but

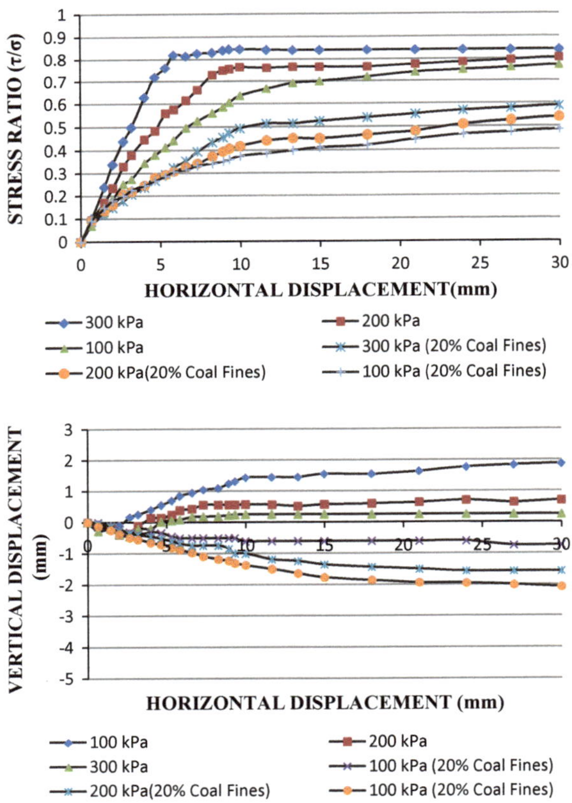

Fig. 6. The stress and deformation curves of compacted coal fouled aggregates in dry state

not completely eliminated by the dry coal fines. Detailed tests on the change in strength parameters due to increase in moisture content of the coal fines reported by Tutumluer et al. (2009) indicated significant decrease in friction angle. The stability of compacted aggregates is a function of friction and deformation. The direct shear induced vertical strain, showed dilatancy at 4% fouling and minimal vertical compressive strain with a maximum value of (2/60) at 20% fouling, thus direct shear interlocking stability was not significantly impaired when the particles are dry.

3.4 Effect of Coal Fines at OMC on Gravel

The effect of mixture of coal fines and gravel that were compacted at OMC on stress strain behaviour and friction angles was also investigated. The stress strain behaviour and stress ratio mobilized by the fouled gravel material due to applied normal stress of 100–300 kPa in direct shear was shown in Fig. 7. For the degree of fouling investigated, no defined maximum peak stress ratio was indicated. At 4% coal fouling, the stress strain behaviour is general strain hardening elasto plastic response at high normal stress, however unlike the dry coal dust; there is no evidence of yielding at lower normal stress of 100 kPa. At 20% fouling the stress strain response is also a straining hardening response. The stress ratio at 4 and 20% fouling tends to a constant value at large imposed horizontal displacement. Increased degree of fouling resulted in decreased mobilized stress ratio for the range of normal stress applied. The effect of increase in the amount of coal fines on the angle of friction of the granitic stones was shown in Tables 2 and 3. As the percentage of wet coal fines increases to 20%, the friction angle of the granitic stones decreased towards the friction angle of the coal fines at the optimum moisture content, implying that the granite–granite intergranular friction is significantly reduced and almost completely eliminated by the wet coal fines. Dilatant direct shear induced vertical strain was indicated by 4% degree of fouling and

Table 2. Intergranular Friction Angles of coal fouled aggregates compacted to different moisture content

% Fouling	Coal (OMC)	CDFA (dry)	CDFA (OMC)
0	6	40	40
4	6	39	39
8	6	37	37
12	6	33	32
16	6	29	22
20	6	27	18

Table 3. Atterberg limits of coal materials

Material	LL	PL
Fresh coal	16	13

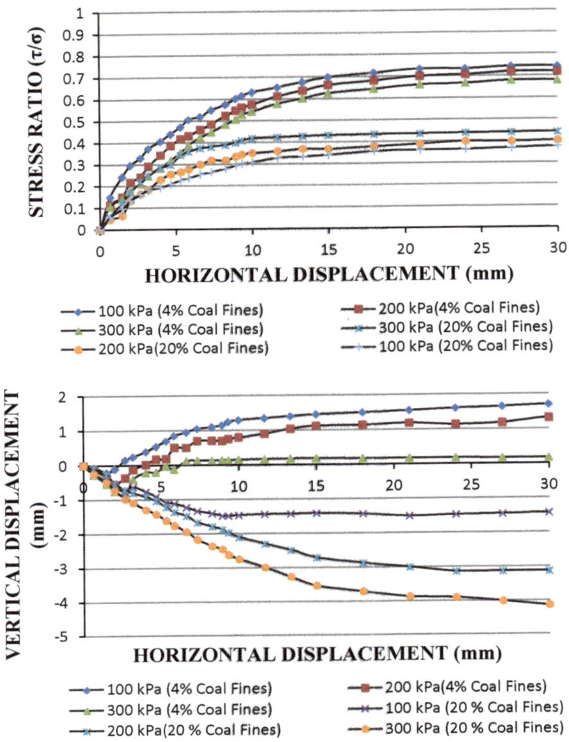

Fig. 7. The stress and deformation curves of coal fouled aggregates compacted at OMC

significant vertical compressive strain with a maximum value of (4/60) at 20% fouling, which is twofold the value indicated by the dry mix. Thus the direct shear interlocking stability can be significantly impaired due to significantly reduced friction angle and increased direct shear induced compressive vertical strain.

3.5 Interface Shear Strength at OMC

The effect of different constitutions of gravel and coal fines that were compacted at OMC on the interface direct shear stress and deformation behaviour was presented in Fig. 8 and Tables 2 and 3. At low degree of fouling, strain hardening response with well-defined peak stress ratio as well as shear induced dilatant volume change was indicated for the normal stresses applied. At 20% degree of fouling, a weakly defined peak followed by strain hardening behaviour was indicated. This behaviour is similar to the stress strain response of 4% fouled aggregate in dry state. For samples compacted at OMC, the shear induced compression of 20% degree of fouling due to interface shear was lower than the value due to direct shear. Figure 8 showed the strength envelopes of coal fouled aggregates (dry and at OMC), soil concrete interface and coal fines at OMC. The DSIFA of the clean aggregate was 50° and Fig. 9 shows that for fouling up to 20%, the mobilized DFISA was higher than the mobilized DSFA. The DSIFA of the clean aggregate was 50°. A general trend of decrease in interface friction angle with

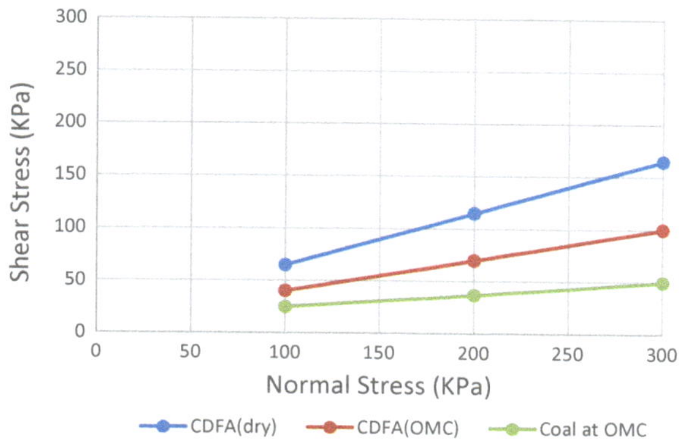

Fig. 8. Strength envelopes of coal fouled aggregates (dry and at OMC), soil concrete interface and coal fines at OMC

increase in coal fines at OMC was exhibited. In particular, a reduction of DSIFA from 50° to 25° or 50% reduction in friction angle, due to increased in degree of fouling by 20% was indicated. In the dry state, the friction angle decreased significantly while the vertical strain increased slightly with increase in fines, thus direct shear interlocking stability was not significantly decreased when the coal particles are dry. In the wet state, reduced stress ratio and increased shear induced vertical compressive strain resulted in significantly decreased intergranular and aggregate concrete slipper interlocking stability, as the friction angle of the fouled aggregate tend to the friction angle of the wet coal material.

Thus although surface depression due to shear induced vertical compression may not be significant, 50% reduction in concrete—aggregate interface friction angle due to 20% coal fouling can significantly impair rail line stability.

4 Intergranular Friction Angles Degradation Ratios in Relation to Fouling Indices

The different degree of fouling investigated were 0–20% of coal fines per dry mass of aggregates. The fines (>0.425 mm) were smaller than 9.5 mm which is the maximum diameter of fouling particles used for the classification of different degrees of fouling represented by the percentage of fouling (PF) (% fouling). In order to classify the different degrees of fouling investigated (0–20%) in terms of PF, were percentages of fouling were adjusted, i.e. 20% fouling was adjusted to (20/120 * 100) = 16.6%. In relations to the PF Classification in Table 3, 16.6% represent a moderately fouled material.

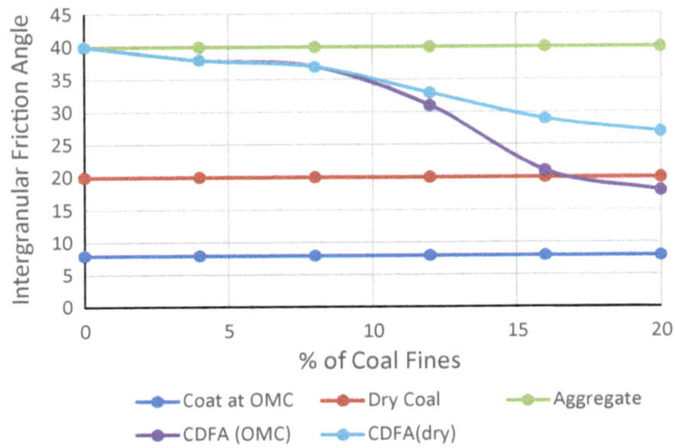

Fig. 9. The change in interlocking friction angles due to coal fouling

However in order to classify the different degrees of fouling investigated in terms of RBFR, the specific gravities of coal (1.25) and Gravel (2.61) were used in Eq. (3). Thus 20% fouling i.e. (2.61/1.25 * 20) = 41.76 (42). In relations to the RBFR classification in Table 3, 42% represent a fouled material.

The rate of degradation of intergranular and interface friction angle due to coal fouling (PF, RBFR) and moisture content, can be represented by the regression Eqs. (3) and (4)

$$\frac{\phi - \phi_f}{\phi} = C_1[F] + C_2[W_1] \tag{3}$$

F is degree of fouling represented by PF or RBFR

C_1 is the coefficient of the rate of fouling

C_2 is the coefficient representing the reduction in friction angle due to change in the moisture content of the fouled aggregate.

$$W_1 = \frac{\phi_{CDFA(OMC)}}{\phi_{CDFA(DRY)}}$$

$$W_2 = \frac{\phi_{CDFA(DRY)}}{\phi_{CSIFA(OMC)}}$$

The regression equation of the degradation ratio of intergranular friction angle due to different amount of fouling classified in terms of PF and RBR and change in moisture content of the fouled aggregate was represented by Eqs. (4) and (5)

$$R = \frac{\phi - \phi_f}{\phi} = 0.241 - 0.253[w] + 0.0158[PF] \tag{4}$$

$$R = \frac{\phi - \phi_f}{\phi} = 0.241 - 0.253[w] + 0.006[RBFR] \tag{5}$$

The critical degree of fouling below which the degradation ratio is zero [Fc] was represented by Eq. (6)

$$F_c = \frac{C}{C_f}. \tag{6}$$

C is the constants of Eqs. (4), (5), i.e. 0.241,
C_f is the coefficients of PF or RBFR.

If the fouling material contains insufficient fine particles, the ballast voids cannot be fully filled. In this case, the ballast bed will sustain drainage and resilience capacities to an acceptable extent. On the other hand, if there are more fines available to fill the voids, a greater PF will result with unsatisfactory drainage and resilience capacities of the ballast Indraratna et al. (2011). However, PF cannot differentiate between types of fouling material including the differences in specific gravity (e.g. coal compared to soil and pulverized rock). RBFR is the ratio of the volumes of the fouling material and void volume of ballast aggregates. Therefore, in comparison with FI, the magnitude of RBFR can better represent the degree of fouling by materials of different specific gravities and thus a more accurate parameter for the tracking of in situ fouling.

5 Conclusions

Coal drives the bulk of sub saharan economy. Rail transportation over large distances in semi arid climate distributes fine particles of coal on rail foundations. Road transportation of coal by trucks is also common. The effect of increasing amount of fine coal dust on the shear strength parameters of Nelspruit granitic gravel investigated in a medium size direct shear apparatus show that the stress strain behaviour of the fouled granitic material can be strain hardening or weak strain softening elasto plastic response depending on sample moisture content and magnitude of applied normal stresses.

The effect of fouling on direct shear particle–particle interlocking stability was dependent on the moisture content of the coal. In the dry state, the friction angle decreased significantly while the vertical strain increased slightly with increase in fines, thus direct shear interlocking stability was not significantly decreased when the coal particles are dry. In the wet state, reduced stress ratio and increased shear induced vertical compressive strain resulted in significantly decreased intergranular and aggregate concrete slipper interlocking stability, as the friction angle of the fouled aggregate tend to the friction angle of the wet coal material.

The stability of bound and unbound granular materials is governed by friction angles and shear induced strain. The overall implications of the result of this study is that there is a high likelihood of the stability of ballasts being significantly impaired in wet seasons due to increased degree of coal dust fouling and therefore more stringent screening should be recommended during the rainy seasons.

Aggregates that were exposed to 20% fouling which is 16.6% in PF Classification represent a moderately fouled material. However aggregates that were exposed to 20%

fouling or 42% in RBFR classification represent a fouled material. Because 20% fouling result in 50% reduction in CSIFA for fouled aggregates compacted at OMC, RBFR was thus more sensitive to the changes in CSIFA than PF.

A 20% coal dust fouling can threaten the stability of ballasts especially in rainy seasons and air based cleaning should be implemented at frequency that can ensure that dust accumulation up to 20% fouling never occurs in service.

Acknowledgements. The field work and fabrication of shear box was funded by Rand Coal Engineering PLc. The author is grateful.

References

Das, S.K., Yudhbir, S.: Geotechnical characterization of some Indian fly ashes. J. Mater. Civ. Eng. **17**(5), 546–556 (2005)

Gentzis, T.: Geomechanical properties and permeability of coals from the Foothills and Mountain regions of Western Canada. Int. J. Coal Geol. **69**(3), 153–164 (2007)

Indraratna, B., Salim, W.: Mechanics of ballasted rail tracks—a geotechnical perspective. Taylor and Francis, London (2005)

Indraratna, B., Khabbaz, H., Salim, W.: Geotechnical properties of ballast and the role of geosynthetics in rail track stabilisation. Ground Improv. **10**, 91–101 (2006)

Indraratna, B., Su, L., Rujikiatkamjorn, C.: A new parameter for classification and evaluation of railway ballast fouling. Can. Geotech. J. **48**, 322–326 (2011)

Janardhanam, R., Desai, C.S.: Three-dimensional testing and modeling of ballast. J. Geotech. Eng. **109**(6), 783–796 (1983). https://doi.org/10.1061/(ASCE)0733-9410(1983)109:6(783)

Jasinge, D.: Mechanical properties of reconstituted Australian black coal. J. Geotech. Geoenviron. Eng. **135**(7), 980–985 (2009)

Karim, A., Salgado, R., Lovell, C.W.: Building embankments of coal combustion fly ash-bottom ash mixtures. In: Proceedings of 48th Highway Geology Symposium, pp. 66–74. Knoxville, TN, May 1997

Kim, B., Prezzi, M., Salgado, R.: Geotechnical properties of fly and bottom ash mixtures for use in highway embankments. J. Geotech. Geoenviron. Eng. **131**(7), 914–924 (2005)

Kim, B.J., Yoon, S.M., Balunaini, U.: Determination of ash mixture properties and construction of test embankment—part a. Joint Transportation Research Program, Final Report, FHWA/IN/JTRP-2006/24! Purdue University, W. Lafayette, Indiana, pp 345–352 (2006)

Marto, A., Mahir, A.M., Lee, F.W., Yap, S.L., Muhardi. K.: Morphology, mineralogy and physical characteristics of Tanjung Bin coal ash. In: Proceedings of 4th International Conference on The Recent Advanced in Materials, Minerals and Environment (RAMM) and 2nd Asian Symposium on Material & Processing (ASMP), 1–3 June 2009, Pulau Penang, Malaysia, pp. 34–42 (2009)

Pen, L.W., Powrie, A., Zervos, S., Ahmed, S., Aingaran, S.: Dependence of shape on particle size for a crushed rock railway ballast. Granular Matter **15**(6), 849–961 (2013)

Rottcha, C.: Geotechnical properties of discarded coal from rail sidings. BIng Thesis, (Unpublished) University of Johannesburg, Civil Engineering Department (2011)

Selig, E.T., Waters, J.M.: Track geotechnology and substructure management. Thomas Telford Services Ltd., London (1994)

Sevi, A., Ge, L.: Cyclic behaviors of railroad ballast with the parallel gradation scaling framework. J. Mater. Civil. Eng. **24**(7), 797–804 (2012)

Thys, J.: Mining and population indices in South Africa. Report on Jobs in Provincial Sectors. MP GV Gazzette 0012. Public Works (2003)

TMH: Standard methods of testing road construction material. In: Technical Methods for Highways, vol. 1. Pretoria, South Africa (1996)

TRH 4: Department of Transport. Structural Design of Flexible Pavements for Interurban and Rural Roads. Technical Recommendation for Highways, pp. 1–101, Pretoria, South Africa (1994)

Tutumluer, E., Bombrow, W., Huang, H.: Effect of coal dusts on railroad strength and stability. University of Illinois Lectures. In: 8th International Conference on the Bearing Capacity of Roads Railways and Airfields. June 29–July 2, Champaign Illinois USA (2009)

Explicit Analysis for the Ground Response Behavior Due to the Advancing Excavation of Tunnel in Anisotropy Stress Field

Yu-Lin Lee(✉) and Yi-Xian Xin

Department of Civil Engineering, Chung-Hua University, Hsinchu, Taiwan
rosalee@chu.edu.tw

Abstract. The purpose of this research is that the analytical solutions of non-linear behavior of ground in a deep circular tunnel under anisotropy stress field are derived, and the explicit analysis method (EAM) which bases on the convergence confinement method, the principle of convergence loss, and the simple calculation spreadsheet is particularly proposed. This study includes that introducing an incremental procedure of confinement loss into the analytical solutions derived under anisotropy stress field, establishing the direct calculation logic and flow chart, using a calculation spreadsheet to simply calculate and draw figures to describe the distribution of stress/displacement around tunnel. The results obtained by the explicit analysis and by the finite element analysis are approximately consistent in those curves described in the convergence-confinement method (CCM). The explicit analysis method proposed is easy and directly to simulate the ground response behavior due to the advancing excavation of tunnel.

1 Introduction

The convergence-confinement method (CCM) is a useful theoretical tool for designing support for underground excavations in rock. It consists of three fundamentals: the ground reaction curve (GRC), the support characteristic curve (SCC), and the longitudinal deformation profile (LDP). The ground reaction curve exhibits the increasing trends of radial displacement as the internal support pressure of the tunnel decreases; it plays an important role in determining when the supports are to be installed and how stiff the supports should be. The idea of the ground reaction curve of the rock mass was first presented by Pacher (1964) as a major design component behind the New Austrian tunneling method. The method was further improved by many researchers (Panet and Gellec 1974; Panet 1995) and was proposed later by the French Association of Tunnel and Underground Space (AFTES 1983; AFTES et al. 2001) for application in designs, becoming a useful preliminary tool in the rational design of tunnels.

Many researchers have been working on this method ever since, attempting to develop mathematical and empirical expressions for the ground reaction curve under different behavior hypotheses. The researches on the ground reaction curve from Hoek and Brown (1980), Brown et al. (1983), Brady and Brown (1993), Panet (1995), Wang (1996), Carranza-Torres and Fairhurst (1999, 2000), Oreste (2003), Sharan (2003,

B. Fatahi et al. (eds.), *Sustainable Design and Construction for Geomaterials and Geostructures*, Sustainable Civil Infrastructures,
https://doi.org/10.1007/978-3-319-95753-1_20

2005), Carranza-Torres (2004), Park and Kim (2006), Guan et al. (2007), Serrano et al. (2011), and Alejano et al. (2011) are all applicable to a tunnel that is excavated under hydrostatic stress state conditions. However, only a few studies concerning tunnels excavated under anisotropy stress fields have been conducted (Detournay and John 1988; Panet 1986; Ravandi and Rahmannejad 2013; Shen and Barton 1997; Wong and Kaiser 1991). Originally, the concept of confinement loss λ in the analysis of the convergence-confinement method was proposed by Panet et al. (1982, 1995), and also recommended by AFTES (1983), AFTES et al. (2001) for the purpose of engineering practice design.

The convergence-confinement method is based on four important hypotheses (Oreste 2009): (1) The cross-section of a deep tunnel is circular, (2) The initial stress state is hydrostatic and constant with depth, (3) The rock mass is continuous, homogeneous, and isotropic, and (4) The problem is axis-symmetrical plane strain, and the stress field is planar. In this study, he analytical solution and numerical analysis were used to examine the changes in the ground reaction curve at the tunnel crown, an inclination of 45°, and the side-wall according to various initial in situ stress conditions, and were ultimately used to investigate the influence of the in situ stress ratio K_o (where $K_o = \sigma_h/\sigma_v$, and it relates the horizontal stress σ_h to the vertical stress σ_v) on the ground reaction curve, the stress path at the intrados of the tunnel, and the distribution of stresses/displacements on the cross-sections of a circular tunnel.

2 Analytical Solution for Stresses in the Elastic Region and the Plastic Region

2.1 Stresses in the Elastic Region

The CCM is based on the analysis of the stress and strain state that develops in the rock around a tunnel. The stresses at the periphery of the tunnel are obviously altered due to the advance excavation of the tunnel face. Therefore, the variation of stresses can be mainly interpreted by the stress gradient which is a difference between far-field stress and near-field stress around a tunnel. By using the hypothesis of increments in the numerical analysis, the increment of stresses can be taken into account of the confinement loss λ as a fraction of the stress gradient. This is an important principal concept used by the convergence-confinement method (Lee 1994; Lee et al. 2008, 2009). The variation of stresses can be described by the following representations:

(1) Stresses in far-field (initial in situ stresses, $r \rightarrow \infty$, $\frac{1}{r} \cong 0$):

Before opening the tunnel, the stresses in the rock mass are in equilibrium and are the same as the lithostatic stresses of a non-hydrostatic type ($K_o \neq 1.0$) and constant in the surrounding medium of the tunnel. Hence, the initial in situ stresses in the rock mass can be shown as shown in Fig. 1 and indicated by the following equations:

$$\sigma_r^i = \frac{\sigma_v}{2}[(1 + K_0) + (1 - K_o)\cos 2\theta] \tag{1}$$

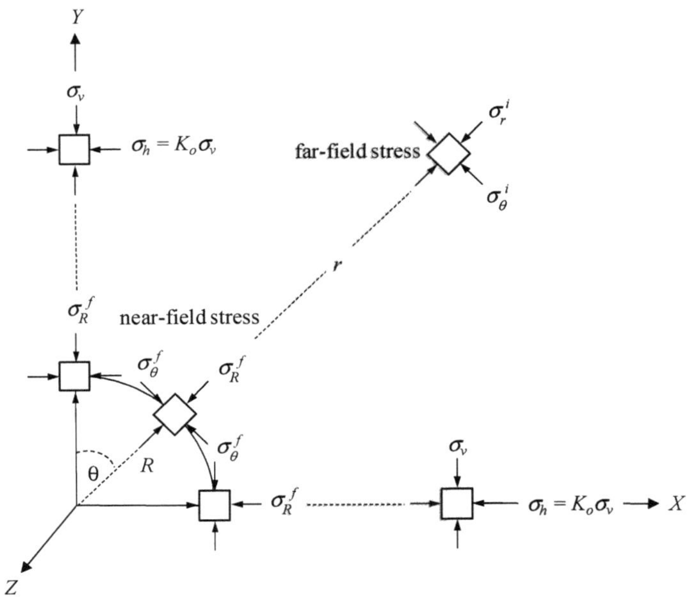

Fig. 1. Schematic illustration of stress variation around a circular tunnel from far-field to near-field in the cylindrical coordinates (r, θ, Z)

$$\sigma_\theta^i = \frac{\sigma_v}{2}[(1+K_0) - (1-K_o)\cos 2\theta] \tag{2}$$

where K_o is the in situ stress ratio or far-field stress ratio, σ_r^i and σ_θ^i are the initial radial stress and the initial circumferential stress in the polar coordinates (r, θ), respectively.

(2) Stresses in near-field (internal boundary stresses or final stresses, $R \leq r < \infty$):

After opening the tunnel, the stresses at any point around the tunnel in the rock mass are altered and can be obtained by the Kirsch's solutions (Kirsch 1898) as follows:

$$\sigma_R^f = \frac{\sigma_v}{2}\left[(1+K_o)\left(1-\frac{R^2}{r^2}\right) + (1-K_o)\left(1-4\frac{R^2}{r^2}+3\frac{R^4}{r^4}\right)\cos 2\theta\right] \tag{3}$$

$$\sigma_\theta^f = \frac{\sigma_v}{2}\left[(1+K_o)\left(1+\frac{R^2}{r^2}\right) - (1-K_o)\left(1+3\frac{R^4}{r^4}\right)\cos 2\theta\right] \tag{4}$$

where R, σ_r^f and σ_θ^f are excavated radius, the final radial stress and the final circumferential stress around the tunnel proximity, respectively. Assuming that the stress gradient due to the advance excavation of the tunnel face is a difference between far-field stress and near-field stress, a difference of the radial stress and of the circumferential stress can be expressed by the following equations, respectively.

$$\bar{\sigma}_r = \frac{-\sigma_v}{2}\left[(1+K_o)\frac{R^2}{r^2} + (1-K_o)\left(4\frac{R^2}{r^2} - 3\frac{R^4}{r^4}\right)\cos 2\theta\right] \tag{5}$$

$$\bar{\sigma}_\theta = \frac{\sigma_v}{2}\left[(1+K_o)\frac{R^2}{r^2} - 3(1-K_o)\frac{R^4}{r^4}\cos 2\theta\right] \tag{6}$$

Using the hypothesis of increments in the numerical analysis, the increment of stresses can be expressed by the multiplication of the stress gradient with the confinement loss λ which is a fraction of the stress gradient, and can be obtained as follows:

$$\Delta\sigma_r = \lambda\bar{\sigma}_r = \frac{-\lambda\sigma_v}{2}\left[(1+K_o)\frac{R^2}{r^2} + (1-K_o)\left(4\frac{R^2}{r^2} - 3\frac{R^4}{r^4}\right)\cos 2\theta\right] \tag{7}$$

$$\Delta\sigma_\theta = \lambda\bar{\sigma}_\theta = \frac{\lambda\sigma_v}{2}\left[(1+K_o)\frac{R^2}{r^2} - 3(1-K_o)\frac{R^4}{r^4}\cos 2\theta\right] \tag{8}$$

As mentioned above, the confinement loss is a scalar which serves to apply a portion of the confinement-removal forces in order to simulate the eventual proximity of the tunnel faces, and which has a value of between 0 and 1. To maintain the simplicity of the equations, one can use the variables $k_1 = 1 + K_o$ and $k_2 = (1 - K_o)\cos 2\theta$, and substitute them into the previous equations. Then the increment of stresses ($\Delta\sigma_r$ and $\Delta\sigma_\theta$) can be obtained as

$$\Delta\sigma_r = \lambda\bar{\sigma}_r = \frac{-\lambda\sigma_v}{2}\left[k_1\frac{R^2}{r^2} + k_2\left(4\frac{R^2}{r^2} - 3\frac{R^4}{r^4}\right)\right] \tag{9}$$

$$\Delta\sigma_\theta = \lambda\bar{\sigma}_\theta = \frac{\lambda\sigma_v}{2}\left[k_1\frac{R^2}{r^2} - 3k_2\frac{R^4}{r^4}\right] \tag{10}$$

Finally, the variation of stresses due to the advance excavation of the tunnel face ($0 \leq \lambda \leq 1$) at a certain point of tunnel proximity (r, θ) can be represented by the following:

$$\Delta\sigma_r = \lambda\bar{\sigma}_r = \frac{-\lambda\sigma_v}{2}\left[k_1\frac{R^2}{r^2} + k_2\left(4\frac{R^2}{r^2} - 3\frac{R^4}{r^4}\right)\right] \tag{11}$$

$$\Delta\sigma_\theta = \lambda\bar{\sigma}_\theta = \frac{\lambda\sigma_v}{2}\left[k_1\frac{R^2}{r^2} - 3k_2\frac{R^4}{r^4}\right] \tag{12}$$

By substituting Eqs. (1) and (9) into Eq. (11) for the radial stress, and Eqs. (2) and (10) into Eq. (12) for the circumferential stress, the following new expressions can be obtained:

$$\sigma_r = \frac{\sigma_v}{2}\left[k_1\left(1-\lambda\frac{R^2}{r^2}\right)+k_2\left(1-\lambda\left(4\frac{R^2}{r^2}-3\frac{R^4}{r^4}\right)\right)\right] \tag{13}$$

$$\sigma_\theta = \frac{\sigma_v}{2}\left[k_1\left(1+\lambda\frac{R^2}{r^2}\right)-k_2\left(1+3\lambda\frac{R^4}{r^4}\right)\right] \tag{14}$$

It should be noted that these equations must be satisfied by the boundary conditions proposed by the convergence-confinement method and also can be verified by the following interpretations: (1) When $\lambda = 0$, it means that the tunnel is not excavated yet, and stresses in the rock mass are the same as the initial in situ stresses or far-field stress (e.g. $r \to \infty$), and (2) When $\lambda = 1$, it means that the tunnel is already excavated, and stresses in the rock mass are the same as the Kirsch's stresses or near-field stress (e.g. $R \leq r < \infty$).

2.2 Stresses in the Plastic Region and the Plastic Radius

In the more general case of rock with elastic-perfectly plastic behavior (Mohr-Coulomb failure criterion), the confinement loss in elastic limit situation λ_e appears when the elastic limit of the rock is reached, for a certain radial stress σ_r and with the decreasing of r, and when the stress state reaches the limit conditions defined by the strength criterion. Such a value of r is called the plastic radius R_p which is not only a function of the peak strength parameters of the rock mass, but also is a function of the confinement loss in elastic limit situation λ_e. When the confinement loss λ continuously increases, plastic radius R_p increases, but the radial stress decreases in the plastic region; therefore this radial stress is obviously a function of the plastic radius R_p ($R \leq r \leq R_p$). For this reason, the radial and circumferential stresses in the plastic region can be obtained by combining the equations of failure criterion and the equations of equilibrium and shown as the followings.

$$\sigma_r = \frac{\sigma_v}{K_p - 1}\left\{[(k_1-k_2)\lambda_e - k_2]\left(\frac{r}{R_p}\right)^{K_p-1} - 2N\right\} \tag{15}$$

$$\frac{\sigma_\theta}{\sigma_v} = \frac{1}{K_p - 1}\left\{[(k_1-k_2)\lambda_e - k_2]K_p\left(\frac{r}{R_p}\right)^{K_p-1} - 2N\right\} \tag{16}$$

where the coefficient of passive lateral pressure K_p, the stability number N, the confinement loss in elastic limit situation λ_e and the plastic radius R_p which can be expressed by the following equations, respectively.

$$K_p = \tan^2\left(45^\circ + \frac{\phi}{2}\right) \tag{17}$$

$$N = \frac{\sigma_c}{2\sigma_v} \tag{18}$$

$$\lambda_e = \frac{(K_p - 1)k_1 + (K_p + 1)k_2 + 4N}{(K_p + 1)k_1 + (K_p - 3)k_2} \tag{19}$$

$$\frac{R_p}{R} = \left[\frac{(k_1 - k_2)\lambda_e - k_2}{\frac{1}{2}\left[(K_p + 1)k_1 + (K_p - 3)k_2\right]\lambda_e - \frac{1}{2}(K_p - 1)(k_1 + k_2)\lambda - k_2}\right]^{\frac{1}{K_p - 1}} \tag{20}$$

It should be noted that the confinement loss in elastic limit situation λ_e is a function of the peak strength parameters of the rock mass (c, ϕ), the location of polar coordinates (r, θ), and the initial in situ stress (σ_v, K_o).

3 Analytical Solution for Displacements in the Elastic Region and the Plastic Region

3.1 Displacements in the Elastic Region

In the elastic region, the radial and circumferential displacements, u_r and u, can be obtained by combining the equations of the constitutive law and the equations of compatibility. The relationship can be expressed as

$$2G\frac{du_r}{dr} = (1 - v)\Delta\sigma_r - v\Delta\sigma_\theta \tag{21}$$

$$2G\left(\frac{u_r}{r} + \frac{1}{r}\frac{du_\theta}{d\theta}\right) = (1 - v)\Delta\sigma_\theta - v\Delta\sigma_r \tag{22}$$

By integrating these equations, the radial and circumferential displacements in the elastic region can be expressed as

$$\frac{2G}{\sigma_v}\frac{u_r}{R} = \frac{\lambda}{2}\left(\frac{R}{r}\right)\left\{k_1 + k_2\left[4(1 - v) - \left(\frac{R}{r}\right)^2\right]\right\} \tag{23}$$

$$\frac{2G}{\sigma_v}\frac{u_\theta}{R} = \frac{\lambda}{2}\left(\frac{R}{r}\right)\left\{k_2\left[2(1 - 2v) + \left(\frac{R}{r}\right)^2\right]\tan 2\theta\right\} \tag{24}$$

where v is the Poisson's ratio and G is the shear modulus of the rock mass.

3.2 Displacements in the Plastic Region

In the plastic region, the total radial and circumferential strains, ε_r and ε_θ, can be decomposed into elastic and plastic parts as

$$\varepsilon_r = \varepsilon_r^e + \varepsilon_r^p \tag{25}$$

$$\varepsilon_\theta = \varepsilon_\theta^e + \varepsilon_\theta^p \tag{26}$$

where the subscripts e and p represent the elastic and plastic parts, respectively.

Additionally, with the purpose of determining the displacement field in the plastic region, a plastic flow rule is needed. By assuming that the elastic strains are relatively small in comparison to the plastic strains and that a non-associated flow rule is valid, the plastic parts of the radial and circumferential strains may be related for the plane strain condition.

$$\varepsilon_r^p + K_\psi \varepsilon_\theta^p = 0 \tag{27}$$

where $K_p = \tan^2\left(45^\circ + \frac{\psi}{2}\right)$, ψ is the dilation angle of the rock. In the plastic region, by combining the equations of the constitutive law and the equations of compatibility leads to the following differential equation:

$$\frac{du_r}{dr} + K_\psi \frac{u_r}{r} = f(r) \tag{28}$$

where

$$f(r) = \varepsilon_r^e + K_\psi \varepsilon_\theta^e \tag{29}$$

or

$$f(r) = \frac{1}{2G}\left[(K_\psi - vK_\psi - v)\Delta\sigma_\theta + (1 - v - vK_\psi)\Delta\sigma_r\right] \tag{30}$$

The differential equation Eq. (28), may be solved by engineering mathematic with the homogeneous solution and the particular solution, and by using the boundary condition for the radial displacement u_r in the plastic region. Finally, the closed-form analytical solution for the radial displacement in the plastic region can be simplified as

$$\frac{2G}{\sigma_v}\frac{u_r}{R} = \lambda_e\left(\frac{r}{R}\right)\left[C_1 + C_2\left(\frac{r}{R_p}\right)^{K_p-1} + C_3\left(\frac{R_p}{r}\right)^{K_\psi+1}\right] \tag{31}$$

where

$$C_1 = -\frac{(1-2v)}{(K_p-1)}\left[\frac{(K_p+1)k_1 + (K_p-3)k_2}{2}\right] \tag{32}$$

$$C_2 = \frac{1 + K_\psi K_p - v(K_p+1)(K_\psi+1)}{(K_p-1)(K_\psi+K_p)}\left[(k_1-k_2) - \frac{k_2}{\lambda_e}\right] \tag{33}$$

$$C_1 + C_2 + C_3 = \frac{k_1 + k_2(3 - 4v)}{2} \tag{34}$$

4 Explicit Procedure (EAM)

The explicit procedure used by this study is also called an incremental procedure which can realize and transform the analytical solutions to an executable computation that can be directly calculated by using a simple calculation spreadsheet (e.g., the Microsoft Excel format). The particular procedure proposed by this paper is called the explicit analysis method (EAM), which can deal with using the confinement loss as an incremental step to simulate the effect of advance excavation of the tunnel face, calculating the stresses and displacements in each step, and drawing the ground reaction curve, the stress path at the intrados of the tunnel, and the distribution of stresses/displacements on the cross-sections of a tunnel. The steps of calculation for the EAM are described by the following representations.

(1) Input data: This relates to the data of the initial is-situ stress, geometry of the tunnel, material properties, and unsupported distance.

(2) In order to estimate the confinement loss λ_z at a certain distance z from the tunnel working face, one can use the following function as a chosen advance effect function of tunnel excavation. Therefore, λ_z can be determined.

$$\lambda_z = \lambda_d + (1 - \lambda_d)\left[1 - \left(\frac{m}{m + z/R}\right)^2\right] \tag{35}$$

(3) Dividing the confinement loss λ_z by n segments, the incremental step of λ can be expressed as

$$\Delta\lambda = \frac{1}{n}(1 - \lambda_z)\frac{(k_1 + k_2)}{2} \tag{36}$$

(4) Calculating each step value of λ as

$$\left.\begin{array}{ll} \lambda_{i+1} = \lambda_z & i = 0 \\ \lambda_{i+1} = \lambda_i + i\Delta\lambda & i = 1 \sim (n-1) \end{array}\right\} \tag{37}$$

(5) Attaining the final value $\lambda_n = \lambda_{i+1}$.

(6) According to the Eq. (19), to estimate the confinement loss in the elastic limit situation λ_e.

(7) If $\lambda_{i+1} < \lambda_e$, it means that the stress state is in the elastic region, and that the radial and circumferential stresses/displacements can be calculated with Eqs. (13), (14) and (23).

(8) If $\lambda_{i+1} \geq \lambda_e$, it means that the stress state is in the plastic region, and that the plastic radius R_p can be calculated with Eq. (20). Once one obtains this value R_p,

the procedure automatically substitutes into Eqs. (15), (16), and (31) for the radial and circumferential stresses/displacements, respectively.

(9) Recording the calculated data, it relates to the ground reaction curve (u_r, σ_r), the stress path (σ_r, σ) and the distribution of stresses/displacements (σ_r, r), (σ, r), and (u_r, r).

(10) When $i < n - 1$, repeat steps (5) through (10).

(11) When $i = n - 1$, the process is not repeated, and the data from each step is recorded.

(12) Drawing the ground reaction curve, the stress path at the intrados of the tunnel, and the distribution of stresses/displacements on the cross-sections of the tunnel.

5 Comparison Between Explicit Analysis and Numerical Analysis

In order to investigate the results obtained by the analytical solutions which are realized by the explicit procedure (EAM), it is necessary to try to find the numerical analysis, or finite element method (FEM), which can provide rigorous results for comparison. In this study, the mesh prepared of finite element analysis includes that the analysis boundary supported by rollers is twenty times the tunnel excavation radius, using three group elements (excavation, lining and rock mass), 658 elements (540 Q8 elements and 118 T6 elements) and 1971 total nodes; and these conditions are shown in Fig. 2.

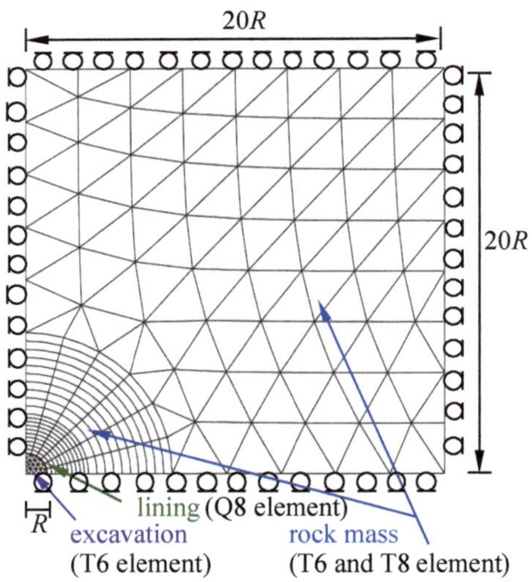

Fig. 2. Finite element mesh and boundary condition

The input data of calculation considered the in situ stress ratio, K_o= 0.8 and 1.2, is shown in Table 1. The comparison of results calculated by the EAM and FEM includes the stresses and displacements at the intrados of the tunnel and the distribution on the cross-sections of the tunnel.

Table 1. Input data of calculation

Parameter	Value
In situ stress ratio, K_o	0.8, 1.2
Vertical stress, σ_v (MPa)	1.0
Tunnel radius, R (m)	5.2
Elastic modulus, E (MPa)	300.0
Unit weight, γ (MPa/m)	0.02
Poisson's ratio, ν	0.25
Cohesion, c (MPa)	0.1
Friction angle, ϕ (°)	30.0
Dilation angle, ψ (°)	30.0

5.1 Stresses/Displacements at Intrados of Tunnel

The behavior of rock mass at the intrados of the tunnel is altered by the advance excavation of the tunnel face, and may be interpreted by the ground reaction curve and the stress path at any point of the intrados of the tunnel. In this paper, the dimensionless radial displacement, $2Gu_r/R_v$, and stresses normalized by the vertical stress, $_v$, in the elastic region and plastic region are plotted in Figs. 3, 4, 5, 6, 7, 8, 9, 10, 11, 12, 13, 14 and 15.

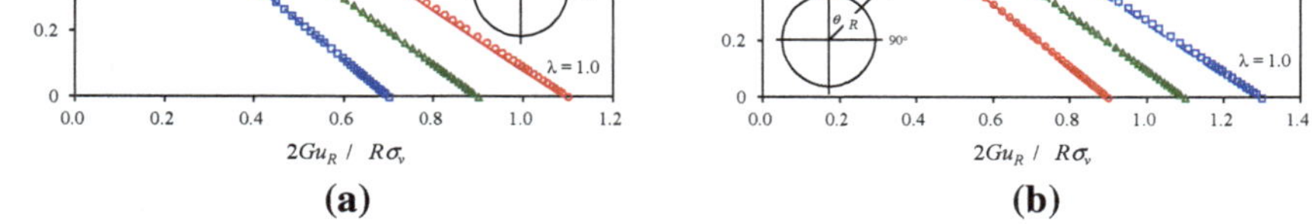

Fig. 3. Comparison of the ground reaction curves between the explicit analysis (EAM) and numerical analysis (FEM) in elastic media under anisotropy stress state conditions **a** $K_o = 0.8$ and **b** $K_o = 1.2$

The comparison of GRCs in the elastic part (linear elasticity) is illustrated in Fig. 3, which shows that in a case of $K_o < 1.0$, the radial displacement at the crown of the tunnel ($\theta = 0°$) is greater than that at an inclination of 45° ($\theta = 45°$), and that at the

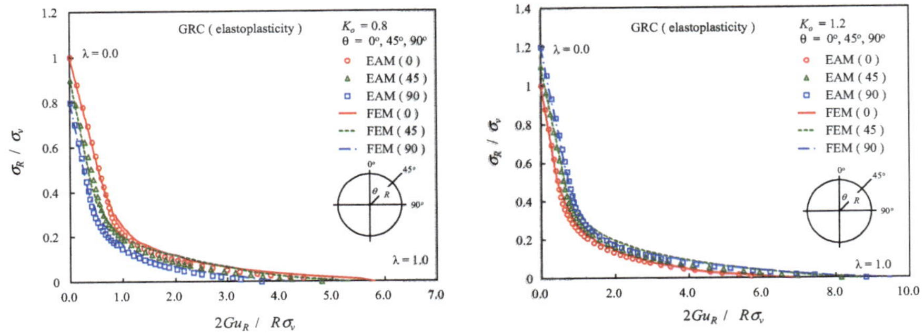

(a) **(b)**

Fig. 4. Comparison of the ground reaction curves between the explicit analysis (EAM) and numerical analysis (FEM) in elastic-perfectly plastic media under anisotropy stress state conditions **a** K_o = 0.8 and **b** K_o = 1.2

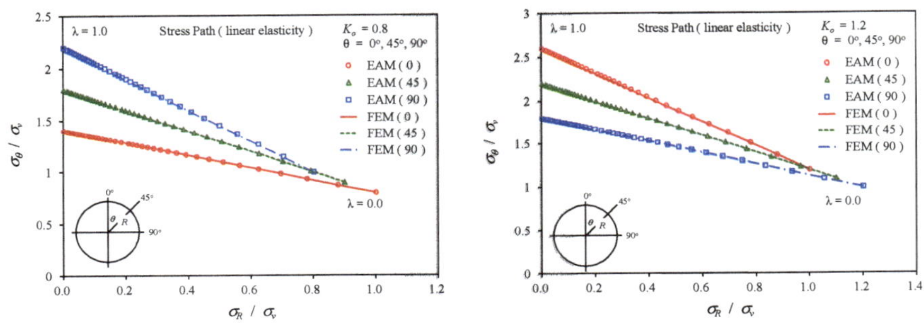

(a) **(b)**

Fig. 5. Comparison of the stress paths at the intrados of the tunnel between the explicit analysis (EAM) and numerical analysis (FEM) in elastic media under anisotropy stress state conditions **a** K_o = 0.8 and **b** K_o = 1.2

side-wall ($\theta = 90°$). In contrast, in a case of $K_o > 1.0$, the radial displacement at the side-wall is greatest, while that at an inclination of 45° is second greatest, and that at the crown is lowest. In addition, it can be found that the results of GRC simulated by the EAM and FEM are obviously coincident from $\lambda = 0$ to $\lambda = 1$.

After that, the comparison of GRCs in the elastic-perfectly plastic part is shown in Fig. 4, in which the observable appearance of the influence of in situ stress ratio in this part of GRC are similar to those investigated in the elastic part, but the radial displacements obtained by the EAM and FEM are slightly different in terms of the maximum values, and the trends of GRCs simulated among these methods at different locations of the intrados of the tunnel are very consistent from $\lambda = 0$ to $\lambda = 1$.

In addition, the stress path at the locations of the intrados of the tunnel in the elastic part is shown in Fig. 5. It can be seen that the results of the stress path as simulated by the EAM and FEM indicate the same values and similar paths. As for the stress path in

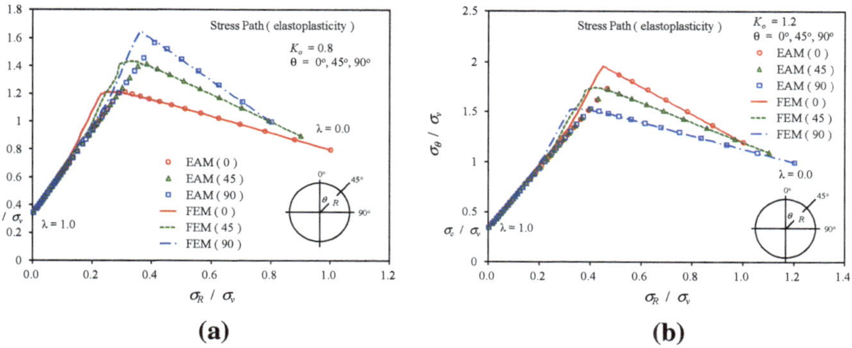

Fig. 6. Comparison of the stress paths at the intrados of the tunnel between the explicit analysis (EAM) and numerical analysis (FEM) in elastic-perfectly plastic media under anisotropy stress state conditions **a** $K_o = 0.8$ and **b** $K_o = 1.2$

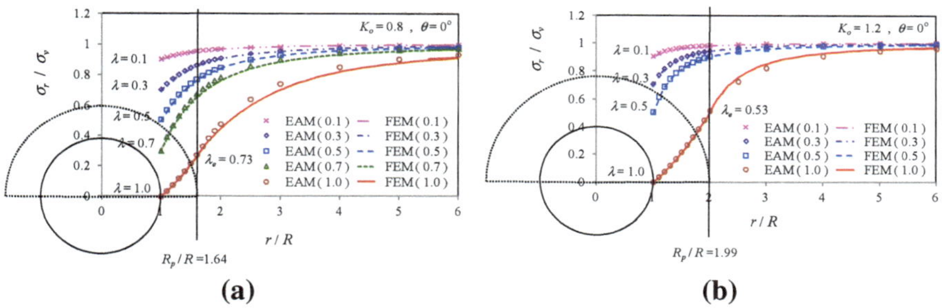

Fig. 7. Comparison of the distribution of radial stress at the crown ($\theta = 0°$) on the cross-sections of the tunnel between the explicit analysis (EAM) and numerical analysis (FEM) in elastic-perfectly plastic media under anisotropy stress state conditions **a** $K_o = 0.8$ and **b** $K_o = 1.2$

Fig. 8. Comparison of the distribution of circumferential stress at the crown ($\theta = 0°$) on the cross-sections of the tunnel between the explicit analysis (EAM) and numerical analysis (FEM) in elastic-perfectly plastic media under anisotropy stress state conditions **a** $K_o = 0.8$ and **b** $K_o = 1.2$

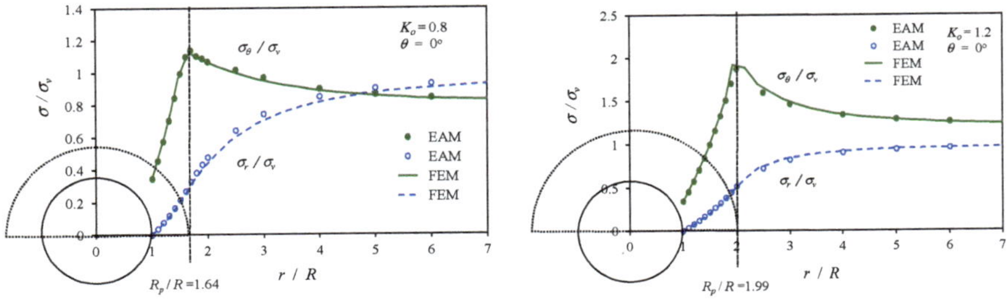

(a) (b)

Fig. 9. Comparison of the distribution of stresses at the crown ($\theta = 0°$) on the cross-sections of the tunnel between the explicit analysis (EAM) and numerical analysis (FEM) in elastic-perfectly plastic media under anisotropy stress state conditions **a** $K_o = 0.8$ and **b** $K_o = 1.2$

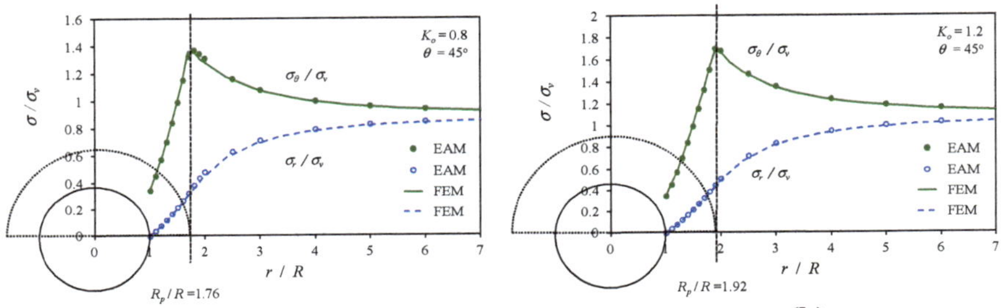

(a) (b)

Fig. 10. Comparison of the distribution of stresses at an inclination of 45° ($\theta = 45°$) on the cross-sections of the tunnel between the explicit analysis (EAM) and numerical analysis (FEM) in elastic-perfectly plastic media under anisotropy stress state conditions **a** $K_o = 0.8$ and **b** $K_o = 1.2$

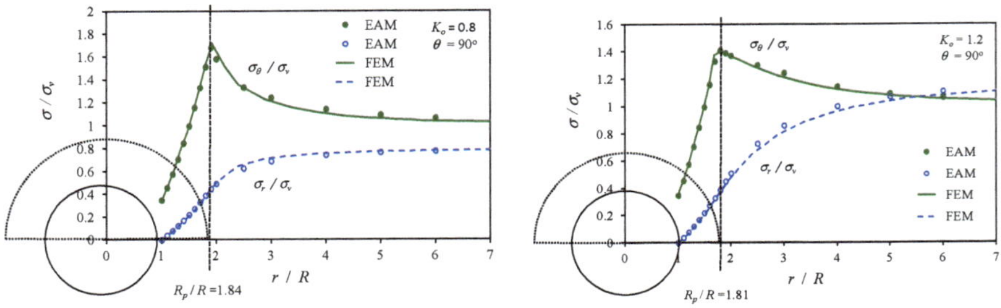

(a) (b)

Fig. 11. Comparison of the distribution of stresses at the side-wall ($\theta = 90°$) on the cross-sections of the tunnel between the explicit analysis (EAM) and numerical analysis (FEM) in elastic-perfectly plastic media under anisotropy stress state conditions **a** $K_o = 0.8$ and **b** $K_o = 1.2$

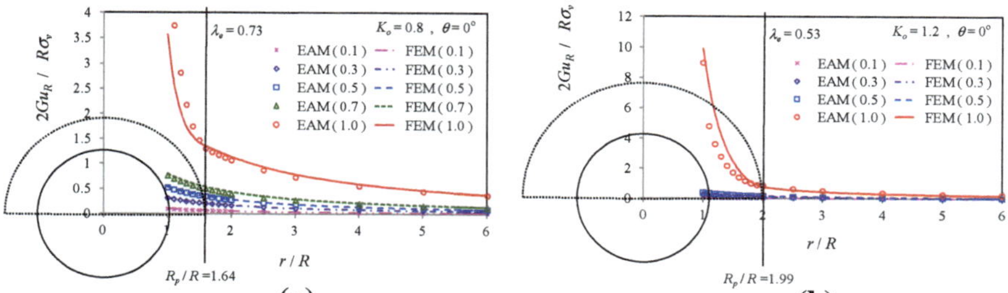

(a) **(b)**

Fig. 12. Comparison of the distribution of radial displacement at the crown (θ = 0°) on the cross-sections of the tunnel between the explicit analysis (EAM) and numerical analysis (FEM) in elastic-perfectly plastic media under anisotropy stress state conditions **a** K_o = 0.8 and **b** K_o = 1.2

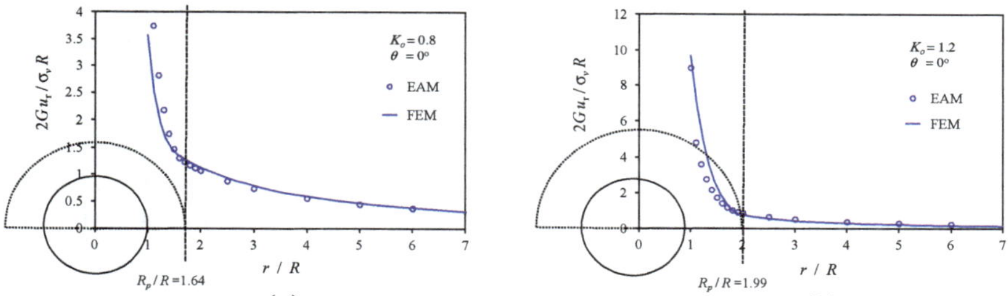

(a) **(b)**

Fig. 13. Comparison of the distribution radial displacement at the crown (θ = 0°) on the cross-sections of the tunnel between the explicit analysis (EAM) and numerical analysis (FEM) in elastic-perfectly plastic media under anisotropy stress state conditions **a** K_o = 0.8 and **b** K_o = 1.2

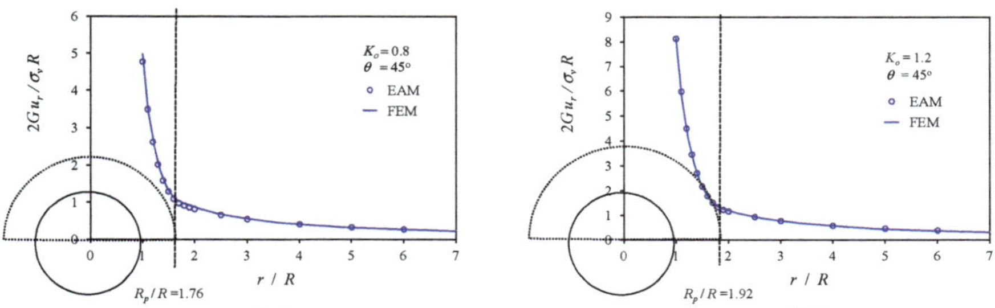

(a) **(b)**

Fig. 14. Comparison of the distribution of radial displacement at an inclination of 45° (θ = 45°) on the cross-sections of the tunnel between the explicit analysis (EAM) and numerical analysis (FEM) in elastic-perfectly plastic media under anisotropy stress state conditions **a** K_o = 0.8 and **b** K_o = 1.2

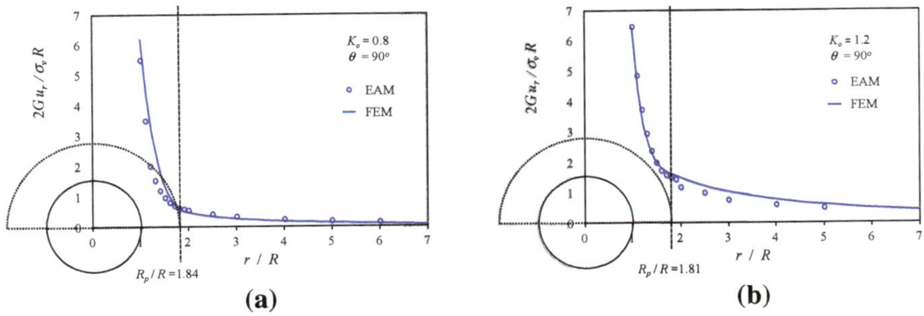

Fig. 15. Comparison of the distribution of radial displacement at the side-wall ($\theta = 90°$) on the cross-sections of the tunnel between the explicit analysis (EAM) and numerical analysis (FEM) in elastic-perfectly plastic media under anisotropy stress state conditions **a** $K_o = 0.8$ and **b** $K_o = 1.2$

the elastic-perfectly plastic part shown in Fig. 6, it displays very suitable trends for the stress path as simulated by the EAM and FEM. This stress path may follow along the line of Mohr-Coulomb failure criterion, finally attaining the last value of σ_c/σ_v.

Finally, it should be noted that the previously obtained results clearly give an important sign that the stresses/displacements at different locations of the intrados of the tunnel exhibit different characteristic trends (e.g. $\theta = 0°$, $45°$, and $90°$); however, the stresses/displacements are caused with the same value of confinement loss which corresponds to a distance to tunnel working face.

5.2 Distribution of Stresses/Displacements on the Cross-Sections of Tunnel

Surrounding a circular tunnel excavation in an anisotropy stress state, one is interested in the distribution of stresses/displacements on the cross-sections of the tunnel. According to the results obtained for the distribution of stresses at the specified locations of the tunnel (e.g. $\theta = 0°$, $45°$, and $90°$) and in the cases of K_o= 0.8 and 1.2 (Figs. 7 and 8), the variation of stresses due to the advance excavation of the tunnel face can be described and verified by the following five stress states:

(1) As $\lambda = 0$ ($r \to \infty$), the stresses are in the initial in situ stress state,
(2) As $0 < \lambda < \lambda_e$ ($R_p < r < \infty$), the stresses are in the elastic region, and the radial and circumferential stresses increase with the increase of the confinement loss, and both stresses are symmetrically separated along the horizontal axis r/R,
(3) When $\lambda = \lambda_e$ ($r = R_p$), the plastic radius appears and stresses are at the elastic-plastic interface. The radial stress begins to change the curvature (Fig. 7), and the circumferential stress attains the maximum value (Fig. 8).
(4) When $\lambda_e < \lambda < 1$ ($R < r < R_p$), the stresses are in the plastic region, and this leads to both the radial stress and circumferential stress being decreased steeply,
(5) Until $\lambda = 1$ ($r = R$), the radial stress becomes zero and the circumferential stress is equal to the uniaxial compression strength of the rock.

In addition, from the point of view of continuity of stresses at the elastic-plastic interface, the stresses are proved and continuously connected in this interface. The comparison of the distribution of stresses between the EAM and FEM presents not only an approximate value, but also is consistent with the trends shown in Figs. 9, 10 and 11. In Figs. 9 and 11, the distribution of stresses at the crown and side-wall in the cases of K_o= 0.8 and 1.2 is just contrary to each other. However, in Fig. 10 the distribution of stresses at an inclination of 45° shows the same trends that were obtained in the hydrostatic stress state field (K_o= 1.0).

Lastly, the variation of radial displacement, u_r, due to the advance excavation of the tunnel face can be also described and verified by the following five states of displacement (Fig. 12): (1) As $\lambda = 0$ ($r \rightarrow \infty$), u_r is equal to zero in the initial in situ stress state, (2) As $0 < \lambda < \lambda_e$ ($R_p < r < \infty$), the u_r is in the elastic region, and gradually increases with λ, (3) When $\lambda = \lambda_e$ ($r = R_p$), the plastic radius appears and u_r is at the elastic-plastic interface, (4) When $\lambda_e < \lambda < 1$ ($R < r < R_p$), u_r is in the plastic region, and leads u_r to increase quickly, and (5) Until $\lambda = 1$ ($r = R$), u_r becomes a maximum value. Finally, the radial displacement is also proved and continuously connected in this elastic-plastic interface. The comparison of the radial displacement distribution between the EAM and FEM is very consistent with the trends shown in Figs. 13, 14 and 15.

6 Conclusions

The following main conclusions can be drawn from the present study:

(1) A theoretically consistent closed-form analytical solution was derived for the elastic-perfectly plastic analysis of a circular tunnel in a rock mass governed by Mohr-Coulomb failure criterion and subject to an anisotropy in situ stress.
(2) The procedure of the explicit analysis method (EAM) is particularly proposed to deal with using the confinement loss as an incremental step to simulate the effect of advance excavation of the tunnel face, calculating the stresses and displacements in each step, and drawing the ground reaction curve, the stress path at the intrados of the tunnel, and the distribution of stresses/displacements on the cross-sections of the tunnel.
(3) The agreement between the finite element results and the proposed closed-form solutions with the explicit procedure was found to be excellent in elastic-perfectly plastic media.
(4) The variation of stresses due to the advance excavation of the tunnel face can be mainly interpreted by the stress gradient which is a difference between far-field stress and near-field stress around a tunnel and can be obtained by means of the hypothesis of increments in the numerical analysis. The increment of stresses can be considered with the confinement loss as a fraction of the stress gradient.
(5) The confinement loss λ serves to apply a portion of the confinement-removal forces in order to simulate the eventual proximity of the tunnel faces and is just a positive scalar, an increment value, or a function for the effect of advance excavation of the tunnel face.

References

AFTES: Recommandations pour l'emploi de la méthode convergence-confinement. Tunnels et Ouvrages Souterrains **59**, 119–138 (1983)

AFTES, Panet, M., et al.: Recommendations on the Convergence-Confinement Method. Association Française des Tunnels et de l'Espace Souterrain, pp. 1–11 (2001)

Alejano, L.R., Rodriguez-Dono, A., Alonso, E., Fdez-Manín, G.: Ground reaction curves for tunnels excavated in different quality rock masses showing several types of post-failure behaviour. Tunn. Undergr. Space Technol. **24**, 689–705 (2011)

Brady, B.H.G., Brown, E.T.: Rock Mechanics for Underground Mining. Chapman & Hall, London (1993)

Brown, E.T., Bray, J.W., Ladanyi, B., Hoek, E.: Ground response curves for rock tunnels. J. Geotech. Eng. ASCE **109**, 15–39 (1983)

Carranza-Torres, C.: Elasto-plastic solution of tunnel problems using the generalized form of the Hoek-Brown failure criterion. Int. J. Rock Mech. Min. Sci. **41**(3), 480–491 (2004)

Carranza-Torres, C., Fairhurst, C.: The elasto-plastic response of underground excavations in rock masses that satisfy the Hoek-Brown failure criterion. Int. J. Rock Mech. Min. Sci. **36**, 777–809 (1999)

Carranza-Torres, C., Fairhurst, C.: Application of the convergence-confinement method of tunnel design to rock masses that satisfy the Hoek-Brown failure criterion. Tunn. Undergr. Space Technol. **15**(2), 187–213 (2000)

Detournay, E., John, C.M.S.: Design charts for a deep circular tunnel under non-uniform loading. Rock Mech. Rock Eng. **21**, 119–137 (1988)

Guan, Z., Jiang, Y., Tanabasi, Y.: Ground reaction analyses in conventional tunnelling excavation. Tunn. Undergr. Space Technol. **22**, 230–237 (2007)

Hoek, E., Brown, E.T.: Underground Excavations in Rock. The Institution of Mining and Metallurgy, London (1980)

Kirsch, G.: Die theorie der Elastizität und die bedürfnisse der festigkeitslehre. Zeit Ver Deut Ing J. **42**, 797–807 (1898)

Lee, Y.L.: Prise en compte des non-linéarité de comportement des sols et des roches dans la modélisation du creusement d'un tunnel. Thèse, École Nationale des Ponts et Chaussées (1994)

Lee, Y.L., Lin, M.Y., Hsu, W.K.: Study of relationship between the confinement loss and the longitudinal deformation curve by using three-dimensional finite element analysis. Chin. J. Rock Mech. Eng. **27**(2), 258–265 (2008)

Lee, Y.L., Hsu, W.K., Lin, M.Y.: Analysis of the advancing effect and the confinement loss by using deformation measurement in tunneling. Chin. J. Rock Mechan. Eng. **28**(1), 39–46 (2009)

Oreste, P.P.: Analysis of structural interaction in tunnels using the convergence-confinement approach. Tunn. Undergr. Space Technol. **18**, 347–363 (2003)

Oreste, P.: The convergence–confinement method: roles and limits in modern geomechanical tunnel design. Am. J. Appl. Sci. **6**(4), 755–771 (2009)

Pacher, F.: Deformationsmessungen in Versuchsstollen als Mittel zur Erforschung des Gebirgsverhaltens und zur Bemessung des Ausbaues. Felsmechanik und Ingenieursgeologie Supplementum **IV**, 149–161 (1964)

Panet, M., Guellec, P.: Contribution a l'étude du sousténement derrière le front de taille. In: Proc. 3rd Cong. Int. Soc. Rock Mechanics, Part B, Denver, vol. 2, pp. 1130–1134 (1974)

Panet, M., Guenot, A.: Analysis of convergence behind the face of a tunnel. In: Proceedings of International Symposium Tunnelling'82, IMM, London, pp. 197–204 (1982)

Panet, M.: Calcul du souténement des tunnels à section circulaire par la method convergence-confinement avec un champ de contraintes initiales anisotrope. Tunnels et Ouvrages Souterrains **77**, 228–232 (1986)

Panet, M.: Le Calcul des Tunnels par la Méthode de Convergence-Confinement. Presses de l'Ecole Nationale des Ponts et Chaussées, Paris (1995)

Park, K.H., Kim, Y.J.: Analytical solution for a circular opening in an elastic–brittle–plastic rock. Int. J. Rock Mech. Min. Sci. **43**, 616–622 (2006)

Ravandi, E.G., Rahmannejad, R.: Wall displacement prediction of circular, D shaped and modified horseshoe tunnels in anisotropy stress fields. Tunn. Undergr. Space Technol. **34**, 54–60 (2013)

Serrano, A., Olalla, C., Reig, I.: Convergence of circular tunnels in elastoplastic rock masses with non-linear failure criteria and non-associated flow laws. Int. J. Rock Mech. Min. Sci. **48**, 878–887 (2011)

Sharan, S.K.: Elastic–brittle–plastic analysis of circular openings in Hoek-Brown media. Int. J. Rock Mech. Min. Sci. **40**, 817–824 (2003)

Sharan, S.K.: Exact and approximate solutions for displacements around circular openings in elastic-brittle plastic Hoek-Brown rock. Int. J. Rock Mech. Min. Sci. **42**, 42–549 (2005)

Shen, B., Barton, N.: The disturbed zone around tunnels in jointed rock masses. Int. J. Rock Mech. Min. Sci. **34**(1), 117–125 (1997)

Wang, Y.: Ground response of a circular tunnel in poorly consolidated rock. J. Geotech. Eng. ASCE **122**(9), 703–708 (1996)

Wong, R.C.K., Kaiser, P.K.: Performance assessment of tunnels in cohesion less soils. J. Geotech. Eng. ASCE **117**(12), 1880–1901 (1991)

Comprehensive Optimization of Support System for Conventional Tunnelling

Kyung-Nam Kang, Won-Suh Sung, and Ki-Il Song(✉)

Department of Civil Engineering, Inha University, 100, Inha-ro, Nam-gu, Incheon, Republic of Korea
ksong@inha.ac.kr

Abstract. Support system such as rockbolt, shotcrete and steel sets are important primary components for conventional tunnelling due to their influence on tunnel stability. Selection of appropriate and cost effective support system and proper installation of tunnel support system are important factors for tunnel construction. RMR and Q-system have been used to classify the rock mass and to design the tunnel support system in empirical manner. However, these methods are time consuming due to iterative design procedure and difficult to verify their validity. In this study, differential evolution algorithm is introduced to optimize tunnel support system design. DEA constructed in this study is based on both theoretical and numerical methods. The suggested optimum design scheme provides cost effective and safe support system for conventional tunneling.

1 Introduction

Tunnel excavation generates the redistribution of the underground stress, resulting in a new biaxial stress and the original tri-axial stress states; the displacement of the excavation surface then occurs inside the excavation cavity due to extinction of underground stress. To prevent the generated displacement, materials such as rockbolt, shotcrete, and steel sets are installed. At present, six standard support patterns are generally determined depending on the use of tunnel and the rock mass classification in Korea (Korean National Tunnel Standard Specification 2015). These materials are important primary support for the tunnel due to their influence on tunnel stability. Therefore, not only the choice of appropriate and cost effective materials is an important factor for tunnel stability design, but also application of proper optimal design method for the tunnel support materials is required.

Many researches on optimal design for construction have been carried out by different scholars. Yin et al. (2000) conducted a study on the optimal placement of stiffeners in the design of the tunnel supports. He used the geometrical optimization method considering the stress condition of the original soil and proposed that the optimum design method can reduce 50% of the tunnel deformation. Ryu et al. (2012) studied optimization of concrete lining design and proposed a reasonable estimation method for relaxation load. Nguyen et al. (2015) argued that empirical methods in general use are not necessarily optimal and that improved results can be obtained using the proposed optimization techniques.

B. Fatahi et al. (eds.), *Sustainable Design and Construction for Geomaterials and Geostructures*, Sustainable Civil Infrastructures,
https://doi.org/10.1007/978-3-319-95753-1_21

Genetic algorithms (GA) is one of the techniques used for optimal design, especially when the objective function is not continuous and there are many variables in the optimization problem. GA converge the optimal solution through mutation, crossover, and selection over generations. Yet, as the number of variables increases, the analysis time become longer. To overcome these limitations, an optimization of tunnel support design using differential evolution algorithm (DEA) (Storn and Price 1997) is suggested in this study.

2 Empirical and Theoretical Design Methods

Tunnel design method can be classified into empirical method, theoretical method, and numerical method. The empirical method is based on rock classification methods such as Rock Mass Rating (Bieniawski 1989) and Q-system (Barton 2002), and Convergence-Confinement Method (CCM) can be considered as theoretical method (Carranza-Torres and Fairhurst 2000). Using various numerical methods such as Finite Element Method (FEM), Finite Difference Method (FDM), Discrete Element Method (DEM), the performance of support system can be evaluated precisely during tunneling. In this study, empirical (RMR), theoretical (CCM) and numerical (FDM) methods are combined and components related to tunnel support system is optimized with DE algorithm.

Unal (1983, 1992) presented an empirical equation for calculating rock load with RMR,

$$p_r = \frac{100 - RMR}{100} \times \gamma_t \times b \tag{1}$$

where γ_t is the unit weight (kN/m^3) and b is the span of the tunnel.

The CCM proposed by Carranza-Torres and Fairhust (2000) refers to the principle of stabilizing the tunnel by controlling the displacement generated after the support is installed in the back of tunnel. In this method, tunnel behavior is represented by the longitudinal deformation profile (LDP), the ground reaction curve (GRC) and the support characteristic curve (SCC) (Fig. 1). LDP displays displacement before and after the plane turns in the longitudinal direction from the tunnel excursion plane, assuming that the support is not yet installed. In GRC, it is observed that the displacement increases when the tunnel internal pressure (p_i) decreases from the initial stress (σ_0). Furthermore, SCC shows that the supporting pressure (p_s) acting on the support material increases with the internal displacement of the tunnel increase. Such CCM has a limitation that can be used in a circular tunnel condition with K = 1.0. CCM or safety factor by rock load can be calculated using Eq. (2).

$$\text{FS} = \frac{p_s}{p_r} \geq FS_{allowable} \tag{2}$$

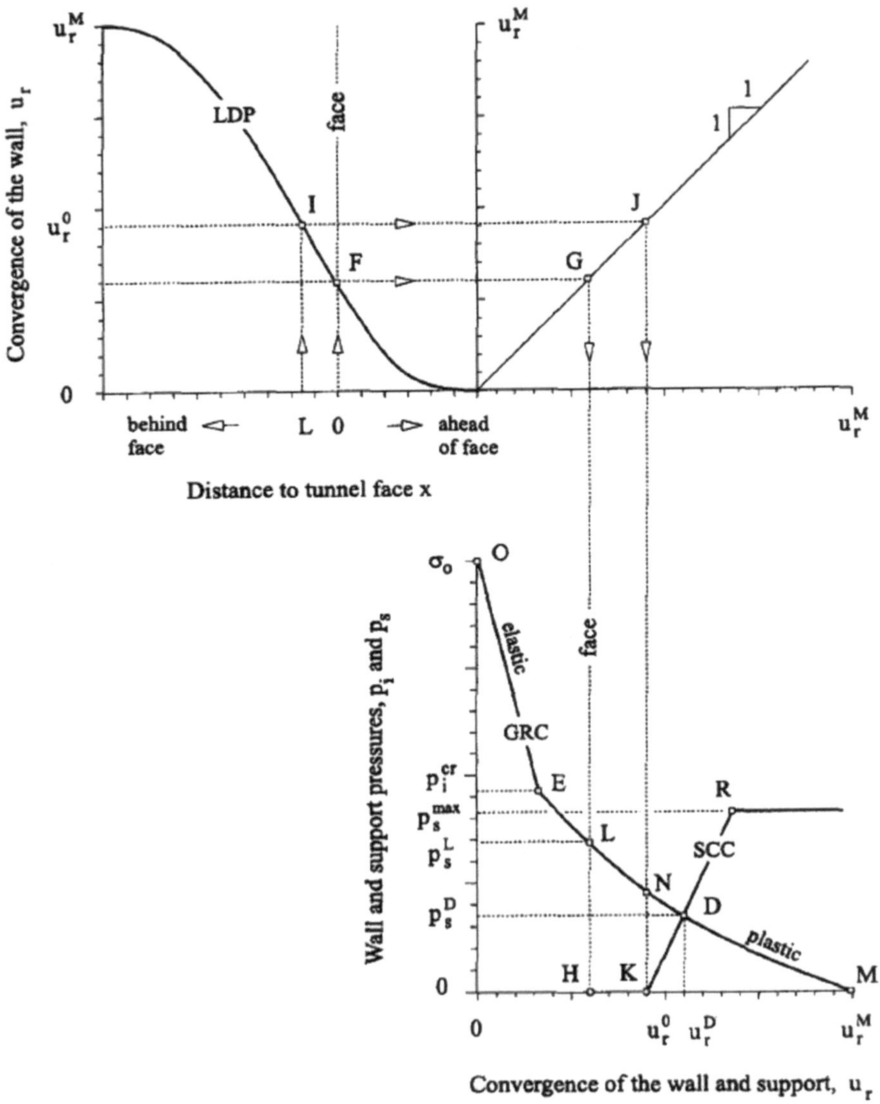

Fig. 1. Schematic representation of the convergence-confinement method (Carranza-Torres and Fairhurst 2000)

where, p_s is maximum supporting pressure of support. p_r is the maximum supporting pressure of support or rock load and $FS_{allowable}$ is the allowable factor of safety ranging from 1.5 to 3.0, but usually with a value of 1.5.

Axial stress of rock bolts and bending stress of shotcrete can be obtained by numerical analysis using Flac3D. According to KR C-12030 of KR Code 2012, the allowable axial stress of the rockbolts of SD350 and D25 is 175 MPa, and the allowable bending stress of the steel fiber shotcrete is 4.5 MPa or more. Each allowable factor of safety is 2.0 and 1.0, respectively.

3 Optimization Method—Differential Evolution Algorithm (DEA)

From the three different vectors randomly selected in the group $X_G = [x_{1,G}, x_{2,G}, \ldots, x_{NP,G}]$, next generation of mutant vector $(v_{i,G+1})$ is generated as shown in Eq. (3) for each of the transformation target vectors,

$$v_{i,G+1} = x_{r1,G} + F \cdot (x_{r2,G} - x_{r3,G}) \tag{3}$$

where NP is the population size and F is a mutation constant with a value between 0 and 2. Equation (4) represents the crossing step. The parent vector is crossed with a mutant vector to generate a trial vector. Here, rand means that the vector to be used for the change is randomly selected, and CR is a mating constant having a value between 0 and 1. I_{rand} is a random integer of [1, 2, … , D].

$$v_{ji,G+1} = \begin{cases} v_{ji,G+1} & if\ rand_j \leq CR\ or\ j = I_{rand}, \\ v_{ji,G} & if\ rand_j > CR\ or\ j \neq I_{rand} \end{cases} \tag{4}$$

$$x_{i,G+1} = \begin{cases} u_{i,G+1} & if\ f(u_{i,G+1}) \leq f(x_{i,G}), \\ x_{i,G} & otherwise \end{cases} \tag{5}$$

Equation (5) represents the selection step. The target vector $(x_{i,G+1})$ is compared with the trial vector $(u_{i,G+1})$. In the next generation, individuals of superior traits are selected. Therefore, each trial vector will deliver an excellent genetic trait in the next generation through comparison with the parent vector. Through this process, the best individuals in the population are retained in the next generation. This process is repeated until the specified end criterion is reached. Figure 2 shows the process of Eq. (4), which is a mutation vector generation equation in a two-dimensional plane.

4 Optimization of Tunnel Support System

Analysis cases are describes in Table 1. The length, longitudinal spacing, transversal spacing of rockbolt and shotcrete thickness are considered as design parameters. For the numerical analysis, a circular tunnel is modeled 25 m below the ground surface. Rockbolt was modeled with a cable element and the shotcrete with shell element. The section used in the analysis is shown in Fig. 3, and the rock mass properties and DEA properties are summarized in Table 2. RMR, γ_t (Unit weight), and material properties of the support were used by Park and Kang (2010). Also this article cited Trueman's article (1988) demonstrating the relationship between RMR and cohesion/friction angle to calculate values. ν (Poisson's ratio) is also calculated with the relationship proposed

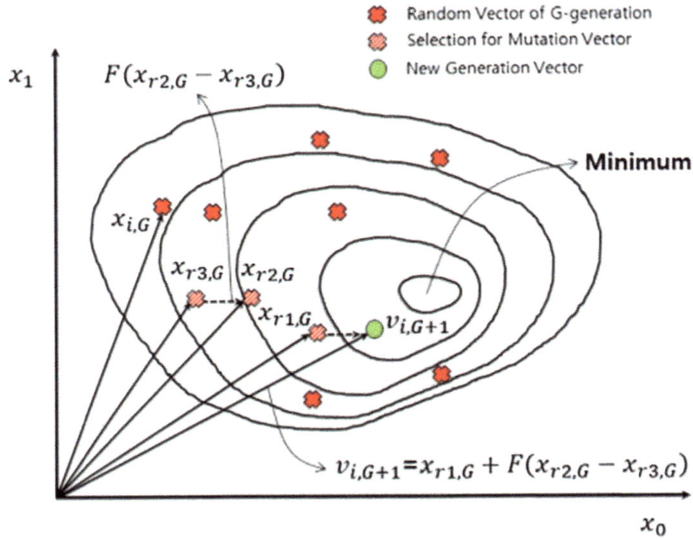

Fig. 2. Procedure of differential evolution algorithm in 2D space (Storn and Price 1997)

Table 1. The design parameters for tunnel support system

Items		Conditions
Rockbolt	Length (m)	3, 4, 5
	Longitudinal spacing (m)	1, 2, 3
	Transversal spacing (m)	1, 2, 3
Thickness of shotcrete (cm)		5, 8 10, 15, 20

by Aydan et al. (1993). Table 2 was applied to Mohr-Coulomb criterion for Flac 3D modeling. The model was implemented with condition that the excavation and supporting beyond 10 m is finished, to analyze the optimization of the support system during construction. The analysis conditions are safety factor for RMR and for CCM of 1.5 or more, and the axial stress of rock bolt and the bending stress of shotcrete are evaluated.

DEA is an algorithm that can search real numbers. However, since the design condition is an integer, random number of real number condition is divided into 3–5 sections to select design condition. Analysis results are summarized in Table 3.

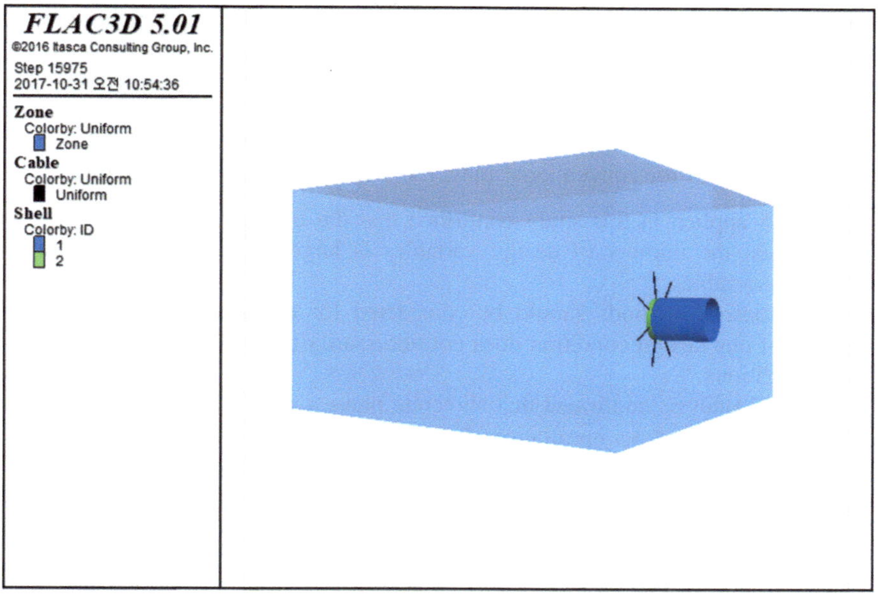

Fig. 3. The feature of 3-dimensional model

Table 2. Rock and DEA parameters used in the numerical analysis

Rock mass condition		DEA	
RMR (Rock Mass Rating)	50	Mutant constant	0.8
γ_t (Unit weight)	25 kN/m^3	Crossover constant	0.8
c (Cohesion)	3.0 MPa	Number of population	10
Φ (Friction angle)	30.0°	Max iteration	20
ν (Poisson's ratio)	0.25	Range of target values	Please refer Table 1

Table 3. The results of optimization

Items		Optimized value
Rockbolt	Length (m)	5
	Longitudinal spacing (m)	3
	Transversal spacing (m)	3
Thickness of shotcrete (cm)		10
FS of RMR		6.76
FS of CCM		1.62
FS of axial stress		8.02
FS of bending stress		1.30

5 Conclusions

In this study, differential evolution algorithm is introduced to optimize tunnel support system design. DEA constructed in this study is based on both theoretical, emprical and numerical methods. The suggested optimum design scheme provides cost effective and safe support system for conventional tunneling. The conclusion is as follows:

1. The DEA applied in this study confirmed that the optimal solution can be found even when the number of design variables is large and there is no correlation between variables.
2. Various design methods should be considered for tunnel design. Because satisfaction of one design condition does not necessarily imply satisfaction of the other two conditions.
3. Numerical analysis confirmed that shotcrete plays a significant role in the stability of tunnel; therefore, its installation should be carefully considered.

Acknowledgements. This work was supported by the National Research Council of Science & Technology (NST) grant by the Korea government (MSIP) (No. CRC-16-02-KICT).

References

An, J.S., Kim, B.C., Moon, H.K., Song, K.I., Su, G.S.: DEA optimization for operating tunnel back analysis. J. Korean Tunn. Undergr. Space Assoc. **18**(2), 183–193 (2016)

Aydan, Ö., Akagi, T., Kawamoto, T.: The squeezing potential of rocks around tunnels; theory and prediction. Rock Mech. Rock Eng. **26**(2), 137–163 (1993)

Barton, N.: Some new Q-value correlations to assist in site characterisation and tunnel design. Int. J. Rock Mech. Min. Sci. **39**(2), 185–216 (2002)

Carranza-Torres, C., Fairhurst, C.: Application of the convergence-confinement method of tunnel design to rock masses that satisfy the Hoek-Brown failure criterion. Tunn. Undergr. Space Technol. **15**(2), 187–213 (2000)

Choi, S.O., Shin, H.S.: Stability analysis of a tunnel excavated in a weak rock mass and the optimal supporting system design. Int. J. Rock Mech. Min. Sci. **41**, 876–881 (2004)

Hoek, E., Brown, E.T.: Practical estimates of rock mass strength. Int. J. Rock Mech. Min. Sci. **34** (8), 1165–1186 (1997)

Kim, et al.: A study on the optimization technique for the plan of slope reinforcement arrangement of soil-nailing in tunnel portal area. J. Korean Tunn. Undergr. Space Assoc. **18** (6), 569–579 (2016)

Nguyen, T., Ghabraie, K., Tran-Cong, T., Fatahi, B.: Improving rockbolt design in tunnels using topology optimization. Int. J. Geomech. **16**(1), 04015023 (2015)

Park, C.S., Kang, M.H.: A study on the numerical analysis of a NATM tunnel with consideration of construction procedure and field measurement. Tunn. Undergr. Space **20**(1), 28–38 (2010)

Storn, R., Price, K.: Differential evolution—a simple and efficient heuristic for global optimization over continuous spaces. J. Global Optim. **11**(4), 341–359 (1997)

Trueman, R.: An evaluation of strata support techniques in dual life gateroads. Doctoral dissertation, University College (1988)

Risk Assessment on Ground Uplift and Pressure Control During Grouting Against Running Water

Zhuo Zheng(✉), Rentai Liu, and Qingsong Zhang

Geotechnical and Structural Engineering Research Center, Shandong University, Jinan 250061, Shandong, China
413708318@qq.com

Abstract. A theoretical model is proposed to estimate the flow patterns of grout suspensions during grouting process with the influence of flowing groundwater. A comprehensive analysis is conducted on the proposed model and the laboratory test results of previous studies. It is concluded that: the temporal and spatial variation of grouting penetration can be analyzed with ideal fluid hypothesis, while the information of grouting pressure can be acquired through real-time monitoring. Thus, the whole information during grouting process is acquired. The pressure of grout in upstream direction is larger than in downstream direction, and the danger of ground uplift is higher. Combining the uplift mechanism of overlying rock mass and grouting penetration and pressure distribution rule, the allowable injection pressure and the risk for overlying rock mass uplift can be estimated. The proposed model provides theoretical basis for risk assignment for overlying rock mass uplift during grouting process.

Keywords: Grouting against running water · Flow pattern · Pressure control
Risk assessment

1 Introduction

Grouting is one of the most widely used methods to control groundwater disasters during geotechnical engineering by sealing the pathway of groundwater and reduces the conductivity of the rock mass (Zhang et al. 2010; Zhang et al. 2012; Li et al. 2014). Controlling on the injection pressure is essential for a successful grouting process, which the grout suspension can spread into the target area and then solidify. However, the grouting process is significantly influenced by flowing groundwater as the resistance is much higher along the upstream direction than the downstream direction. Moreover, the viscosity of grout suspension should be increased in case of being washed away by flowing groundwater. Thus, a higher injection pressure is needed to increase the momentum of grout suspension. On the other hand, the pressure and velocity of groundwater should not be too high. An excessive high injection pressure may cause grout suspension penetrate into a very remote area along the downstream direction, which is useless for sealing the pathway of groundwater in the target area. Furthermore, for shallow buried engineering construction, an excessive high pressure

B. Fatahi et al. (eds.), *Sustainable Design and Construction for Geomaterials and Geostructures*, Sustainable Civil Infrastructures,
https://doi.org/10.1007/978-3-319-95753-1_22

may cause large deformation and even failure of overlying rock mass. Thus, a balance should be made in order to achieve the best grouting effect.

Recently, several researchers have experimentally and theoretically investigated the penetration mechanism of grouting against flowing water. In most cases, the fracture is simplified as a single flat plate. Zhang (2011) developed a modeling experiment system for single flat fracture grouting against flowing groundwater and proposed the U shape penetration mechanism. Liu et al. (2012) studied the penetration mechanism of cement based grout under both static and flowing groundwater conditions, and analyzed the flow pattern and pressure distribution characteristics. Some other researchers studied the grouting penetration process with different constitutive model of grout suspension and obtained a series of conclusions (Zhan et al. 2011; Zhan and Sui 2011).

However, few studies have considered the grouting pressure distribution and the uplift mechanism of overlying rock mass during the grouting process. In the following sections, theoretical analysis and laboratory tests are conducted to obtain a basic recognition on pressure distribution and penetration process of grouting. Next, a lower limiting value for uplifting criterion is proposed from a conservative perspective. Finally, a controlling method for grouting pressure is proposed based on the criterion, which provide theoretical basis for risk assignment for uplifting of covering rock mass during grouting process.

2 Fundamental Mechanism of Overlying Rock Mass Uplift

There are two main factors influencing the uplift of overlying rock mass during grouting process: one is the rheological properties of the grout suspension, and the other one is the mechanical properties of the rock mass and the surrounding stress condition. For cases of shallow buried depth, the crustal stress field is mainly caused by the dead load of the overlying rock mass. Thus, fractures with an almost horizontal inclination angle are mostly unconstrained and thus most likely to deform under the load of grouting pressure. When the grouting pressure acting on the fracture surface is larger than the constraining force from the surrounding rock mass, a plastic deformation and uplift may occur, as shown in Eq. (1):

$$F_1 \leq F_2 \tag{1}$$

Here, F_1 represents the grouting pressure acting on the fracture surface. F_2 represents the reaction force loading on the rock mass when it is moving upward. As the grouting parameters vary with time and location, such as pressure, viscosity and velocity of the grout suspension, the values of both F_1 and F_2 are not constant during grouting process. With increase of grouting time, the penetration distance of grout suspension keeps growing, thus the load bearing area of the overlying rock mass increases. In the meantime, constrains also increase since a larger area is influenced by grouting. To sum up, calculation on the uplift force needs to be combined with analysis of flow pattern of grout suspension during process of grouting against flowing groundwater. Thereby, fluid pressure at different grouting time can be obtained and potential uplifting area of overlying rock mass can be determined. Finally, evaluation criteria can be established.

For the simplest case, the influence of flowing groundwater on the penetration process and the pressure distribution is neglected. Assuming the fracture is horizontal, grout suspension penetrates uniformly along all direction, as is shown in Fig. 1. The shearing strength of the rock mass is also neglected.

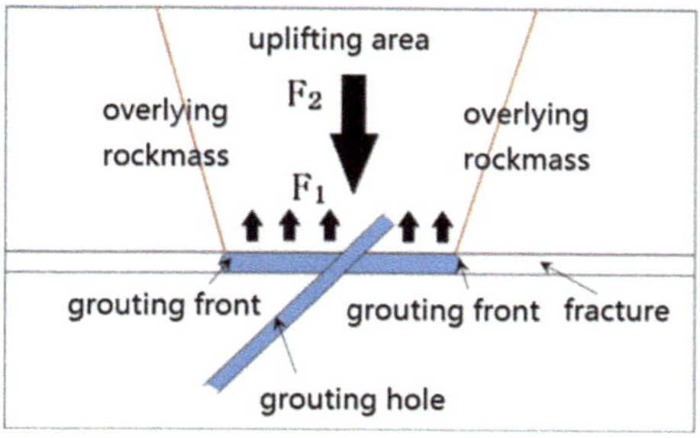

Fig. 1. Uplift area induced by grouting pressure

Thus, F_2 can be represented as:

$$F_2 = \rho \cdot g \cdot h \cdot \pi \cdot I^2 \cdot k_1 \tag{2}$$

Here, ρ is the density of the overlying rock mass. g is acceleration of gravity. h is the distance between the fracture and ground surface. I is grout penetration distance. k_1 is the influence of the inclination angle of uplifted overlying rock mass, and its value is 1 when the inclination angle is 0°. According to Littlejohn and Bruce (1975) and Yap and Rodger (1984) β can be determined as 45°, and Eq. (2) is rewritten as:

$$F_2 = \rho \cdot g \cdot h \cdot \pi \cdot I^2 \cdot \left(1 + \frac{h}{I} + \frac{1}{3}\left(\frac{h}{I}\right)^2\right) \tag{3}$$

3 Penetration Rules When Grouting Against Flowing Groundwater

The mathematical solution for the variation process of grout suspension penetration form is a highly nonlinear problem. However, for an engineering problem, errors in a small range are allowable, so an approximate solution is sufficient for design of engineering project. By regarding the grout suspension and groundwater as ideal fluid, the process of grouting against flowing groundwater can be simplified to irrotational flow of incompressible fluid. Thus, it can be analyzed with potential field theory. The fracture can be regarded as an infinite plane, by neglecting the boundary effect. Then, this problem can be further simplified to a superposition of uniform flow field and point source flow field, as shown in Fig. 2.

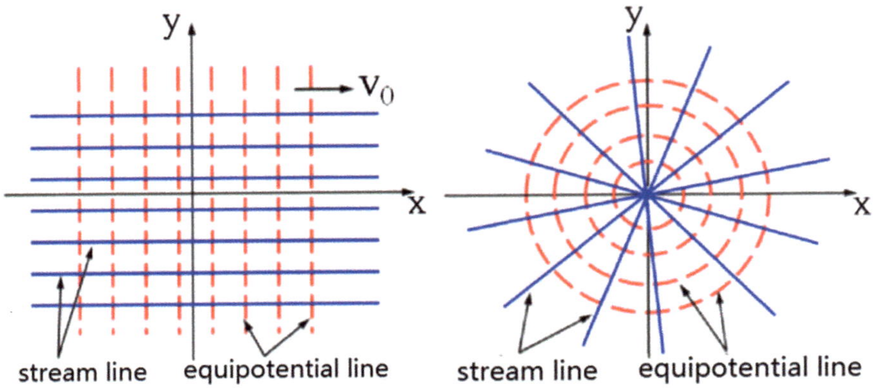

Fig. 2. Velocity potential of uniform flow field and point source flow field

The velocity potential can be written as:

$$\Phi_l = U_0 x + C_1 \tag{4}$$

$$\Phi_R = \frac{Q}{2\pi} \ln\sqrt{x^2 + y^2} + C_2 \tag{5}$$

The velocity potential after superposition can be written as:

$$\Phi_l = U_0 x + \xi \cdot \frac{U_0}{2\pi} \ln\sqrt{x^2 + y^2} + C \tag{6}$$

Equation (6) is the potential function of the grouting penetration against flowing groundwater, which considers the above simplification. Here, ξ is the ratio between injection flow rate and water velocity.

$$\xi = \frac{Q}{U_0} \tag{7}$$

The corresponding stream function can be deduced from Eq. (6) and written as:

$$\Psi = U_0 y + \xi \cdot \frac{U_0}{2\pi} \arctan\left(\frac{y}{x}\right) + C' \tag{8}$$

Figure 3 shows the stream lines of Eq. (8) by setting ξ as 4π m/s. The stream lines are colored by blue.

A further analysis on the distribution of the stream lines showed that: stream lines starting from the point source are confined within a envelop line, which is also the boundary between the stream lines starting from the point source and from infinity. Thus, it can be concluded that the envelope line represents the extreme penetration distance of grout suspension. The equation of the envelop line is written as:

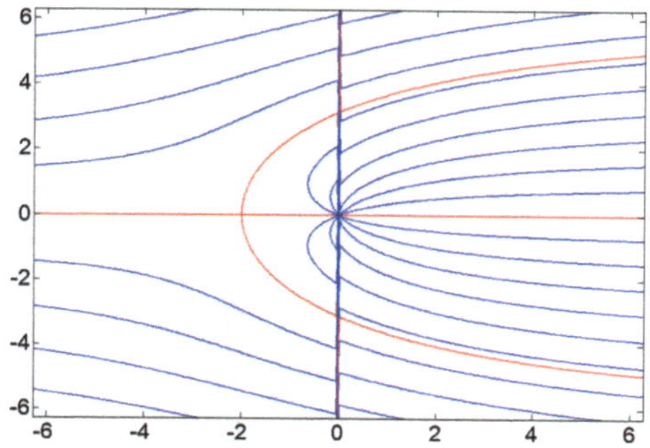

Fig. 3. Streamline of grout penetration with running water

$$\begin{cases} y + \frac{\xi}{2\pi}\arctan\left(\frac{y}{x}\right) = 0, & (y \neq 0)(x \leq 0) \\ y + \frac{\xi}{2\pi}\arctan\left(\frac{y}{x}\right) = \pm\frac{\xi}{2} & (x > 0) \end{cases} \tag{9}$$

From Eq. (9), the extreme penetration distance of grout suspension against the water flowing direction and the extreme width can be calculated, as shown in Fig. 3.

Assuming the radius of grouting hole is r_0, the temporal and spatial variation of the grout suspension can be represented in the form of differential equation as:

$$\begin{cases} \frac{dx}{dt} = U_0 + \xi \cdot \frac{U_0 x}{2\pi(x^2 + y^2)} \\ \frac{dy}{dx} = \xi \cdot \frac{U_0 y}{2\pi(x^2 + y^2)} \end{cases} \tag{10}$$

Solution of Eq. (10) shows the fluid field change and spatial and temporal variation of penetration process of grout suspension. Calculation results are compared with those from laboratory test, which was conducted previously in our study (Li et al. 2016a, b), as shown in Figs. 4 and 5.

From Fig. 4, it can be seen that the flow pattern becomes more and more anisotropic as grouting time increases. In the upstream direction, the penetration rate drop down rapidly, and the grout front converges towards the line of the extreme penetration scope. In the downstream direction, the grout front keeps moving forward and the penetration width keeps growing as grouting time increases. The velocity of grout suspension decreases gradually and finally converges to the velocity of groundwater.

By comparing Figs. 4 and 5, it can be seen that the flow patterns are similar, which proves the rationality of the ideal fluid assumption. For constant injection flow rate condition, the simplification on the viscosity of grout and groundwater only influences the pressure distribution. The theoretical calculation cannot accurately estimate the pressure distribution, but the result of spatial and temporal variation of penetration area is basically accurate. Another difference caused by ideal fluid assumption is the water resistance abilities, which means whether the flow pattern is more like a circle or an ellipse, but this influence is small and can be neglected.

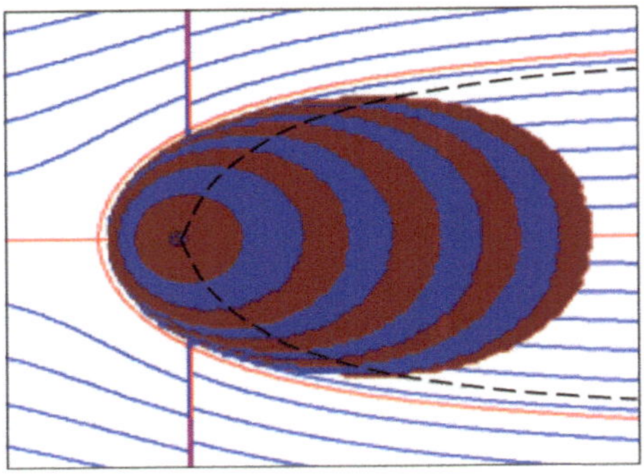

Fig. 4. Temporal and spatial variation of grout penetration

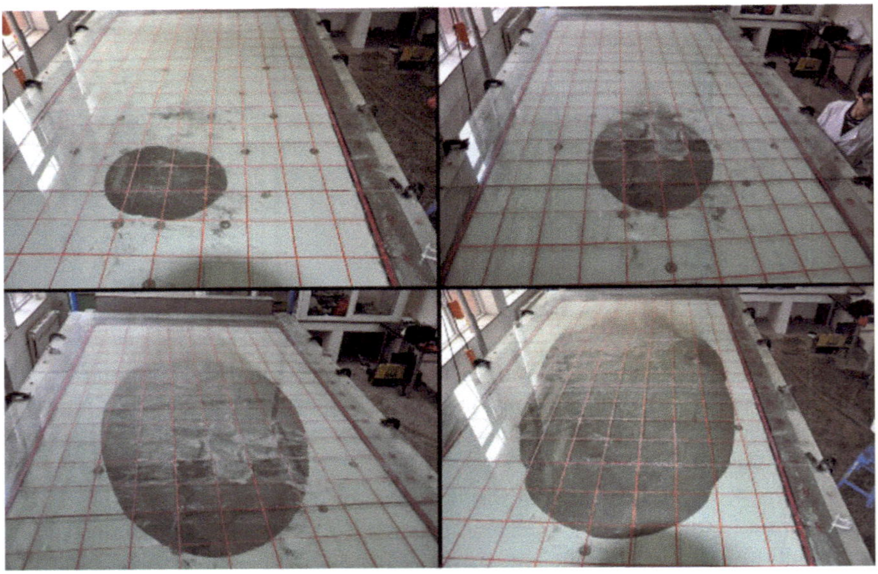

Fig. 5. Experiment pictures of grout penetration form

4 Pressure Distribution During Grouting Process

According to the proposed theoretical model, the loaded area of overlying rock mass can be determined by calculating the penetration distance at different times. In order to calculate the whole uplifting force, the pressure distribution of grout suspension is also needed. Here, a lower limit value of grouting pressure is determined according to theoretical analysis and laboratory test from a conservative perspective.

For ideal fluid, the pressure distribution can directly deduced from the velocity field through Cauchy-Lagrange integral equation (Blazek 2005).

$$\frac{\partial \Phi}{\partial t} + \frac{1}{2}|\nabla \Phi|^2 + \frac{p}{\rho} = C(t) \quad (11)$$

The ideal fluid assumption neglected the viscosity of the grout, which causes big difference of pressure distribution comparing with reality. Thus, the rheological property of grout suspension should be sufficiently considered. Liu (2012) established the pressure distribution equation for Bingham fluid in single flat fracture with static water under constant grouting flow rate condition, as shown in Eq. (12):

$$p = -\left(\frac{3}{2}\tau\right)r - \frac{6q\mu(t)}{\pi b^3}\ln r + C \quad (12)$$

Here, p represents the pressure at grouting hole, π represents the yield stress of the fluid, r represents the distance from the center of grouting hole to any point within the penetration scope, q represents the grouting flow rate; b represents the width of the fracture, μ(t) represents the viscosity of the fluid as a function of time, C represents a integration constant.

According to mass conservation equation, r and q obeys the following relationship:

$$r_t = \sqrt{\frac{qt}{\pi b}} \quad (13)$$

Using boundary conditions of $r = r_t$ and $p = p_w$, pressure distribution equation is obtained from Eq. (12):

$$p(r,t) = \frac{3\pi}{b}\left(\sqrt{\frac{qt}{\pi b}} - r\right) + \frac{6q\mu}{\pi b^3}\ln\left(\frac{1}{r}\sqrt{\frac{qt}{\pi b}}\right) + p_w \quad (14)$$

In the equation, the first two terms represents the contribution to pressure dissipation from yield stress and viscosity of the fluid, respectively. It can be seen that when the influence of viscosity is ignored, the pressure decreases linearly along penetrating direction.

Under the influence of flowing groundwater, the pressure distribution of grout suspension shows some similarity with that of static water condition: Along upstream direction, grout suspension is constrained by groundwater pressure all the times during the penetration process. The stress state is similar to that of static water conditions. Thus, the pressure distribution is also similar. On the other hand, pressure distribution along downstream direction is very different from that of static water condition and is much more complicate, which can only be analyzed with experimental method.

In our previous studies (Li et al. 2016a, b), laboratory tests were conducted on grouting process in single flat fracture, and pressure distribution along the axis of symmetry was obtained, as shown in Fig. 6. It can be observed that a high injection rate leads to a high grouting pressure. The pressure is highest around the injection hole and is lower in further place. Along downstream direction, the pressure drops nonlinearly, and this trend becomes more obvious as grouting time increases. In other words,

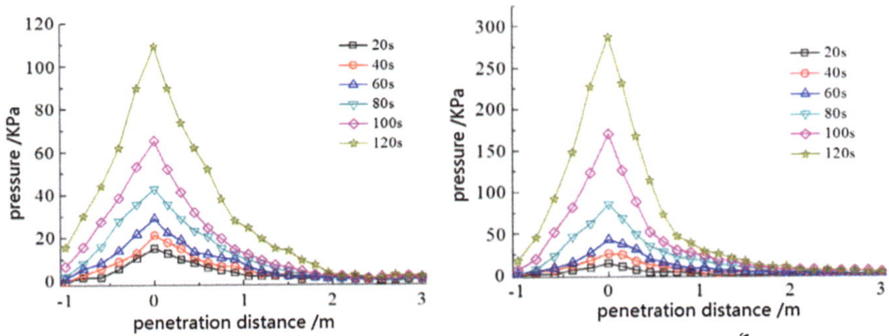

(a) U_0=0.3m/s, ξ=0.33 (b) U_0=0.3m/s, ξ=0.66

Fig. 6. Grouting pressure distribution along longitudinal axis with different grouting time

although the pressure influencing area is larger, but its magnitude is lower at any point compared with that in upstream area. In the very remote place of the downstream side, the uplift force can be neglected even if the grout has reached. Thus, the risk for uplift is higher in upstream area. By assuming that the pressure distribution in downstream area is equal to that in upstream area, the whole uplift force can be calculated as a conservative value.

5 A New Criterion for Uplift of Overlying Rock Mass and Allowable Injection Pressure

During grout penetration process, the grout viscosity is most influential on grouting pressure distribution and is a crucial factor for the success of grouting process. The requirement for grout viscosity in grouting design is that: it should remain at a low value at the initial stage, which is beneficial for grout fluid reaches further area. When the grout suspension approaches the boundary of the groundwater pathway, it increases quickly to improve the ability to resist flowing groundwater. According to our previous study (Liu et al. 2012), for cement-silicate grouting material, when the volume ratio of these two components is 1:1, 2:1 and 3:1, the corresponding viscosity-time relationship for each case can be represented as:

$$\mu(t) = 0.003182t^{2.23} + 0.04 \tag{15}$$

$$\mu(t) = 0.000843t^{2.69} + 0.08 \tag{16}$$

$$\mu(t) = 0.01864t^{2.07} + 0.16 \tag{17}$$

On the basis of all the analysis above, the risk of uplift of overlying rock can be calculated. Firstly, field investigation should be carried out to acquire the occurrence of the fractures and the velocity of groundwater. Then, penetration area is calculated according to Eqs. (10) and (11). Next, F_1 can be calculated combining with Eq. (12)

Also, F_2 can be calculated from the area of grouting penetration and resisting force per unit area according to Eq. (2). Finally, the admissible value of injection pressure can be calculated according to Eq. (1).

Assuming here is a horizontal fracture. Its depth is 6 m. The density of overlying rock mass is 2.5 kg/m^3. The grout material is cement-silicate mixture with volume ratio of 1:1. The velocity of groundwater is 0.1 m/s and ξ is 2π. The penetration distance and the value of F_1 and F_2 at different times can be estimated. The calculation results are shown in Fig. 7. As can be seen, both F_1 and F_2 increases with grouting time, but the increasing rate of F_2 is much higher. The maximum permissible value of the injection pressure is determined by the intersection point of the two curves. If the groundwater is not successfully sealed with the maximum injection pressure, grouting method should be altered to control the injection pressure by reducing injection rate and the viscosity of the grout, or injecting the grout intermittently. The value of F_1 and F_2 are then recalculated until the groundwater can be successfully sealed.

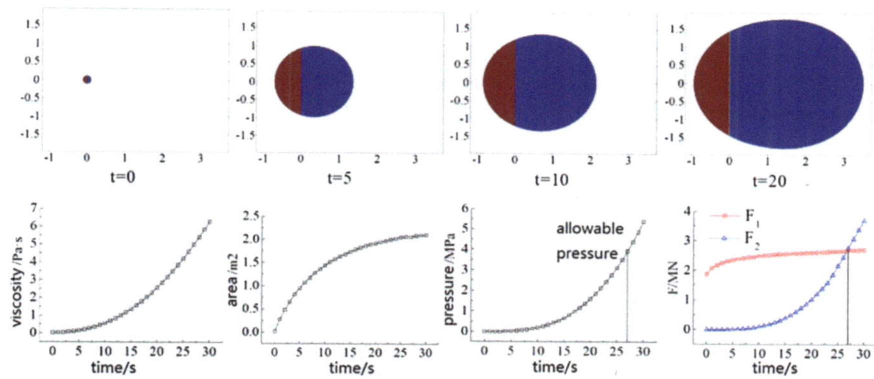

Fig. 7. Calculation results of grouting flow pattern and grouting uplift criterion

6 Discussion

In reality, the distributions of fractures encountered in geotechnical engineering construction are often complicated. However, uplift of rock mass is most likely to happen if there is an almost horizontal fracture with a much larger aperture than other fractures. This case is regarded as the most unfavorable condition. The proposed method mainly aims at evaluate the risk of rock mass uplift in this situation. For a fracture with an inclination angle, the calculated uplifting force should be modified with a reduction factor and a higher allowable injection pressure may be derived. The proposed method may also be applicable for fracture network which is composed of one large opening fracture and many small fractures, by considering the grout suspension mainly penetrates in the large fracture and neglecting the influence of the smaller ones. However, the rationality for the simplification and the specific calculation method still need further study.

7 Conclusion

(1) A theoretical model for uplifting of overlying rock mass during grouting process is proposed and a method to determine the allowable injection pressure is developed, on the base of uplift mechanism of overlying rock mass, and grouting penetration and pressure distribution rule. The proposed model provides theoretical basis for risk assignment for covering rock uplift during grouting process.
(2) When estimating the uplifting force and constraining force during grouting process, the temporal and spatial variation of flow patterns of grout suspension can be analyzed with the ideal fluid assumption. Under the constant injection flow rate condition, the error aroused by this approximation can be neglected, while the information of pressure distribution can be acquired by real time monitoring in the field. A combination of these results provides the whole information during the grouting process.
(3) The pressure distribution along the upstream direction can be approximately regarded as the same as that under static water condition. Along the downstream direction, grouting pressure at any point is lower than that of the corresponding point in upstream area. Therefore, the risk for uplift of the overlying rock mass is higher. During calculation, the pressure distribution in the downstream direction can be regarded the same as that of upstream direction, and the calculation result is considered as a conservative value.

Acknowledgements. This work is support by the national key research and development plan of China (2016YFC0801600).

References

Blazek, J.: Computational Fluid Dynamics: Principles and Applications, 2nd edn. Elsevier, Amsterdam (2005)
Li, S., Liu, R., Zhang, Q., et al.: Protection against water or mud inrush in tunnels by grouting: a review. J. Rock Mech. Geotech. Eng. **8**(5), 753–766 (2016a)
Li, S., Zhang, W., Zhang, Q., et al.: Protection against water or mud inrush in tunnels by grouting: a review. Rock Mass Mech. **35**(03), 744–752 (2016b)
Littlejohn, G.S., Bruce, D.A.: Rock anchors: state of the art-Part 1. Ground Eng. **8**(5), 25–32 (1975)
Liu, J., Liu, R., Zhang, X., et al.: Penetration law model test and numerical simulation of cement fracture grouting. Chin. J. Rock Mech. Eng. **12**, 2445–2452 (2012)
Liu, R.: Study on penetration and plugging mechanism of quick setting cement based slurry in underground dynamic water grouting and its application [Ph.D. thesis]. Shan Dong University, Jinan (2012)
Yap, L.P., Rodger, A.A.: A study of the behavior of vertical rock anchors using the finite element method. Int. J. Rock Mech. Min. Sci. Geomech. Abstr. **21**(2), 47–61 (1984)
Zhan, K., Sui, W.: Design of model test system for grouting into a single fracture with flowing groundwater. Res. Explor. Lab. **10**, 19–23 (2011)
Zhan, K., Sui, W., Gao, Y.: A model for grouting into single fracture with flowing groundwater. Rock Mass Mech. (06), 1659–1663 (2011)

Zhang, Q., Han, W., Li, S., et al.: Comprehensive grouting treatment for water gushing analysis in limestone breccias fracture zone. Chin. J. Rock Mech. Eng. **31**(12), 2412–2419 (2012)
Zhang, X. Study on mechanism of slurry penetration and sealing at the process of underground engineering moving water grouting and its application (2011)
Zhang, X., Li, S., Zhang, Q., et al.: Filed test of comprehensive treatment method for high pressure dynamic grouting. J. China Coal Soc. **35**(08):1314–1318 (2010)

Impacts of Nanotechnology Adoption on Sustainable Construction

Ayodeji E. Oke(✉), Clinton O. Aigbavboa, and Kgothatso Semenya

Department of Construction Management and Quantity Surveying, Faculty of Engineering and the Built Environment, University of Johannesburg, Johannesburg, South Africa
{aoke, caigbavboa}@uj.ac.za, kgothatsosemenya@gmail.com

Abstract. The architectural, engineering and construction industry plays a significant role in the development of any country but, the industry have been experiencing a number of challenges relating to safety, environmental issues and general performance especially as a result of continuous adoption of traditional and conventional construction materials. In a view to minimize these challenges and achieve sustainable development, this study examines the effects of the adoption of nanotechnology principle to traditional construction materials. A review of existing literature materials on nanotechnology, sustainable construction and construction materials results in identification of 12 major impacts of nanotechnology on construction materials. In designing the questionnaire for the study, these factors were subsequently adopted and the research instruments were distributed to construction professionals, that is, quantity surveyors, engineers, construction managers, project managers and architects, in private and public establishments in Gauteng region of South Africa. The adoption of nanotechnology will not only help in reducing waste and provide new means of innovation but will also aid reforming of existing environmental remediation techniques. However, the adoption may also result in adverse effects on environment and human health if not properly managed by experts and concerned regulatory agencies. This indicates that as beneficial as the concept of nanotechnology may be to the performance of architectural, engineering and construction projects, care should be taken by contractors or developers, construction professionals and regulatory agencies, to understand and apply the principle in accordance to standards and regulations to avoid any potential negative results.

Keywords: Construction materials · Nanotechnology · Project success Sustainable construction · Sustainable development

1 Introduction

According to Chong (2004), current advancements which are essential drivers of the twenty first century and the new economy include microelectronics, biotechnology, data innovation and nanotechnology as well as the modifying capacity, empowering and supporting related base frameworks and materials. Nanotechnology is often

B. Fatahi et al. (eds.), *Sustainable Design and Construction for Geomaterials and Geostructures*, Sustainable Civil Infrastructures,
https://doi.org/10.1007/978-3-319-95753-1_23

depicted as an empowering innovation and major advancement, it is required to prompt various creative improvements in different fields of innovation and ranges of utilization in the public arena as well as in the commercial environment. The concept of nanotechnology has the potential for expansive changes that will in the long run influence all territories of life. Such changes will without a doubt have solid repercussions for the environment and general society, and carry with them not just the fancied and expected impacts. This may include such things as developments as enhancements to items, procedures and materials; financial development; new employments for talented specialists; help for the earth; and further strides toward economical business, additionally unforeseen and undesirable symptoms and outcomes (Steinfeldt et al. 2007).

The applications of nanotechnology are gaining ground in almost all engineering fields. To the extent its application is worried with development, the branch itself manages different sorts of materials. The properties of different materials perhaps improved by the applications (Rao et al. 2015), and furthermore, produce systems towards materials and whose composition and parts reflect novel and overall improved synthetic, physical and organic properties, phenomena and procedures due to their nanoscale size (Tegart 2004). According to Wegner et al. (2005), regarding structural biomaterials, the desire functionality and properties with regards to materials may be realized in biomaterials, through the production of next generation materials with great promises of undeniably great serviceability and hyper-performance. Furthermore, these products will exert strength properties now only seen with composite materials having their base composition as carbon. In addition, even in environments with high moisture content, these new hyper-performance bio-products will be capable of longer service.

Zucker and Darby (2005) state that nanoscale science and technology has every one of the reserves of the sort of leap forward transformative advancement in which cascades of important scientific discoveries create the technological opportunities that transform existing industries and create new ones. Furthermore, nanotechnology is anticipated that would represent a noteworthy extent of technological advancement and economic growth over the next several decades. This study therefore examines the overall positive effects of adopting nanotechnology in the quest of achieving sustainable development goals in the construction industry.

2 Nanotechnology and Sustainable Construction

The potential impacts of applying nanotechnology to traditional construction materials are vast and vary from human, societal, environmental, etc. Due to the application of this we discover new materials termed nanomaterials and nanocomposites. Karn and Bergeson (2009) suggest that there is an overemphasis on applying nanotechnology towards producing financial advantages without considering unfriendly societal and ecological effects.

According Ofori (1998), the construction industry lingers behind different segments in its reaction to the issues of nature. In the developing nations, this issue gets even less consideration in spite of the fact that it is of basic significance. Work on the environmental ramifications of development action in developing nations ought to be embraced as an issue of direness. Considering these asset and skill requirements which these nations confront, accentuation ought to be on counteractive action. Furthermore, with consideration being brought to the related weakness of the private sector in developing countries, the government's role as a construction client would be of great importance and significance. Therefore, construction companies and related specialists ought to be encouraged and provided with a continuous mandate to ceaselessly search for inputs and methods for working which will minimize the negative effect of construction activity on the natural environment (Ofori 1993).

Nanomaterials and nanocomposites with one of a kind physical and substance properties are progressively being utilized in the construction industry to enable novel applications. However, with all the desired applications considered, being threatened continuously with the opportune concern about their potential (unintended) impacts to the environment and human health (Lee et al. 2009). According to the Zhu et al. (2004), the impact of different nanomaterials on the indigenous habitat is fervently in nanotechnology and ecological research, furthermore various on-going examinations concentrate on the vulnerability in regards to the potential impacts of materials that exist on the nanoscale with properties that are not quite the same as when utilizing the material on a smaller scale or full scale. Some work in such manner demonstrates that the potential impacts might be insignificant (Lin et al. 2004). Furthermore, Roco and Bainbridge (2005) state that nanotechnology can be used to remediate waste and contamination, create frameworks and materials that utilization assets most effectively, reuse contamination into crude materials, and guarantee security and manageability of new materials.

The incorporation of nanomaterials in construction is expected to enhance the most important characteristics and properties of traditional construction materials (for example, strength, durability, and lightness) (Sobolev et al. 2016), offer new collateral functions such as energy-saving, self-heating, and anti-fogging (Zhu et al. 2004), and give principle parts to upkeep instruments, for example, structural wellbeing sensors (Zhang et al. 2006). As far as the previous favorable circumstances of nanomaterials, nanotechnology in development was chosen as one of 10 focused on uses of nanotechnology ready to determine the developing world's most concerning issues (ARI-News 2005). In any case, numerous cases in advanced history show the unintended ecological effects of at first encouraging advances, including the ponder arrival of "valuable" chemicals, for example, DDT, which was utilized to control intestinal sickness and other water-borne maladies yet was later observed to be cancer-causing to people and poisonous to a few feathered creature animal groups (Turusov et al. 2002).

As the generation and utilization of nanomaterials increases, so does the likelihood of their discharge in the earth, which builds the potential for antagonistic impacts on human and ecological wellbeing (Lee et al. 2009). Lee et al. (2013) further states that the utilisation of nano-materials in construction brings in a high possibility of hazardous effects to ecological receptors and humans considering the fact that nano-materials and those incorporated in construction materials will be released through various channels during their entire life cycle (disposal/recycling, demolition, construction and manufacturing). Consequently, to utilise the prospective benefits of nanotechnology in construction, vast efforts are needed to create relevant strategies to alleviate the release of nano-materials into environment and provide procedures to manage their environmental risks with regards to their usage in construction related activities. Lee et al. (2013) states that even though a number of studies have attested to the toxic impacts of nano-materials, there is still a knowledge gap that remains considering the risk to the environment of nano-materials linked with their exposure pathways for the likely receptors (humans and ecosystems).

Environmental degradation due to unsustainable methods for production and other human exercises uncovered the whole world population to increased risks. Innovative techniques utilizing nano-built materials and devices can be conveyed for the expulsion of polluting molecules in air, water and soil. Therefore, cleaner producing procedures and strategies, applying nano-scale systems, could likewise contribute to reduced levels of ecological contamination, particularly in the creating scene where global benchmarks are frequently not protected. Arsenic in soil and water for example, is a generally diffused issue in a few areas of the developing world (Bürgi and Pradeep 2006). A simple, cheap but effective nano-scale technique to remove arsenic involves titanium dioxide nanoparticles (Pena et al. 2005). Nanomaterials have been appeared to be viable in evacuating metal particle pollution. A more extensive use of such innovations outfitting revelations in nanotechnology could have a constructive and more extensive effect on the wellbeing conditions and regular environment of a great many individuals.

Zhu et al. (2004) states that nanomaterials have indicated huge potential in securing the earth's life and improving sustainability, with variety of nanoparticles having been studied for use in chemical processing to reduce waste or replace toxic materials, and for treatment or remediation of pollutants in the environment. According to Ibrahim et al. (2016) one of the most capable approaches towards the reformation of techniques for the remediation of the environment is nanotechnology. Hanus and Harris (2013) further emphasise that nanotechnology has the potential to reduce the environmental impact and energy intensity of structures, as well as improve safety and reduce the related financial requirements/costs associated with infrastructure.

The advent of nanotechnology may bring new materials and new means of innovation altogether, however, its potential risks seem to pose a serious threat towards the health of any living organism. The current study examines the positive impact of nanotechnology adoption on development of infrastructural projects in the construction industry.

3 Research Methodology

This article evaluates the influence of nanotechnology concept on traditional construction materials in the South African construction industry. To achieve the objective, quantitative design approach was adopted through the administration of questionnaires on relevant stakeholders in the industry. The targeted groups of respondents for the study include engineers, architects, project managers, construction managers as well as quantity surveyors that are involved in the design and execution of construction projects in Gauteng Province of South Africa. To obtain necessary and relevant information, purposive sampling was employed to sample 64 respondents including chemical engineers with experience of construction activities and concept of nanotechnology as applicable to the construction industry.

The questionnaire adopted is a close-ended and multiple-choice type comprising of two sections. The first part was designed to obtain information regarding the characteristics of respondents while the second part dwells majorly on the objective of the study, which is concerned with various potential health, environmental and societal impacts of nanotechnology on construction materials with a view to achieving sustainable development. A five point Likert scale (1 = strongly disagree; 2 = disagree; 3 = neutral; 4 = agree; 5 = strongly agree) were adopted for this aspect of the questionnaire. The Likert scale were further employed to compute mean item score (MIS) and standard deviation (SD), which were used to determine and rank the significance of the impacts of nanotechnology.

4 Findings and Discussion

During the period of the survey, 64 questionnaires were administered on the respondents while 55 were retrieved. 52 of them were finally used for analysis because the remaining 3 of the retrieved instruments were not correctly and completely filled.

Table 1 shows the respondents' ranking of potential health, environmental and social impacts of applying nanotechnology to traditional construction materials in Gauteng, South Africa. It revealed that reform environmental remediation techniques

Table 1. Impacts of nanotechnology on sustainable construction

	MIS	SD	R
Reform environmental remediation techniques	3.96	0.91	1
Reduce waste	3.83	0.76	2
Provide new means of innovation	3.81	0.97	3
Replace toxic materials	3.79	0.87	4
Aid in environmental protection	3.69	0.83	5
Enhance sustainability	3.67	0.98	6
Improve the health of structures	3.63	0.79	7
Contributes to reduced levels of environmental pollution	3.62	0.75	8
Reduce material environmental impact	3.62	0.91	8
Remove disease causing chemicals in materials	3.56	0.8	9
Increases the potential for adverse effects on environmental health	2.98	1.21	10
Increases the potential for adverse effects on human health	2.92	1.12	11

MIS Mean Item Score; *SD* Standard Deviation; *R* Rank

was ranked first with a MIS of 3.96 and a SD = 0.91; reduce waste was ranked second with a MIS of 3.83 and a SD 0.76; provide new means of innovation was ranked third with a MIS of 3.81 and a SD = 0.97; replace toxic materials was ranked fourth with a MIS of 3.79 and a SD = 0.87; aid environmental protection was ranked fifth with a MIS of 3.69 and a SD = 0.83; enhance sustainability was ranked sixth with a MIS of 3.67 and a SD = 0.98; improve the health of structures was ranked seventh with a MIS of 3.63 and a SD = 0.79; these were ranked eighth: contributes to reduced levels of environmental pollution with a MIS of 3.62 and a SD = 0.75, and, reduce material environmental impact with a MIS of 3.62 and a SD = 0.91; remove disease causing chemicals in materials was ranked ninth with a MIS of 3.56 and a SD = 0.80; increases potential for adverse effects on environmental health was ranked tenth with a MIS of 2.98 and a SD = 1.21, and increases the potential for adverse effects on human health was ranked eleventh with a MIS of 2.92 and a SD = 1.12. The Cronbach's Alpha (α) for this section became 0.830.

The findings with regard to the utilized Likert scale and the obtained Mean Item Scores reveal that for the potential health, environmental and societal impacts of applying nanotechnology to traditional construction materials, ranked from first to the ninth the respondents agree with the impacts except for the last two impacts, the tenth and eleventh, in the revealed findings were the respondents reflect a slight disagreement with the impacts in question. This conclusion is brought forth by the interpretation of the MIS with an assumption that for a MIS below or equal to 3.55 the respondents either reflect that they strongly disagree, disagree or a neutral capacity towards the question. Furthermore, for a MIS beyond 3.55, the respondents reflect that they agree and strongly agree with the question.

The findings are in one accord with the study carried out by Roco and Bainbridge (2005) stating that the applications of nanotechnology are to be used to remediate waste and contamination, deliver frameworks and materials that utilization assets all the more productively, reuse contamination into crude materials, and guarantee wellbeing and maintainability of new materials. However the findings with regards to the carried out survey are not in agreement with the studies carried out by Turusov et al. (2005) and Lee et al. (2013) which both attest to the unintended health, environmental and societal impacts of such applications through initially promising technologies. These studies state that as the generation and utilization of nanomaterials increment, so does the likelihood of their discharge in nature, which builds the potential for unfriendly consequences for human and ecological wellbeing.

5 Conclusion and Recommendations

The study has explored various potential health, environmental and societal impacts of applying nanotechnology to traditional construction materials. Literature revealed that there are potential health, environmental and social impacts of applying nanotechnology to traditional construction materials. These impacts include: increment of the potential for adverse effects on human health, reduce waste, builds the potential for antagonistic impacts on ecological wellbeing, contributes to reduced levels of environmental pollution, remove disease causing chemicals in materials, aid in environmental protection, enhance sustainability, replace toxic materials, reduce material environmental impact, improve the health of structures, provide new means of innovation and reform environmental remediation techniques.

The findings from the questionnaire study obtained from respondents are that the major potential impacts of nanotechnological applications to traditional materials include the reformation of environmental remediation techniques, reduction of waste, replacement of toxic materials and aid in continuous environmental protection in Gauteng Province, South Africa. Therefore, it can be said that the objective of the research study was met based on the findings from the questionnaire survey. Although there is not much available literature on the potential health, environmental and societal impacts of applying nanotechnology to traditional materials more research and development is required. Human and environmental challenges form major parts of the ground need for sustainability, therefore if studies exist that show the possibility of such applications being harmful, more research and continuous experiments are required to obtain the desired possible outcomes of the particular applications.

The value of a research study lies in its beneficiaries. The people who stand to benefit from this research are mainly the construction professionals; educating themselves about on current and emerging technologies that aim to bring positive change and aid to the industry. This research study can also be of great benefit to construction companies who have structure developed to carry out researches on emerging technologies that will aid company innovation and later investment. However, further study can be extended to the adoption of nanotechnology for other construction materials. More so, qualitative, historical and other forms of quantitative data can also be adopted to further examine the importance of nanotechnology not only for construction materials, but for general use in the construction industry.

References

ARI-News: Nanotechnology in Construction—One of the top ten answers to world's biggest problems, May 3 (2005)

Bürgi, B.R., Pradeep, T.: Societal implications of nanoscience and nanotechnology in developing countries. Curr. Sci. **90**(5), 645–658 (2006)

Chong, K.P.: Nanotechnology in Civil Engineering Nanoscience Foundation, Arlington (2004)

Hanus, M.J., Harris, A.T.: Nanotechnology innovations for the construction industry. Prog. Mater Sci. **58**(7), 1056–1102 (2013)

Karn, B.P., Bergeson, L.L.: Green nanotechnology: straddling promise and uncertainty. Nat. Resour. Environ. **24**(2), 9–23 (2009)

Lee, J., Lee, S., Chung, E., Reyes, V.C., Mahendra, S.: Nanomaterials in Civil Engineering. In Springer Handbook of Nanomaterials, pp. 1039–1062. Springer, Berlin (2013)

Lee, J., Mahendra, S., Alvarez, P.J.J.: Potential environmental and human health impacts of nanomaterials used in the construction industry. In: Nanotechnology in Construction, vol. 3, pp. 1–14. Springer, Berlin (2009)

Lin, T., Wang, H., Wang, H., Wang, X.: The charge effect of cationic surfactants on the elimination of fibre beads in the electrospinning of polystyrene. Nanotechnology **15**(9), 1375 (2004)

Ofori, G.: Research on construction industry development at the crossroads. Constr. Manag. Econ. **11**(3), 175–185 (1993)

Ofori, G.: Sustainable construction: principles and a framework for attainment-comment. Constr. Manag. Econ. **16**(2), 141–145 (1998)

Pena, M.E., Korfiatis, G.P., Patel, M., Lippincott, L., Meng, X.G.: Adsorption of As(V) and As (III) by nanocrystalline titanium dioxide. Water Res. **39**, 2327–2337 (2005)

Rao, N.V., Rajasekhar, M., Vijayalakshmi, K., Vamshykrishna, M.: The future of civil engineering with the influence and impact of nanotechnology on properties of materials. Proc. Mater. Sci. **10**, 111–115 (2015)

Roco, M.C., Bainbridge, W.S.: Societal implications of nanoscience and nanotechnology: maximizing human benefit. J. Nanopart. Res. **7**(1), 1–13 (2005)

Sobolev, K., Lin, Z., Flores-Vivian, I., Pradoto, R.: Nano-engineered cements with enhanced mechanical performance. J. Am. Ceram. Soc. **99**(2), 564–572 (2016)

Steinfeldt, M., Von Gleich, A., Petschow, U., Haum, R.: Nanotechnologies, Hazards and Resource Efficiency. Springer, Heidelberg (2007)

Tegart, G.: Nanotechnology: the technology for the twenty-first century. Foresight **6**(6), 364–370 (2004)

Turusov, V., Rakitsky, V., Tomatis, L.: Dichlorodiphenyltrichloroethane (DDT): ubiquity, persistence, and risks. Environ. Health Perspect. **110**(2), 125–128 (2002)

Wegner, T.H., Winandy, J.E., Ritter, M.A.: Nanotechnology opportunities in residential and non-residential construction. In 2nd international symposium on nanotechnology in construction, Bilbao, Spain, vol. 8 (2005)

Zhang, W., Suhr, J., Koratkar, N.: Carbon nanotube/polycarbonate composites as multifunctional strain sensors. Jpn. Nanosci. Nanotechnol. **6**(4), 960–964 (2006)

Zhu, W., Bartos, P.J.M., Porro, A.: Application of nanotechnology in construction—summary of a state-of-the-art report. Mater. Struct. **37**(273), 649–658 (2004)

Zucker, L.G. & Darby, M.R.: Socio-Economic Impact of Nanoscale Science: Initial Results and Nanobank (No. w11181). National Bureau of Economic Research (2005)

Author Index

B. Fatahi et al. (eds.), *Sustainable Design and Construction for Geomaterials and Geostructures*, Sustainable Civil Infrastructures,
https://doi.org/10.1007/978-3-319-95753-1

MIX
Papier aus verantwortungsvollen Quellen
Paper from responsible sources
FSC® C105338

If you have any concerns about our products,
you can contact us on
ProductSafety@springernature.com

In case Publisher is established outside the EU,
the EU authorized representative is:
Springer Nature Customer Service Center GmbH
Europaplatz 3, 69115 Heidelberg, Germany

Printed by Libri Plureos GmbH
in Hamburg, Germany